Second Edition

Sustainability
for the 21st Century
Pathways, Programs, and Policies

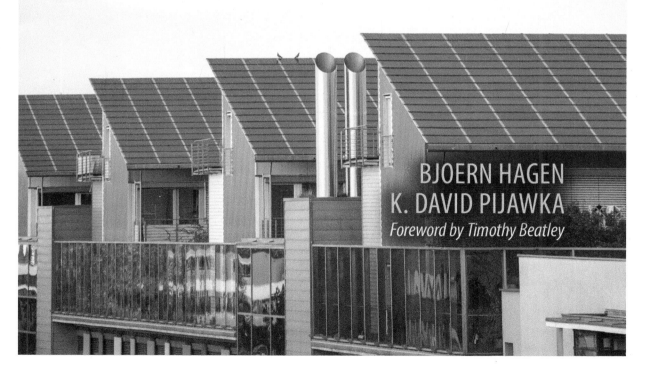

BJOERN HAGEN

K. DAVID PIJAWKA

Foreword by Timothy Beatley

Kendall Hunt
publishing company

Contents

Foreword   *vii*

About the Editors   *xi*

Acknowledgements   *xiii*

Overview

Chapter 1   Pathways to a Sustainable Future: A Look Inside   1
K. David Pijawka and Bjoern Hagen

Chapter 2   An Intellectual History of Sustainability   17
Craig Thomas

Urban Sustainability

Chapter 3   Earliest Urbanism   43
Charles Redman

Chapter 4   What Should Sustainable Cities Look Like?
Programs, Policies, and Initiatives   53
Bjoern Hagen and Ariane Middel

Chapter 5   Urban Sustainable Design   87
Lauren Allsop

Chapter 6   Examining Urban Sustainability through Urban Models   111
Subhrajit Guhathakurta

Chapter 7   Resilience: An Innovative Approach to Social,
Environmental, and Economic Urban Uncertainty   127
Stephen Buckman and Nelya Rakohimova

Urbanization Issues

Chapter 8   Understanding Climate Risk: Mitigation, Adaptation,
and Resiliency   145
Bjoern Hagen

Chapter 9   Sustainable Agricultural Systems for Cities   173
Rimjhim M. Aggarwal, Carissa Taylor, and Andrew Berardy

Chapter 10   Urban Ecology, Green Networks and Ecological Design   195
Edward Cook

Chapter 11   Managing Water and Its Use: The Central Issue
for Sustaining Human Settlements   213
Ray Quay

Chapter 12   Sustainable Transportation   233
Aaron Golub and Jason Kelley

Chapter 13   Energy in the Sustainable City   255
Martin J. Pasqualetti and Meagan Ehlenz

Chapter 14   The Urban Heat Island Effect and Sustainability Science:
Causes, Impacts, and Solutions   273
Darren Ruddell, Anthony Brazel, Winston Chow, Ariane Middel

Moving Forward

Chapter 15   Can We Teach an Education for Sustainable
Development? Measuring the Fourth "E"   295
Chad P. Frederick and K. David Pijawka

Chapter 16   China's Pursuit: Smart Sustainable Urban Environments   307
Douglas Webster, Feifei Zhang, Jianming Cai

Chapter 17   Sustainability in Indian Country: A Case
of Nation-Building Through
the Development of Adaptive Capacity   333
Judith Dworkin

References   353
Contributor Biographies   385
Index   393

Foreword

Facing the Urban Age with a Passionate Optimism

Tim Beatley, University of Virginia

There is an understandable tendency these days to be pessimistic about the future. To be sure, a planet of 10 Billion or more, with signs of serous climate change, severely constrained availability of water and food, and limited success to date at tackling global poverty, to name a few, are not trends that inspire confidence. As the planet continues its high-speed charge in the direction of cities, it remains to be seen whether this new global era of the urban will usher in new abilities to solve these many challenges, or simply exacerbate them and make them intractable.

While the challenges facing the world today seem daunting, I frequently find myself feeling optimistic and excited about the possibilities, especially as a city planner or perhaps because I am a city planner. I am frequently heard saying that I can't remember a more exciting time to be in the urban planning and design disciplines, as I look back at a quarter-century career.

How cities and urban life will be designed and planned may be the single most important task ahead, and the book you are about to read and the insights, knowledge, and case examples it offers will help to steer the way. Sustainability, we all agree, will need to be the essential goal for everything we do from here on. It is no longer something optional, no longer simply lofty language, but an essential lens and metric against which we judge how well we are doing. You will need the depth of understanding and knowledge the editors and chapter authors so expertly provide here. You will also need to be inspired, hopeful, and optimistic that your work in the future can make a discernible and significant difference.

Why be optimistic? I believe there are many reasons. There are new urban and environmental sensibilities afoot for one. The emergence of the local food movement, for instance, and the fact that many more people seem to want to know where their food comes from, how it was produced, and what impacts its production might have. And the fact that many want to be directly involved in the growing of some or all of that food. Cherishing and celebrating the unique flavors, origins, provenance of food has led to a global slow food movement that has now expanded to embrace slow wood, slow fish, even *slow cities*.[1] Slow is a pejorative word in many cultures and societies, of course, but it is the emphasis on quality of living, savoring experiences, and on deepening relationships between people and the communities and landscapes in which they live that is encouraging.

There are also promising new alliances that will aide in advancing community sustainability. The emergence of a new collaborative spirit between public health and urban planning, and new understanding of the importance of design for

---

[1] http://cittaslowusa.org/

healthy cities and places—which actively foster walking and biking—is encouraging. Physicians, nurses, and public health professionals of all sorts are seeking out partnerships with city planners. And in planning, we understand the powerful new frame that health and well–being can provide and the promising and potent new collaborations with medicine that can help to further elevate the importance of urban design and planning.

We also know much more now than we did about cities and what it will take to create livable, sustainable, healthy places. The last decade has seen an explosion of new research findings and scholarship, providing essential insights about the kinds of urban conditions and design strategies that will be needed to ensure that we make progress on advancing global health and happiness. There is much new research, for instance, on the important role that nature plays in our lives, and that we are likely to be happier, healthy, and lead more meaningful lives when we have daily connections to nature (even inside buildings). This has given rise to movements to return to nature (what we have been calling *biophilic urbanism or biophilic cities*).[2]

With the emergence of citizen science and crowd sourcing through smart phone technology, there are new ways that average citizens are participating and engaging in the life of places they care about. With the world poised to go from 3.9 billion smart phone subscriptions in 2016, to more than 6.8 billion by 2022,[3] the opportunities are tremendous to connect individuals not only to each other but to cultivate new understandings of communities and environments in which they live. It also offers new pathways for activism and cultivating connections with the nature around us. For example, how will the world change with 6.8 billion people using smart phone apps like *i-birds* and *i-trees*?

We have many new and inspiring design and planning ideas, enabling us to re-imagining cities. A decade or two ago, I would have had a hard time imaging buildings, or neighborhoods, that produce all or nearly all of their own energy from renewables, or high-rise apartment towers where sky-forests grow (such as *Bosco Verticale* in Milan). And while there are many project designs and plans that will forever remain at the level of renderings, there have been few times in history where there has been so much creativity and imagination aimed at cities and urban life. And never have we been closer to fully recognizing what livability requires—certainly access to shelter and food, and healthy living and working environments, but increasingly much more—walkable, vibrant community spaces; trees and greenery; community gardens; and abundant nature all around.

Many cities around the world are rising to the fore, setting for themselves ambitious green targets and visions for the future. Several years ago the City of Copenhagen declared its intent to become the world's first Carbon Neutral Capital City. Vancouver, in 2012, declared its intentions to be the Greenest City in the World. For these two cities, these are not empty rhetorical gestures, sound bites, or politics, but serous commitments, backed with detailed actions, plans, and targets.

---

[2] E.g. See Beatley, *Biophilic Cities*, Washington, DC: Island Press, 2011.
[3] Cerwall, P., Jonsson, P., Carson, S., Möller, Barboutov, K., Furuskar, A., Inam, R., Lindberg, P., Ohman, K., Sachs, J., Sveningsson, R., Torsner, J., Wallstedt, K., Gully, V. 2017. *Ericsson mobility report*. June 2017; p. 7. Accessed at https://www.ericsson.com/assets/local/mobility-report/documents/2017/ericsson-mobility-report-june-2017.pdf.

Cities like Copenhagen and Vancouver reflect the important perspective that the production and consumption patterns of global cities, their ecological footprints, extend far beyond their borders and that ethical duties exist to think and act at the global scale. How to harness the economic and political power of global cities on behalf of oceans, for instance (we are on the blue planet after all) is a challenge, yet possible. Working to educate and engage an urban citizenry about distant habitats and organisms (as well as closer ones that may be out of sight, beyond the land's edge) must be a key goal.

A recent article in the journal *Foreign Affairs* declared that "The age of nations is over. The new urban age has begun."[4] Cities will be expected to exert new forms of global leadership moving forward. And here is yet another cause for some optimism. In the absence of national leadership, many cities have already taken the lead in addressing global problems, such as climate change, though obviously much more is needed. National and state governments will continue to be important, of course, but increasingly, it is the city-state that will carry sway and the sphere of influence that will perhaps have the most to do with shaping quality of life and global sustainability. The new era of cities suggests unusual opportunities. Increasingly, cities are negotiating their own city-to-city trade agreements, with a staff and organizational capacity to act unilaterally on the global economic stage. Large cities especially will need to be ready to spearhead global conservation treaties, negotiate and sign them, broker new global agreements regarding resource extraction, and generally exert new leadership on the global stage.

There has been a broadening of the city planning agenda that is also helpful and an increasing recognition that cities must at once consider many different challenges. Along with *sustainability*, there are now a newer suite of complementary words that make up the language of planners and urban designers and managers. A sign of the times that resilience has emerged as a potent word and aspiration in a world where Hurricane Sandy's will become more common, and where cities will become the first responders in periods of drought, heat waves, and disease outbreaks. The news is good here, as foundations like the Rockefeller Foundation, with its 100 Resilient Cities Initiative, have significantly elevated the profile and importance of resilience and tools available for advancing urban resilience. Initially, the Rockefeller Foundation awarded 100 cities funds to hire new Chief Resilience Officers (CRO).[5] And many cities, from New York to Rotterdam to Dhaka, are re-thinking the ways in which they occupy and inhabit the spaces on the edge of the dynamic coastal edge, suggesting new adaptive strategies but more profoundly, new modes of adaptive urban life.

Understanding the ability of coastal cities to move and shift and respond to sea level rise, and planning for this new dynamism, suggests new ways of seeing cities as ecosystems. And there are many other important ways of re-conceptualizing cities with implications for sustainability and livability. Importantly, we can take stock of and comprehensively understand cities in terms of metabolisms, their complex flows of inputs and outputs. Sustainability and resilience will require that we understand these flows—for instance of energy, materials, food, water—and

---

[4] Parag Khanna, "Beyond City Limits," Foreign Policy, 2010.
[5] http://www.100resilientcities.org/

work to modify them, at once shortening supply lines where we can (e.g. growing more food locally and regionally, decentralized renewable energy), reducing the extent of these flows (e.g. designing homes and work spaces that use only tiny amounts of energy, for instance), and working toward a circular metabolism where possible (waste becomes an input to something else).

Partly, this is a bold undertaking to re-imagine the very nature of cities and urban life. We need to plan cities, as architect William McDonough suggests, that act and function more like forests, like natural systems—producing the energy it needs, living off incoming solar energy, providing essential habitat for human and non-human species alike, and recycling and reusing the material outputs (there is no such thing as waste in nature and so should not be in cities either). Understanding cities as places of abundance, bountiful and restorative as much as consumptive and destructive, is an essential mind shift and a more hopeful vision for the future.

Global sustainability will also require us to re-think the nature of economic relationships in an increasingly interconnected planet. No longer can large cities simply assume that massive flows of food, energy, and building materials are derived in benign and non-destructive ways, but there will be a need in the new urban era for city-states to be more engaged and proactive. Cities can and must be leaders in re-tooling these flows to ensure global sustainability, but also global fairness. There remain relatively few examples, unfortunately. New York City's adoption of a policy of transitioning away from the purchasing of tropical hardwoods is an example, but there will be many ways in which urban consumption, especially in the affluent cities of the global North, will need to be accompanied by commitments to aid in the ecological restoration and enhanced health and quality of life of the cities in the Global South. The promise of a new model of global urbanism is that it will harness the power of these metabolic flows and human demands on behalf of both planetary health and human welfare.

I believe, as well, that any compelling model of future urbanism must also take into account the need for daily contact with nature and the natural world. We increasingly recognize that as we advocate and work towards compact, dense, and sustainable cities, these must be *nature-ful* places, places that provide us with a sufficient urban nature diet to be happy, healthy, and lead fully meaningful lives. Nature is not optional but essential. We have grown to understand that we can help make cities more resilient in the face of climate change and the many other shocks we are facing.

Planners and urban designers, then, face serious challenges in creating just, sustainable, resilient, and biophilic cities. But I am optimistic that this can be done, and that life on the urban Earth can be healthful, uplifting, and meaningful as well as respectful of environmental limits.

We can harness the considerable new energy and thinking that exists around cities today (much of which is contained herein), new technologies, research, and new ways of re-imagining the very nature of what a city is. These are exhilarating times in sustainability education and the potential to make a positive difference has never been greater. You will need the skills, knowledge, and inspiration too, and I can think of no better preparation than the articles and chapters in this text. Sustainability, in all its flavors, is an essential pathway to our global future.

About the Editors

K. David Pijawka and Bjoern Hagen

David Pijawka is Professor and Associate Director of the School of Geographical Sciences and Urban Planning at Arizona State University (ASU) and a Senior Sustainability Scientist at the Julie S. Walton *Global Institute of Sustainability*. With other professors he teaches and leads the largest course on sustainability at the university and perhaps globally, *Sustainable Cities*. Professor Pijawka is also affiliated with the Schools of Tranborder Studies, Public Affairs, the Barrett Honors College, and the American Indian Research Institute, among others. He has published extensively on sustainability topics including research on energy related impacts, resiliency in disasters, planning for sustainable neighborhoods, and community well- being. He has also published on topics including transportation risk analysis, socio-economic impacts of nuclear waste storage, risk perceptions of hazards and climate change, neighborhood sustainable design, and environmental impacts on communities especially Environmental Justice studies. Dr. Pijawka has completed numerous planning studies on American Indian communities especially comprehensive plans, visioning and strategic planning. He was the director of a multi - year multi- university Center on the US-Mexico Border Region and developed a recognized indicators initiative on border urban quality of life.

Professor Pijawka has been recognized for his innovative teaching, and in addition to awards, has been invited to lecture all over the world on teaching sustainability. He has given key-note lectures at various universities and international conferences and has received grants from the National Science Foundation, US Department of Transportation, US Environmental Protection Agency, Nuclear Regulatory Commission, Department of Energy, among many others. Dr. Pijawka has been recognized with over 30 awards including an award from the NAACP for his work on Environmental Justice. He has authored over 30 books, monographs, and governmental reports on topics ranging from quality of life indicators to urban sustainability and disaster planning.

Bjoern Hagen holds a M.Sc. in Spatial and Environmental Planning from the University of Kaiserslautern, Germany, and received his Ph.D. in Environmental Design and Planning from Arizona State University. He is currently an Assistant Research Professor in the School of Geographical Sciences and Urban Planning and an Instructor with the School of Sustainability at Arizona State University.

Dr. Hagen has worked on projects focusing on sustainable urban and regional development with the Development Agency of Rhineland Palatinate (Entwicklungsagentur Rheinland-Pfalz e.V) and the redesign of the UNESCO World Heritage site Völklingen Ironworks towards a sustainable future. In recent years, he has conducted research in the areas of climate change mitigation and adaptation, public risk perception, and risk communication. By studying the nature of public perceptions of global climate change in different countries and over time, his research contributes to improving climate change communication efforts to reduce greenhouse gas emissions and increase the adaptive capacity and resiliency of urban environments. Dr. Hagen is also conducting research focusing on the social-cultural dimensions of sustainable and resilient cities. One current study focuses on the efforts of the city of Freiburg, Germany to implement sustainable development patterns by studying two local neighborhood developments.

Acknowledgments

This book would not have been possible without the feedback from students in Arizona State University's *Sustainable Cities* course over the past decade on the topics covered in the class and approaches used in meeting sustainability learning objectives—as well as the commitment of faculty involved in teaching and discussing this broad–based interdisciplinary and introductory course focused on cities. Most of the textbook chapters were written specifically for this book after discussions about content coverage, perspective, structure, cases, and particularly, on meeting educational outcome measures. One challenging question was asked of each author: what are the concepts, cases, and ideas that you are conveying in your chapter that will elucidate the current problems in urban sustainability and how they may be solved over the next decades? Major thanks are given to those very special faculty members who are committed to sustainability education and who went the extra mile to write these chapters. These professors include: Anthony Brazel, Darren Ruddell, Winston Chow, Rimjhim Aggarwal, Ariane Middel, Martin Pasqualetti, Meagan Ehlenz, Charles Redman, Subhrajit Guhathakurta, Aaron Golub, Jason Kelley, Douglas Webster, Jianming Cai, Edward Cook, Judith Dworkin, and Ray Quay.

The book also benefitted from chapters written from recent dissertations by new faculty and specialists making innovative advances in their topics in this volume. These include subject matter on resiliency, sustainable development, history of sustainability, urban environmental policy in China, and sustainability education. Thanks for these important contributions: Andrew Berardy (Arizona State University), Stephen Buchman (University of South Florida), Chad Frederick (University of Louisville), Nelya Rakhimova (University of Dresden, Germany), Carissa Taylor (Arizona State University), Feifei Zhang (Arizona State University), and Craig Thomas (Salem State University).

Thanks to Professor Timothy Beatley, from the University of Virginia, who wrote the book's Foreword and provided material from his own book *Biophilic Cities* (Island Press) as an insert to one of the chapters. Dr. Beatley has had a continuous and close association with the sustainability program at Arizona State University, lecturing in the *Sustainable Cities* course and at the Julie Ann Wrigley Global Institute of Sustainability as well as interacting with ASU's Barrett Honors College students on his Global *Biophilic Cities Initiative* where ASU is a partnership institution.

A large part of working with the authors on the book occurred in Germany at the University of Kaiserslautern with Bjoern Hagen. Together, we carried out a study on the social and well–being sustainability dimensions of Freiburg, Germany that appears in the book. Much thanks to Kaiserslautern for hosting us and to the people in Freiburg who generously gave of their time to support our studies on social sustainability. Thanks to Kevin McHugh, Nabil Kamel, Christopher Boone, Judith Dworkin, Katherine Crewe, Gloria Jeffery, and Kathrin Hab for their insightful reviews. Much thanks to the students and faculty in the Chinese universities and research centers for their high level of engagement and interaction, and especially to Douglas Webster, Jianming Cai, and Feifei Zhang for producing a compelling chapter on the move toward sustainability in Chinese cities.

Significant credit for this book and the course goes to Michael Crow, President of Arizona State University, who had the vision to invest in education programs, transdiciplinary research, and community engagement in sustainability. In the first ten years, the *Sustainable Cities* course alone enrolled around 8,000 students. Over the last two year, the course, offered both on ground and online, has engaged around 1,500 students each year. This targeted course, which compromises much of the Education for Sustainable Development (ESD) work in this text, is only one of myriad sustainability education opportunities offered at ASU.

The *Sustainable Cities* course enables students to learn, many for the first time, the compelling promise of sustainability science and its solutions for cities. Imagine, as it happened, the president of one of the largest universities in the US coming in to the course to lecture to a class of 500 plus freshman on the importance of sustainability and a panel of first-year students posing questions on the lecture. There are few places like this.

I am also grateful to Kendall Hunt for selecting this book for publication and working with us on every facet of its development. Lastly our great thanks to Michele D. Roy for editing this edition and going above and beyond to assist with the design and production of this compelling educational tool for 21st century students.

David Pijawka and Bjoern Hagen
June 2017

Chapter 1

Pathways to a Sustainable Future: A Look Inside

K. David Pijawka and Bjoern Hagen

Scope and Context

This textbook has evolved over the course of a decade, beginning with the introduction of a broad-based interdisciplinary course titled *Sustainable Cities* at Arizona State University (ASU). Its content, range of topics, sustainability concepts, and the nature of the questions asked reflect a continual progression of ideas and initiatives that reflect the changes in sustainability itself. The book grapples with topics, issues, and solutions at the urban scale within a global context. Each topic—like parts of a city—connects to other topics and issues and—like the field of **sustainability**—is complex, challenging, and **interdisciplinary**. Each chapter is presented holistically, exploring the urban connections while covering the individual topic's essential dimensions and issues.

  The guiding idea for this book is interconnectivity. That is, each individual aspect of a topic area has a temporal element, a connection between the past, present, and future. It also has geographical, political, and urban connections. For example, to understand and fully appreciate sustainable transportation, we need to understand land use, goals for carbon reduction, urban design, alternative transit systems, and environmental justice. This interconnectivity applies to how the text can be used as well.

  Instructors can combine chapters as units that complement each other. This will help students understand the connections between the various elements. For instance, Chapters 5 (Urban Sustainable Design), 10 (Urban Ecology, Green Networks, and Ecological Design), and 16 (China's Pursuit: Smart Sustainable Urban Environments) can function as an "urban design" unit. Chapter 5 introduces the idea of urban design and its various concepts. Chapter 10 addresses how urban areas both function as an **ecosystem** and impact the ecosystems that they interact with. Chapter 16 completes the unit by demonstrating how China is incorporating the different design and ecosystem aspects covered in the first two chapters. Likewise, Chapters 7 (Resilience: An Innovative Approach to Social, Environmental, and Economic Uncertainty), 8 (Extreme Climate Conditions: Environmental, Social, and Economic Impacts), 11 (Managing Water and Its Use:

Sustainability Sustainability is about making long-term decisions to balance economic, environmental, and social needs and ensure a high quality of life for present and future generations.

Interdisciplinary The field of sustainability offers complex challenges that require an interdisciplinary approach meaning that experts from different fields, with diverse training and knowledge, and distinctive methodological approaches need to work together.

Ecosystem Ecosystems presents a biological system consisting of interacting organisms and their physical environment.

Adaptive Capacity The capacity or ability of a system to adopt to changes in its environment.

Sustainability Science Emerged in the 21st century as a new academic discipline and focuses on the interrelationships between natural and social systems, and with how those interactions affect the challenge of sustainability.

Resiliency Resiliency refers to the ability of a system to cope with and bounce back from negative impacts.

Urban Heat Island (UHI) A global phenomenon generally seen as being caused by a reduction in latent heat flux and an increase in sensible heat in urban areas as vegetated and evaporating soil surfaces are replaced by relatively high impervious, low albedo paving and building materials.

The Central Issue for Sustaining Human Settlements), and 17 (Sustainability in Indian Country: A Case of Nation-Building Through the Development of Adaptive Capacity) could function as an "issues & solutions" unit. The unit would address climate change and water issues, and discuss resilience and **adaptive capacity** as concepts that could be applied to determine potential solutions.

Our objectives are fourfold. The first is to deliver a broad-based textbook on urban sustainability. Our audiences include students, faculty who will become instructors in this field or are interested in current questions and the foundational core of urban sustainability, and the public who look to see how sustainability can be applied to their communities and to them, personally. Another important audience includes those working in municipal, regional, and state governments—community planners, environmental resources personnel, policymakers—and members of professional organizations seeking information on the practice of **sustainability science** and its application to the urban, natural, and built environment.

To fulfill this objective, chapter authors have been carefully selected. They are scholars in, and educators on, their respective specializations. At the same time, they are well informed about the myriad sustainability issues and trends that cut across cities. They are well versed in the interdisciplinary nature of sustainability, approaches used to address research on their topics, and underlying ethical issues. And, as much as possible within the constraints of a chapter, they connect their specific topic to other urban influences.

The second objective of this book is to understand that "modern sustainability" has had its successes and disappointments. From these have arisen new tools and approaches such as scientific urban predictor models, interdisciplinary methodologies, and policy and planning programs that prepare us to tackle the difficult challenges in restoring and transforming cities and communities into resilient and livable places. This book promises to illustrate how cities are using these new methods and approaches. Especially relevant to this objective are Chapter 4 by Bjoern Hagen and Ariane Middel who look at the programs, policies and initiatives that provide a framework for sustainable cities; Chapter 5 by Lauren Allsopp who lays out various concepts for designing sustainable, human-centric urban areas; Chapter 6 by Subhrajit Guhathakurta who gives attention to urban simulation modeling for sustainability; and Chapter 7 by Stephen Buckman and Nelya Rakhimova who explore the concept of **resiliency**. In addition, Chapter 4 offers an insert on the emerging concept of *Biophilic Cities* by Tim Beatley that examines the role that nature plays in cities—socially, psychologically, and economically. It also looks at the importance of ecological restoration and preservation.

In addition, several chapters investigate the various challenges facing cities. Bjoern Hagen contemplates the broader impacts of global climate change in Chapter 8; Ray Quay looks at water management policies and practices in Chapter 11; Aaron Golub and Jason Kelley probe the past, present and future of transportation in Chapter 12; Martin Pasqualetti and Meagan Ehlenz review energy policies and alternative energy sources in Chapter 13; and Darren Ruddell, Anthony Brazel, Winston Chow, and Ariane Middel define the **urban heat island** (UHI) effect and what has been learned about the mechanisms that drive it in Chapter 14. The authors delve into how these issues impact cities and their inhabitants, offer potential options for mitigating future impacts and adapting to those that are inevitable, and present case studies depicting how cities are successfully meeting the challenges. Chapter 16 by Douglas Webster, Feifei Zhang, and Jianming Cai examines the steps that China is taking to meet the challenges presented by its rapid economic growth and related urbanization as well as the role its national and municipal governments play in sustainable development.

Readers will find several chapters that cover topics not typically considered urban issues but that present crosscutting approaches for urban sustainability as well. For example, Chapter 9 by Rimjhim Aggarwal and Carissa Taylor addresses the concepts and practices of agriculture in the urban context. They take a close look at how agriculture has been traditionally relegated to rural areas and the problems this has created for feeding the world's growing urban population. Cook's Chapter 10 on urban ecological design explores the work of landscape ecologists and how it has taken root in urban planning. Stemming from the ecological sciences, this chapter is indispensable for understanding urban sustainability in terms of the value of **ecosystem services** and their benefits for cities.

The third objective is to provide a model or framework of **sustainable development** that includes health and well-being factors, quality-of-life considerations, cultural restoration and sustainability, and the full range of institutional capacity factors for decision making in the context of **coupled human–ecological systems**. Chapter 17 by Judith Dworkin demonstrates an expanded sustainability framework using a combination of innovative health services, adaptive resiliency, and cultural restoration as it is implemented in the Tohono O'odham Nation.

The last objective is to link the book to its educational value for sustainability learning. Chapter 2 by Craig Thomas provides a unique perspective on the rise of today's *wicked problems,* including **socio-ecological systems (SES)**, international sustainability, and what we can learn from the pioneers of sustainable thinking. Insights into the lessons to be learned from the earliest cities are investigated in Chapter 3 by Redman. In Chapter 15 by Frederick and Pijawka, we argue the case for developing and employing sustainable learning outcomes and competencies. As the principal instructors of ASU's *Sustainable Cities* course, the editors framed the text chapters based on their course experience. Many of this text's authors contributed course material, which has been tested for its educational value in the course.

Educational Value

The growth of sustainability as a field of study has been recent, but it is now burgeoning with new conceptualizations, frameworks, methods, and applications. In the last decade, its vocabulary has expanded to include such topics as livability, resiliency, health and culture, adaptation, anticipatory governance, coupled human–ecological systems, and myriad forms of social-participatory processes seeking acceptable solutions to achieve sustainable cities. How sustainability education is delivered is critically important for the field to advance and the science to be accepted and applied.

The authors of this text are academic experts in sustainability science. They are aware of the growing field of education for sustainable development, including the emergent literature on learning outcomes and competencies. Chapter 16 specifically looks at these sustainable education competencies and how learning

Ecosystem services The combined outcomes of an ecosystem that are beneficial to humans and nature including processes such as the production of oxygen, decomposition of waste and the production of clean water.

Sustainable development The most common definition by the Brundtland Commission defines sustainable development as development that meets the needs of the present, without compromising the ability of future generations to meet their own needs.

Coupled human–ecological systems Humans are dependent on natural systems, such as the water cycle or other nutrient cycles. On the other hand, humans alter these cycles by their effect on the natural environment. This combination of dependence and effects are characteristics of human ecological systems.

Wicked Problems Problems that are incredibly difficult to resolve because they are difficult to define, evolve constantly, and have no final solutions. Instead, any resolution most likely generates further issues. These kinds of problems require a new approach to research and decision making.

Socio-ecological systems (SES) Complex, dynamic systems consisting of natural, biophysical and human cultural and social factors that regularly interact across spatial, temporal and organizational scales.

outcomes can be measured. The book also pushes the boundaries by integrating learning competencies such as interdisciplinarity, systems thinking, anticipatory problem-solving approaches, ethical perspectives, and assessments. Other competencies such as sustainable methodologies involving adaptive governance and citizen engagement, collaborative experience, and strategic planning are included as well. The authors engage the reader with knowledge and substance, and demonstrate how urban solutions are being accomplished. Transdisciplinary thinking and practice is threaded throughout the themes and chapters of the book.

Themes

Each chapter introduces its topic as a sustainability issue, provides a history of problems and their causes, and profiles the problems as manifested in cities. Illustrations of practices and solutions tried by various cities are often shown in case studies with a focus on urban programs, strategies, or policies. The authors were asked to provide more than one case study solution for each sustainability problem when possible. The following section briefly describes the major crosscutting themes and explorations in this text.

Measuring a city's sustainability progress: Developing measures to determine sustainability progress over time has been central to sustainability analysis and policy development. Understanding how well a city is performing has become an important governmental and policy objective. One of the ways of doing this is to track results using indicators and making city comparisons. Indicators enable city governments and the public to identify problems and prioritize them in a policy agenda. The objective of Chapter 4 is to inform us of various programs and plans developed by three of the world's most sustainably advanced cities in the areas of transportation, urban design, citizen engagement, protection and enhancement of natural systems, and housing. It explores three of the principal indicators in use today for cities—the green city, resiliency, and livability. Chapter 6 describes urban modeling systems and how they can be used to simulate various urban futures. It shows how models explore numerous variables, their interactions, and potential impacts. Through these explorations, readers learn the criteria needed to fare well on the indicators and how to connect them to city plans.

The history of sustainability: The question—when did sustainability begin— is an important one to pose. In Chapter 2, Thomas takes a unique approach to answering this question. He does not start with the post-1980s United Nations (UN) initiatives on sustainability discourse and the current debates over climate change, which many consider the beginning of sustainability thinking. Rather, he links pre-1980 historical events to underlying concepts of sustainability thinking, identifying the central and essential intellectual discourses on sustainability and its meanings. Of significance to the present discourse on sustainability is the importance of past events such as past environmental disasters, the Dust Bowl, and the rise of suburbanization. The history of early cities and what they can tell us about sustainable systems are discussed in Chapter 3. Of interest is Redman's insights on instances of societal "collapse" occurring at times when environmental resources were seriously diminished. These chapters help the reader place sustainability within a historical frame that can inform current and future directions of sustainability science and urban development.

Social dimensions of sustainability: The original UN **Brundtland**-based "three-pillar" sustainability framework centered on the articulation of social dimensions. Over the last decade, substantial new approaches have advanced the social aspects of sustainability thinking and practice. Social well-being and resiliency indicator discussions are found in the chapters focused on resiliency and livability where (more often than not) the issues are embedded in ideas about community participation, social and neighborhood stability, social justice policies, leadership, and collaboration. For example, Chapter 4 has an insert on the social dimensions of a neighborhood in Freiburg, Germany, one of the most sustainable cities in the world. Preliminary data from new research by the editors confirm the importance of acting on social sustainability goals from the start. From these discussions, readers gain knowledge on the role of citizen engagement in decision making, the importance of social networking, and building community resiliency programs for success.

Among the social well-being issues covered are environmental justice issues. These are threaded throughout the text as seen in discussions of differential impacts of access to transportation, diverse impacts among social groups and places related to climate action planning, disparities among social groups in relation to "food deserts," and spatial differences in terms of heat island impacts across various social and income groups. While environmental justice concerns fluctuate across the urban landscape, the once hidden patterns of injustice, such as neighborhood access to urban parks, are becoming apparent. The 21st century also continues to harbor urban legacy problems of the past, such as the lack of toxic cleanup in poor areas of cities, the staying power of Brownfield problems, and spatial variances in exposure to industrial toxics, especially among children. Addressing these issues throughout the text enables readers to broaden their thinking on what constitutes environmental justice and its role in sustainable urban areas.

Sustainability challenges are now beginning to address health issues and quality-of-life issues as part the social domain as well. Chapter 17 expands our understanding of the history of environmental justice and the importance of sustainability for American Indian nations as we recognize the dire health crisis facing American Indian peoples, the unacceptable high levels of unemployment, and the purposeful destruction of Indian cultures as part of their history. Using a case study of the Tohono O'odham Nation, this chapter argues for a holistic model of sustainability that expands the three-pillar framework to include an additional and unique set of factors that include effective health services, restoration of lost and traditional culture, and adaptive decision-making organizations with emphasis placed on community resiliency and education.

Brundtland Commission Report The source of the modern paradigm of sustainability. It is significant for demanding ethical issues such as gender and economic equity be part of sustainability's definition. It requires a balance of economic, social, and environmental dimensions.

Sustainability: A Trajectory from International Conferences to Local Actions

Nearly three decades have passed since the 1987 Brundtland Commission's report on sustainable development that provided what has become the prevailing framework and definition of sustainable development. This definition includes maintaining our bio-capacity (security of our ecosystems), meeting society's present needs (alleviation of poverty, food security, social justice), and securing intergenerational equity (resources for future generations) (WCED, 1987). These years can now be condensed into a short but vibrant history ranging from setting global environmental targets and measures to reducing global environmental threats to enhancing social resiliency through efforts by the United Nations, individual countries, and particularly urban areas.

Chapter 2 characterizes and contextualizes this history as part of a larger intellectual history of conservation, ecology, and environmentalism. It also reviews many of the principal initiatives and programs stemming from United Nations sustainability conferences and their targets; alternative programs and measures; and agreements aimed at establishing principles and recommendations for action upon which most countries could agree. These attempts to reach a global consensus on action programs and principles have proved to be important in sustainability's staying power as well as a global influence of the initiatives at the local level.

The United Nations Framework Convention on Climate Change (UNFCCC) is part of the *Rio Convention*, one of three global agreements adopted at the Rio Earth Summit in 1992. It entered into force on March 21, 1994 and now has near-universal membership with 197 countries, or Parties to the Convention, ratifying it. The other Rio Conventions are the *UN Convention on Biological Diversity* and the *Convention to Combat Desertification*. It now also incorporates the *Ramsar Convention on Wetlands*. The next step in the global fight against climate change was the UN Kyoto Protocol (1997).

The agreement sought to reduce carbon emissions by 5.2 percent below the emission levels of 1990 by 2012. Now in 2017, we have failed to meet this goal. In fact, the concentrations of carbon dioxide in the atmosphere rose 20 percent faster between 2000 and 2004 than in the 1990s. With the first Kyoto Protocol ending in 2012, a second-phase 8-year amendment, the Doha Amendment to the Kyoto Protocol, was adopted. It was limited in scope as numerous countries were not committed or obligated to implement the protocol. However, local level commitments to utilize renewable energy sources in 2012 alone did show promise by reducing GHGs by 6 percent. The newest global action, The Paris Agreement, entered into force on November 4, 2016.

The agreement builds on the UNFCCC. As of May 8, 2017, 144 of the 197 Parties to the Convention have ratified the agreement. The goal is to strengthen the global response to climate change threats. It supports the Doha Amendment targets and seeks to strengthen countries' capacity for dealing with climate change impacts. To achieve this, the Paris Agreement is implementing financial supports, a new technology framework, and an enhanced capacity-building framework and emphasizing transparency. This is intended to help developing and the most vulnerable countries meet their objectives (UNFCCC, n.d.). While these global agreements set ambitious goals and attempt to coordinate a unified strategy towards combatting climate change, it is at the city level where we see the most progress.

Urban initiatives in sustainability are leading the way for on-the-ground real-world solutions for mitigating and adapting to climate change impacts. Cities are setting design standards to reduce carbon emissions, implementing alternative transportation policies, and providing renewable energy incentives. For example, many larger cities have engaged in local climate action planning, implemented climate change adaptation plans (e.g., Chicago, Denver, New York, San Francisco), and established green building programs, among other actions. Although global carbon emissions targets could not be reached, there have been numerous targets set for renewable energy generation at the local level. These tend to be much more successful than their global counterparts.

The start of the 21st century and subsequent years represent a noteworthy and foundational shift toward the acceptance and actualization of sustainability thinking, actions, and experimentation in cities. Such efforts will only expand and strengthen as new and robust sustainability frameworks are developed, policies are validated through practice, and sustainability planning becomes increasingly embedded in community culture. Yet, as this book demonstrates, development of city programs and policies are not without their obstacles. We identify the problems in need of sustainable solutions, examine each topical area the urban solutions attempted to resolve at conceptual and practical levels, demonstrate what can be done through comparative analysis, and explore how we can improve and enhance city institutional capacities. This is the rationale for the book's title, *Sustainability for the 21st Century: Pathways, Programs, and Policies.*

Global Challenges in the 21st Century

The Brundtland report informed us that we need to think carefully about how we consume our natural resources and the importance of alleviating poverty through sustainable development programs. *Limits to Growth,* published in 1970 by an international team of researchers at the Massachusetts Institute of Technology, demonstrated how the "earth's interlocking resources – the global system of nature in which we all live – probably cannot support present rates of economic and population growth much beyond the year 2100, if that long, even with advanced technology." The 2004 update to the *Limits to Growth: The 30-Year Update* showed us that the human ecological footprint (EF) had indeed exceeded the Earth's **biocapacity**. Each was a pivotal moment in sustainability.

Biocapacity The capacity of an ecosystem to produce and regenerate useful biological materials and absorb waste material generated by humans.

We are now at another pivotal period, at a crossroads in formulating ideas to take us through the coming decades of the 21st century. We are much more aware of the global threats and issues through international conferences and reports, global monitoring and indicator programs, scientific and social science research advances, and emerging sustainability education programs. Yet, the global problems of climate risks, ecological refugees, environmental injustice, declining biodiversity and marine life, lack of safe drinking water for many parts of the world, and food insecurity remain challenging and difficult to resolve. Despite the lack of worldwide consensus and capacities, however, global action on all aspects of sustainability continues.

Over the last decade, the three-pillar framework produced by Brundtland (environment, economy, and social equity) has expanded in practice to include ideas and programs about urban adaptation to disasters, resiliency, anticipatory governance, coupled human–ecological systems, benefits of ecosystem services in cities, and social well-being and livability, which include issues that fall into health and cultural domains. While the advances in urban sustainability are encouraging, there are continuing and unsettling threats questioning our ability to achieve a sustainable future at the urban as well as the global level. These threats or challenges fall into several categories:

a. Substantial and continuous global population growth well into the 21st century, particularly in the least-developed countries;
b. Very high levels of urbanization for developed countries and major increases in urbanization for poorer countries;
c. Increasing emissions of carbon dioxide and higher atmospheric concentrations of GHGs;
d. An ever-expanding global Ecological Footprint;
e. Food insecurity; and
f. Exacerbation of disasters and catastrophes.

The goal for sustainable development is to prevent, or at the very least, significantly diminish, the ongoing depletion and deterioration of natural capital that leads to "overshooting" the globe's biocapacity. These challenges are deep and complex and will adversely impact cities through the 21st century. The question that we attempt to answer here is what do we need to be aware of as we plunge into the second half of the 21st century while searching for sustainable solutions for the world's cities. Society often does not have the experience, know-how or even the capacity to manage these growing, substantial, and intractable future threats. New ideas, skills, education, and institutional and policy tools will need to be advanced in sustainability science to meet these challenges. This is in line with the warning posited in the introduction of *Limits to Growth: The Thirty-Year Update*. Signed by more than 1600 scientists, the note states:

> Human beings and the natural world are on a collision course. Human activities inflict harsh and often irreversible damage on the environment and on critical resources. If not checked, many of our current practices put at serious risk the future that we wish for human society and the plant and animal kingdoms, and may so alter the living world that it will be unable to sustain life in the manner that we know. Fundamental changes are urgent if we are to avoid the collision our present course will bring about. (Meadows et al., 2004)

Population Growth Rates

The question of the connection between population, resources, and scarcity has been a protracted issue since Malthus, and despite contentious debates, population growth continues to pose an underlying threat to the future. In 2013, the world's population stood at 7.5 billion people according to the United Nations *World Population Prospects*. Projections showed an increase to 8.1 billion by 2025, 9.6 billion in 2050, and up to 10.9 billion by 2100 (United Nations, 2013). A more disquieting population scenario offered in the same report estimates much larger population increases based on higher fertility rates, showing 10.9 billion by 2050 and 16.6 billion people by 2100. A worst-case scenario places the earth's population at just under 30 billion by 2100.

These population estimates greatly exceed the earth's carrying capacity, which is estimated at between 6 to 10 billion people. The problem is further complicated by the fact that much of the population growth will take place in the developing world, especially in least-developed countries such as Africa, Indonesia, and India (United Nations, 2013). The United Nations population report suggests that at least 8 of the present "least-developed" countries in the world will be among the 20 most populated countries by 2100. These least-developed countries are also those with existing resource shortages and deteriorating environmental conditions as well as those prone to conflict.

Based on population growth estimations, these areas will confront more serious resources declines and challenges in adequately providing societal needs in employment, education, health services, and basic infrastructure. These populations will also be highly vulnerable in terms of coping with adverse health conditions. For example, several West African countries confronted a rapidly spreading **Ebola virus** epidemic in 2014. The US Centers for Disease Control and Prevention estimate that over 11,300 people died between the start of the outbreak and January 2016 when the last of the countries, Liberia, was declared virus free (42 days without incident). The World Health Organization determined that "There have been more cases and deaths in this outbreak than all others combined" with the most severely affected countries being Guinea, Liberia and Sierra Leone primarily due to degraded health systems and infrastructure from prolonged conflict and instability (WHO, 2016). Despite the threat of population growth relative to environmental resources, there is hope.

> **Ebola Virus** Ebola Virus, or Ebola haemorrhagic fever, is a severe, often fatal virus transmitted to people from wild animals that then spreads from human to human via direct contact with the blood, secretions, organs or other bodily fluids of infected people, and with surfaces and materials contaminated with these fluids.

One of the most populated countries on earth, China, has lessons for other countries undergoing rapid population growth over the next few decades. As Chapter 16 tells us, over 500,000 people die prematurely each year in China because of air pollution. Some attribute the rise in stroke, ischemic heart disease and lung cancer, from 800,000 cases in 2004 to over 1.2 million cases in 2012, to **$PM_{2.5}$** impacts, much of which is generated by coal-burning plants and automobiles (Liu, 2017). To make Chinese cities more sustainable, major societal shifts are underway.

> **$PM_{2.5}$** Particulate Matter - Fine particulate manner in the form of tiny particles in the air that can be a concern to human health and reduces visibility.

China is poised to transform its cities under the 13th National Plan currently being developed. Among the changes being discussed are significant reductions of coal burning to meet energy needs, major mass transit investment and construction, redistribution of industry to lessen the high rate of rural-to-urban migration, and implementation of policies to regulate ecosystem services loss. Chapter 16 explores China's initiatives, including a case study on one of China's model sustainable cities, Hangzhou, as it provides important lessons for similar urban areas.

Urbanization Issues

Urbanization is the process by which cities grow; how fast they grow is known as the urbanization rate. In the decades to come, the projected population growth will combine with rapid urbanization to further impact the most vulnerable countries resulting in pronounced urban vulnerabilities. The largest cities in poorer countries will be the hardest hit as they lack basic infrastructure, viable governance, and possess markedly low levels of community resiliency (Chapter 2). Like the Haiti

> **Urbanization** The process of more and more people moving from rural to urban areas, result in cities are growing in area and population.

disaster several years ago and the Ebola outbreak in West Africa, the levels of system resiliency (including institutional capacities to respond to and tackle the immediate problem and build long-term solutions) are just not present.

High rates of urbanization result from natural population growth, rural-to-urban migration, and country-to-country migration. This rapid growth is often unplanned and unmanaged, with poor infrastructure, underemployment, and mismatched development patterns. In 2008, over 50 percent of the global population resided in urbanized areas of the world for the first time in history. This reflects a rapid climb from 1950, when cities represented only 30 percent of the world's population. Currently, it is at 54.5 percent according to the 2016 United Nations report, *The World's Cities in 2016* (United Nations, 2016). If the estimates of the United Nations are realized, 66 percent of the world will reside in cities by 2100, and many of these cities will be larger than any cities in history. Given our practices today, these cities will likely be highly fragmented and infrastructure deficient with low levels of adaptation and resiliency.

Compounding these problems is the likelihood that such areas will not have the capacity to absorb and respond to (or even prepare for) environmental shocks, such as serious drought and water shortages for agricultural production. Even those countries with lower urbanization rates, like those in Africa, are changing very quickly. By 2100, the United Nations forecasts an urbanization rate of over 55 percent for this region. In these poverty-stricken countries, sustainable development must be directed at reducing vulnerability to avoid what Jared Diamond calls "collapse" (Diamond, 2005). Implementing resilient urban systems at this stage will not be optional. Thus, there is a clear need to find viable solutions over the next decades for urban regions that do not have adequate infrastructure and resources in education, health services, employment, water security, and food systems to meet the needs of their growing populations.

Another major global impact from rapid growth is the structural changes beginning to be observed. For instance, the United Nations estimates that 2.5 billion people will be added to urban populations between now and the next three and one-half decades (United Nations, 2014). This would not be as daunting if urbanization was distributed equally, but it is not. In 1990, around the time of the Brundtland report, there were 10 **megacities**—cities with over 10 million people. In 2016, there were 31. By 2030, we likely will have 41 megacities with most of them growing well beyond the 10 million population mark. Added to the 41 megacities, the United Nations estimates that there will be more than 60 other cities with populations between 5 and 10 million residents during this same period (United Nations, 2016).

Megacities Cities or metropolitan areas with more than 10 million inhabitants.

From a sustainability perspective, this may not be a crisis if all megacities were like Tokyo. It has around 38 million residents, yet is considered a highly resilient city characterized by mass transit and high-density residential districts coupled with consistent high levels of employment. However, most of the world's megacities are located in developing countries without even the basic infrastructure to support their populations. In addition, many megacities are situated along coastlines and may be impacted by climate-related disasters such as rising sea levels and flooding over the next few decades. Adaptation policies to reduce the impacts of climate change will be a necessity for these areas.

These urban changes also impact rural areas. There are signs now indicating that there is going to be growing future food insecurity as rural areas are being depleted of natural resources and soil fertility. Innovative sustainability solutions on how to feed the growing urban populations while protecting the world's rural and agricultural areas are needed. A principal objective of sustainability will involve determining how to reduce these massive vulnerabilities. As urban populations continue to swell and more unprecedented challenges arise, the time to implement sustainability planning and intervention becomes more urgent.

This book informs us that sustainability solutions for highly vulnerable urban regions will require new long-term solutions and the establishment of new, responsive and anticipatory governance to build resilient human–environmental and adaptable infrastructure systems. Developing new urban governance systems over the next few decades will be particularly important to slow the adverse conditions of the most vulnerable regions. Several chapters consider the topic of resiliency in the context of sustainable development. Chapter 7 provides a comprehensive picture of the concepts and practices needed in advancing community resiliency. Augmenting this information is Chapter 17, which addresses the importance of applying resiliency ideas to the health services and cultural restoration of American Indian nations. Indian Nations have been neglected in the past, their lands exploited for their natural resources, and in most cases stripped of their assets, including their indigenous cultures. We posit that sustainability provides a foundation for economic and cultural restoration and advancement.

Greenhouse Gases and Ecological Footprints (EF)

Two additional global issues stemming directly from population growth are **greenhouse gas** (GHG) emissions and the expanding global **Ecological Footprint** (EF). For these two challenges, the beginning of the 21st century is a continuation of past trends as population numbers continue to increase, natural resources diminish beyond their sustainable yield, and regional conflicts disrupt communities and agricultural development. Cities are the largest emitters of GHGs (around 75 percent of all GHG emissions) through automobile fuel consumption, construction processes, and energy generation. They are also the areas most impacted by climate-based natural disasters that result from these issues. Chapter 8 presents an in-depth discussion of GHG emissions.

The Global Footprint Network (GFN) defined EF as a measure of "human appropriation of ecosystem products and services in terms of bioproductive land and sea area needed to supply products" (Global Footprint Network, 2012). In other words, EF is a measure of environmental resource consumption translated into the amount of land required to support that level of consumption. It is often used to compare countries or even cities,' use of natural capital/resources on a per capita basis and is typically conceptualized in **Global Hectares**.

EF has two analytical dimensions. The first is the demand for human consumption measured in land required for food crops, waste disposal, and absorption of carbon dioxide emissions. The second is the regenerative capacity to meet these demands. The availability of regenerative land to meet each demand is termed *biological capacity or biocapacity*. Researchers, policy makers, and resource managers also

Greenhouse Gas (GHG) Gases in the atmosphere that absorb and re-emit solar radiation back to the earth's surface. The four most common greenhouse gases in the atmosphere CO_2, CH_4, halocarbons, nitrous oxide (N_2O).

Ecological Footprint The amount of biologically productive land and water area required to produce all the resources an individual, population, or activity consumes as well as the land area needed to absorb the waste they generate, with consideration for prevailing technology and resource management practices.

Global Hectares A biologically productive hectare, figured on the world average of biological productivity for a given year, that accounts for both the biocapacity of a hectare and the demands on that hectare. A hectare is 1/100th of a square kilometer, 10,000 sq. m., or 2.471 acres.

examine EF to determine trends and rates at which a country's resource base is being used. They track biocapacity against EF to determine if a country or city has an ecological reserve or deficit. For some countries, especially industrialized and developed countries with large populations, the size of their EF and the resulting ecological deficit often extends beyond the boundaries of the country.

In 2013, worldwide EF consumption was estimated to be 2.87 global hectares per capita (GHC) with a biocapacity measure of 1.71 GHC (GFN, 2017). This results in a global deficit of 1.16 GHC or the level of consumption exceeded the earth's biocapacity by 1.16 GHC. In the same year, the US' EF was observed at 8.59 GHC while its biocapacity was 3.78, resulting in a deficit of 4.81 GHC. The country's EF was the sixth highest EF in the world. It trails Luxembourg at 13.09, Qatar at 12.57, Australia at 8.8, Trinidad/Tobago at 8.76, and Canada at 8.76. In terms of deficits, the US is fourth among this group. Luxembourg's deficit is the highest at 11.51, Qatar's is 11.36, and Trinidad/Tobago's is 8.76. Australia and Canada both have biocapacity reserves of 6.87 and 7.42, respectively. Despite these and similar reserves, the global deficit of 1.16 GHC is a concern as it reflects the fact that humans as a group are using 1.16 hectares of resources more than exists on the planet.

GFN also shows the global ecological use in terms of number of earths. In 1961, global ecological resource generation stood at 0.8 earths, meaning we had a biocapacity reserve of 0.2. Basically, we were using resources at a rate that took the Earth 8 months to regenerate. We were "living within our means," albeit barely. In the early 1970s, the deficit reached 1 earth. We were at a tipping point, using ecological resources at a rate that took the Earth a whole year to regenerate. Starting around 1990 and lasting until around 2000, ecological resource consumption stood at 1.2 earths. We had "overshot" the Earth's **carrying capacity**. As of 2017, the **ecological overshoot** reached 1.6 earths (GFN, 2017). Continued EF growth and widening ecological deficit represents a serious and fundamental sustainability challenge for cities and countries, alike.

There are just a few projections of biocapacity deficits. One, showing a deficit of 3.0 earths by 2050, is calculated without planned interventions or "business-as-usual." Those developing these projections acknowledge that we can promote sustainability through interventions and lower the chance of meeting this prediction. For example, we may be able to maintain our current rate of consumption or perhaps even reduce it if significant shifts to renewable energy are made. This is because carbon accounts for ~60% of humanity's EF and is its fastest growing component (GFN, 2017a). Reducing our dependence on fossil fuels will go a long way to moving us towards sustainable consumption. This book shows that pathways to sustainability exist, knowledge and policies are available, and plans can be implemented to transform our cities and, ultimately our countries, toward significantly improved levels of sustainability.

Food Insecurity

With respect to global food insecurity, estimates show a decline of persons living with continuous hunger and undernourishment by 17 percent since 1990. This is certainly a welcome improvement, and it signifies that something

Carrying Capacity The maximum sustainable population an area will support before undergoing environmental deterioration.

Ecological Overshoot The world's ecological deficit caused by human overuse of ecosystem products and services; the process of exceeding the Earth's carrying capacity, leading to environmental deterioration and ecosystems' inability to provide adequate services.

is working in our sustainability efforts. However, it remains a global concern that one in every nine people on earth suffers from chronic malnutrition and hunger (FAO, 2015). Furthermore, malnutrition is distributed unevenly with the largest concentrations located in developing countries, such as Africa and Asia, and areas experiencing continual conflict and political upheaval. A recent international report for these countries estimates an average undernourishment rate of 12.9 percent (FAO, 2015).

There are three primary food insecurity issues facing sustainable development. First, although undernourishment rates are declining, the total number of undernourished people remains monumental at around 794.6 million persons. Second most of the universally hungry are living in dire conditions in underdeveloped regions. To address this component, solutions must address underdevelopment, social inequity, and health discrepancies as well as hunger alleviation. Third, the rate of progress to alleviate undernourishment is relatively slow and has not yet met any of the international targets such as those set by the Millennium Development Goals. These issues are not simply country issues however.

There are glaring concerns in the US and other countries over urban food availability and quality, particularly in central urban areas. Chapter 9 addresses food insecurities and the conditions that bring them about, giving specific attention to urban "food deserts" that result from a lack of fresh, healthy, and affordable food options. Food deserts adversely and disproportionately impact inner city neighborhoods. It is surprising and, rather disturbing, that one of the richest countries in the world is facing a new form of environmental injustice well into the 21st century. Like solutions for underdeveloped regions, solutions for urban areas must take into consideration social justice, health, social well-being, and legacy problems (such as toxics and pollution exposure) as well as food security. Innovative sustainability models need to be advanced that combine community resiliency in terms of well-being, support, leadership, civic engagement, and sustainable neighborhood design and rebuilding.

Disasters and Extreme Events

The first 15 years of the 21st century have left sustainability scholars in a quandary regarding the extreme events or disasters being experienced around the world. The experiences during and after Hurricanes Katrina and Sandy in the US show that we need to improve our knowledge about coastal cities and how to protect them. The 2010 Haiti earthquake that killed between 220,000 and 316,00 people and displaced 1.5 million people is still causing substantial problems including over 55,000 still displaced as of September 2016—almost seven years after the disaster (CNN, 2016). The BP Gulf Coast Oil Spill off the US coast with its possible decades-long ecosystem impacts leads to sobering questions about energy policy, regulation of offshore oil drilling, response capacities, long-term hazard mitigation, and economic restoration. The 9.0 magnitude earthquake and subsequent tsunami and nuclear disaster in Fukushima, Japan in 2011 killed 20,000 people and displaced hundreds of thousands more. Five years after these events, tens of thousands of people, mostly elderly, were still living in temporary housing. Estimates of long-term contamination include 22 million square meters of radioactive waste and more than 800,000 tons of highly reactive water with 400 tons added daily with no plan for disposal. Cleanup is expected to take more than 40 years (CNN, 2016a).

As unfortunate and devastating as these and similar events can be, there is much to learn from understanding their full human–environmental causes and subsequent effectiveness of remediation efforts. For instance, what can these events tell us about improving our institutional capacities to prevent, respond to, restore, and rebuild damaged or destroyed communities? What can they tell us about societal responses to uncertainties? We also need to face the particularly problematic challenge of learning how we can perform better in collaborating with residents in rebuilding communities. Recovery processes to date have been focused primarily on rebuilding infrastructure and housing. Experiences in New Orleans (Hurricane Katrina) and New York (Hurricane Sandy) demonstrate the need to take into consideration community members well-being, participation, engagement, and social justice in remediation efforts. In Chapter 11 by Ray Quay, **"anticipatory governance"** institutional processes for mitigating adverse future shocks in the areas of potential water shortages provide an excellent example of how residents can be engaged.

Much of the current disaster research stems from societal difficulties in long-term rebuilding and the need to reduce human risk and vulnerability. The emergence of resiliency as a principal area within sustainability science has also spurred extensive research. Within sustainability science, disasters are often defined by the interactions between human settlements and environmental systems that either act as catalysts or are disrupted by disasters. **Systems-thinking** enables us to look at these problems through the lens of human–ecological coupled systems. It recognizes that "wicked" problems, such as natural disasters and subsequent events as well as climate-related extreme events, are characterized by high levels of complexity and uncertainty.

To address these complexities and uncertainties, emerging frameworks for sustainability solutions for disasters consist of both **mitigation** and **adaptation** policies. Disaster mitigation seeks to prevent or reduce the impacts and risks of hazards before an emergency or disaster occurs. These efforts are often in response to previous disasters. Adaptation typically takes place after a disaster and involves changes that enable human-ecological systems to function within the new parameters created by the disaster.

In terms of climate-related disasters, mitigation typically applies to the source of the global warming problem, that is, reducing carbon dioxide (CO_2) emissions through alternative energy generation, especially renewables. This is in large part because of the role that CO_2 and other greenhouse gas (GHG) emissions play in changing the Earth's climate. GHGs greatly impact the Earth's energy balance. They act as a **climate forcing** mechanism, thereby impacting the Earth's warming and cooling systems. GHG mitigation targets set in the first Kyoto Protocol (1997) involved the reduction of carbon dioxide emissions by 5.2 percent less than the prevailing 1990 levels. This reduction was never achieved. The solution, as Martin Pasqualetti argues in Chapter 13, is to expand renewable energy generation (solar, geothermal, hydropower, and wind). The US Energy Information Administration (EIA) estimates that 14.9% of US energy in 2017 is generated

Anticipatory Governance The development and analysis of a range of possible scenarios, rather than a forecast or selection of a single scenario. It presents a new model of decision making while dealing with high uncertainties.

Systems-thinking An approach to problem solving that views all parts of a system as a whole and considers how they influence and impact each other.
Mitigation policies Policies that focus on reducing the causes and risks a of potential negative events (i.e. climate change, economic downturns, or health epidemics).

Adaption policies Policies focus on reducing the impacts of negative events that cannot be prevented anymore.

Climate forcing Climate forcing's are environmental processes that influence Earth's climate. They may be external to Earth and its atmosphere or may be internal to the planet and its atmosphere.

from renewables (EIA, 2017). The International Energy Agency (IEA) found that ~22% of worldwide energy generation is from renewables. This is clearly an important step, but not enough for the dramatic GHG reduction needed reduce concentrations to nonthreatening levels.

The second paradigm, adaptation, is required in combination with mitigation. GHGs build and concentrate in the atmosphere long after they are released. This means that the Earth will continue to suffer adverse climate change impacts even if we were to stop generating GHGs immediately. Adaptation assumes that human actions can reduce climate-based impacts through: (a) significant investments in infrastructure to prevent extreme damages and reduce the levels of extent and intensity of impacts; (b) land-use planning policies to reduce vulnerabilities, including special zoning restrictions, open space regulations, flood insurance, and (in some places) wetlands protection; and (c) enhancement of community resiliency measures in terms of decision-making capacities. Chapter 8 addresses both mitigation and adaptation response and policies to climate change. Adaptation responses to reduce risk among various kinds of hazards are discussed also in Chapters 7, 11, and 14.

Two of the disasters identified above, the 2010 Haiti earthquake and 2011 Fukushima nuclear disaster, can provide insight about sustainability in the coming decades. The disaster in Haiti illustrates how the loss of environmental resources, high levels of poverty, and poor health services combined with low community support networks can create significant recovery problems. The Fukushima disaster teaches us about the implications of living with risk and how decisions are often made without being fully aware of the potential ramifications of human–technology–natural systems and human safety. While there is a current interest in replacing fossil fuel energy with nuclear energy to decrease atmospheric GHG, we need to remember the lessons learned from the Chernobyl, Three Mile Island, and now Fukushima nuclear catastrophes. These cases support the need for sustainability to be positioned with sufficient methods, knowledge, and policies to handle these seemingly intractable trade-offs between human safety and development.

Final Thought

In total, this book focuses on sustainability's past, present, and future, using historical review and context. It documents current urban conditions and solutions, and maps out issues, concerns, and possible new approaches that cities can take to ensure a sustainable future. Finally, it urges a broader, more inclusive definition of sustainability and calls for an expanded role for sustainability education.

Chapter 2

An Intellectual History of Sustainability

Craig Thomas

Introduction

According to sustainability scholar David Orr (2002), the term sustainability has become "the keystone of global dialogue about the human future" (p. 145). This sustainability dialogue has contributed to what might be thought of as a new paradigm or "worldview" that looks toward the future with respect to actions in the present. It is well known by its three-pillar framework of "environment, economy, and society." This framework is most often applied to economic development that takes into account both human needs and environmental constraints.

The term "sustainable development" was introduced in international development discourses of the early 1970s, such as during the United Nations Conference on the Human Environment (Balboa, 1973), *Limits to Growth* (Meadows, Randers, Meadows & Behrens, 1972) published the same year, and the United Nations World Commission on Environment and Development's *Brundtland Report* also known as *Our Common Future* (WCED, 1987). By the United Nation's Earth Summit in Rio, often referred to as the Rio Earth Summit, sustainable development had become the catch-all phrase, as it is found in 12 of the Summit's 27 principles (Summit, 1992). Over the past two decades, the concept has enjoyed a surge in public and academic visibility, encompassing wide ranging applicability, such as the valuing of ecosystem services, food security, urban design, building materials and construction, and energy systems.

When discussing the origins of sustainability, most people refer to the three pillars and the *Brundtland Report* as the source. While the *Brundtland Report* was transformational, the concept dates back to the 17th century when Hans Carl Von Carlowitz coined the term "Nachhaltigkeit," or sustainability, in his 1713 work *Silvicultura Oeconomica*. American foundations of the concept reach back to its first use in American forestry practices at the turn of the twentieth century.

Defined simply, sustainability is *the capacity to endure* and has been a main concern of society as far back as we are aware. Sustainability discussions have

Sustainable development Economic growth and development that does not deplete natural resources nor damage ecosystems. The *Brundtland Report*'s (1987) definition is commonly used as a definition of sustainability: "Sustainable development is development that meets the needs of the present, without compromising the ability of future generations to meet their own needs" (WCED, 1987).

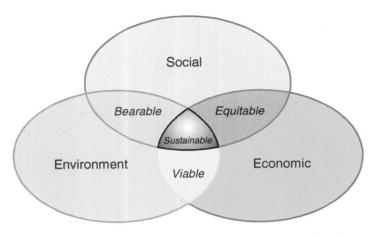

The three pillars were introduced with the 1987 *Brundtland Report* but since have been expanded, interpreted, and reinterpreted in attempts to define *sustainability*

guided societies whenever signs of dwindling resources and collapse make themselves apparent. Thus, sustainabilty can also be defined as the opposite of collapse. Archeologists and geographers like Jared Diamond have explored how soil loss, changing climates, population growth, and the homogenization of natural systems have been a concern for centuries. In *Collapse: How Societies Choose to Succeed or Fail* (2006), Diamond uses the US State of Montana as a case study to explore the "problems of toxic wastes, forests, soil, water (and sometimes air), climate changes, biodiversity losses, and introduced pests" that are reflective of the "types of problems that have undermined pre-industrial societies in the past" (p. 35).

This chapter will begin by examining the nature of socio-ecological problems and the US "Dust Bowl" of the 1930s. It will then look at historical and current influences on sustainability thinking, including major thinkers of the "environmental movement" in the US. Next, it examines the sustainability tradition as framed by the United Nations (UN) and other international conferences designed to address sustainability problems and define sustainability. Finally, it offers an alternative tradition posed by naturalists and ecologists to enhance our understanding of the UN tradition and shape a contemporary understanding of sustainability.

Dust Bowl One of America's first major socio-ecological disasters caused by drought and over-farming of the Central Plains.

Socio-Ecological System (SES) Problems

Land degradation, loss of biodiversity, unprecedented population growth and consumption, and twenty-first century global traumas/crises such as the Fukushima disaster are just some of the highly complex problems vexing our institutional abilities to anticipate and respond effectively to such problems—especially over the long-term. These global challenges call for a change in the worldview that accounts for the magnitude of myriad complex and integrated **socio-ecological systems** (SES, or "coupled system"). Systems thinkers advocate that what is often divided into 'natural' and human systems be considered a single, complex SES (Redman, Grove & Kuby, 2002, p. 161).

These complex SES problems are often characterized as *wicked* problems—or problems of such complexity that they require the integration of knowledge from many disciplines coupled with experiential and empirical knowledge from various fields (Kates & Parris, 2003). Urban planners and science of design experts Horst Rittel (1930-1990) and Melvin Webber (1920-2006) first used the term *wicked problems* in 1973 to define such seemingly confounding dilemmas in a general theory of planning and to denote the need for a new and significantly broader mode of thinking for their resolution. Examples of wicked problems include

Socio-ecological System A coupled, and inextricably linked, human-natural system with complex and emerging properties that defies understanding using one or more of the traditional academic disciplines such as *economics* or *environmental management*.

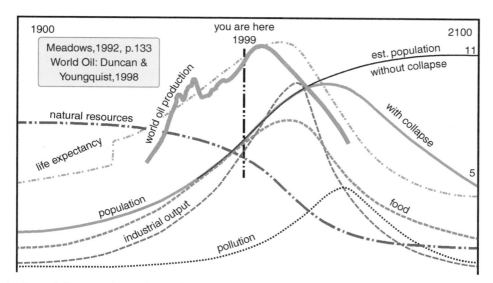

In *Beyond the Limits* (1992) Meadows, Meadows & Randers illustrate intertwined and overlapping problems that require an understanding of SES—coupled human-natural systems—to solve.

global warming and cooling, healthcare, the AIDS epidemic, international drug trafficking, nuclear weapons, nuclear energy and waste, and social injustice. As if wicked problems were not sufficiently daunting, many SES problems have been characterized as *super-wicked* given that the time available to solve them is running out at the global level. Super wicked problems worsen with each passing day of inaction. Solutions require information about the future as well as the present and past. Thus, they have no viable testing ground, meaning that they cannot be solved by trial and error. Every action counts significantly as each has good or bad consequences (Rittel & Webber, 1974).

Coupled socio-ecological systems and related terms such as *resiliency* and *adaptive management* (covered later in this book) originated at least as far back as the 1960s and 1970s, particularly in the work of Canadian ecologist and systems-thinker C. S. Holling, and in findings by international and interdisciplinary groups such as the authors of ***Limits to Growth*** (Meadows, et al., 1972). Holling's work also linked natural and social sciences in new fields like *urban ecology*. In recent decades, American researchers in the field of sustainability science such as Robert Kates and William Clark (1999) have expanded this work.

In addition, important research bodies, like the Proceedings of the National Academy of Sciences (PNAS), have recognized SES as inextricably intertwined with *transdisciplinary* Sustainability Science.

SES constitute the important *wicked problems,* such as the decline of land health in the form of ecosystem diminution, agricultural erosion and depletion, and annual biodiversity losses at a rate estimated at 100 to 10,000 times past extinction rates (Wilson, 2002), that Sustainability Science addresses. In developing countries facing these immense environmental problems coupled with staggering increases in population and consumption, ecologists and activists like Vandana

The Limits to Growth (1972) A book that suggest we would soon reach the Earth's carrying capacity, and launching the term *sustainable development* into the modern sustainability discourse.

Shiva place SES problems center stage. Along with climate change, these are among sustainability's most urgent problems.

Solving these complex problems requires integrating theory and practice, as well as understanding the linkage of minute, individual actions in the present to collective consequences that may manifest many generations in the future. What the great majority of scientists and citizens alike failed to appreciate at the time of the Industrial Revolution (1760–1840) was that while it "freed people from that land," it was an *evolutionary* action. Not only did it free people from working on the land, it affected a paradigm-shift. The relationship that had existed between human beings and the land for thousands of years changed. Presently, many sustainability scholars speak of that time in history in terms of *decoupling* agricultural and natural systems. They call for a *recoupling* of these systems based on traditional practices that align society with its natural and local environmental limits.

The next section examines the nature of SES through the historic lens of the "Dust Bowl" as it took place in the 1930s in the United States and is reflective of a significant SES collapse resulting from this decoupling.

America's First Major SES Problem: the "Dust Bowl"

To illuminate the now common problems of salinization and desertification taking place all over the world, this section begins with a striking example of the devastation that these issues can cause. The 1930s Dust Bowl still resonates as one of the first and largest SES problem in the history of the US. Massive clouds of dirt, silt, and important minerals laid down over thousands of years of geological history filled the sky for hundreds of miles, blanketing communities across 11 states.

The worst of these "black blizzards" occurred on May 9, 1934. As the devastating cloud moved east, it gathered more and more soil. Historians estimate that over 350 million tons of dirt formed a wall approximately 10,000 feet high. This wall would eventually make its way across the country to the Atlantic Ocean, coating everything in its path. Ten million tons of soil was dumped on Chicago alone (Worster, 1979). On May 11th, it hit New York. It had traveled over 2000 miles away in just two days. See more on the Dust Bowl at http://www.history.com/topics/dust-bowl.

As environmental historian Donald Worster (1979) put it, the Dust Bowl signified "the final destruction of the old Jeffersonian ideal of agrarian harmony with nature" (p. 45). For most of US history, American farmers held a progressive doctrine of reaping ever higher and higher yields. Now "Nature" was striking back, eradicating agricultural communites across the Great Plains and creating over 2.5 million *ecological refugees.*

"Houses were shut tight, and cloth wedged around doors and windows, but the dust came in so thinly that it could not be seen in the air, and it settled like pollen on the chairs and tables, on the dishes," writes the great American novelist John Steinbeck illustrating this event in *Grapes of Wrath* (1939). The novel opens by portraying farmers staring numbly at a wasteland of dying crops. As the farming families looked fearfully to their leaders, believing they and the community could remain "whole," there was still hope in their own work ethic. Indeed, the federal government would come to the rescue. But, day after day, as thousands of tons of topsoil flew into the air, and huge gusts of wind carried their livelihoods away, these families were forced to rethink their way of life.

What caused this SES problem? It was not just the natural strip winds and tornados endemic to the Western prairies and Great Plains of America; nor was it the traditional natural climatic cycles of 50- to 100-year droughts and floods. For decades prior to this event, farming and ranching in this area stripped the land of its natural vegetation and the grasslands that kept the land alive. When droughts came, the wind easily lifted the barren dead soil, carrying it across the country.

Historians generally agree that early Americans embraced a view that misunderstood the ecology of the Great Plains. They assumed that the agricultural principles of the East and its temperate climates could be applied to the region by the mere addition of water. In 1909, Congress passed the Enlarged Homestead Act that granted settlers 320 acres of "dryland farming" and brought thousands of "sod-busters" to the area from 1910–1930. These settlers would irrevocably alter the region's biotic equilibrium. Hundreds of thousands of acres were stripped of their natural vegetation to make room for cash crops, such as wheat and corn. Vast irrigation networks eventually diverted the flows of the major rivers to support this agriculture. Despite these incentives, only a few independent farmers were successful on these dryland farms (Worster, 1979).

Courtesy Library of Congress

Day after day, year after year, for almost a decade in many places of the Great Plains, the "Dust Bowl" brought tragedy to many Americans

While scientists and researchers made great advancements at the time in understanding how ecosystems function, in practice, farmers did not heed warnings to rotate crops and let significant-sized patches lie fallow annually. In other words scientific advances in the ecological sciences did not manifest in regional agricultural policies. In addition, ranchers given 4,000 acre parcels in the original Homestead Act (1862) transformed the land through enormous cattle drives that destroyed many endemic species of plants and animals and eroded the soil. These migrations especially impacted river beds whose flow was already greatly diminished due to expanding settlements and agricultural practices.

Further, both farmers and ranchers viewed the land as subservient to their purposes. This **anthropogenic** viewpoint and corresponding practices ultimately deprived the land of much needed minerals like nitrogen and phosphorus, and simultaneously robbed it of the many benefits of its natural cycles. Mono-cropping, which became the primary farming method, was designed for maximum yields and used ubiquitously. For instance, while it was well-known that legumes and alfalfa could be rotated with wheat to build a more sustainable humus soil, "progressive farming" drove machinery intensive farming practices to increase national production from 112 to 375 million

Anthropogenic of, relating to, or resulting from the influence of human beings on nature (Merriam-Webster)

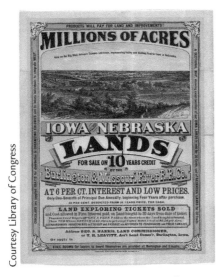

The government encouraged mass migrations west. Settlers often did not have the land and water resources to sustain a farm through the generational and seasonal climate changes

bushels in the three years from 1929–1932 alone (Worster, 1979). With mono-agriculture, new strains of wheat, lack of crop rotation, etc., an agricultural culture emerged that would, in time, wholly lose its resiliency. Nutrients set down over millions of years were quickly depleted by these intensive, widespread farming practices, causing vast quantities of dirt to be loose enough for seasonal winds to easily lift off the ground.

At the turn of the twentieth century, scientific advancements in agricultural machinery, the development of early pesticides and herbicides, and new petro-intensive practices like mass chemical fertilization also contributed to the collapse of the agricultural region. The fragility of biotic communities collided with conservation laws geared toward agricultural and ranching efficiency and the steady increase in the production of yields. Warnings against these methods went unheaded.

It would serve us well to remember, as population ecologist Lester Brown (2011) put it, "the archeological record indicates that civilizational collapse does not come suddenly out of the blue. . . . Economic and social collapse was almost always preceded by a period of environmental decline" (p. 9). While the Dust Bowl was an important lesson that helped the US change its agricultural practices, this same SES problem is now occurring in places like Africa and China. Global drought, flooding, massive human migrations, premature deaths due to poor air quality, and agricultural yields far below those expected due to changing climates are central to today's sustainability discussions. These kinds of problems have plagued humankind for millenia. Thus, important historical analysis can benefit us in understanding the key tenets of sustainability as we will discuss next (Diamond, 2006).

"Maximum Sustainable Yield"

The international concept of sustainability has deep roots that reflect long-held theories of national power in Europe. The dwindling timber supplies following the Thirty Years War (1614–1648) in France and Germany and the seeds of industrialism were inextricably integrated with their treatment of nature. This short background section will describe the contributions of Carl von Carlowitz, Thomas Robert Malthus, and American forester and conservationist Gifford Pinchot who all influenced the intellectual understanding of "**sustainable yield**" as it spread from Europe to America through forestry management and the maintenance of natural capital.

Sustainable Yield the harvesting a given natural resource without diminishing its natural capital or lowering ecosystem productivity.

The origins of sustainability in America begin with the appearance of the concept *sustainable yield*, advanced by the first Chief Forrester of the US Forest Service (1905–1910), Gifford Pinchot. He applied European concepts of good use, conservation, and sustainable yield to develop the first national park system as one that was designed to increase the wealth of the country.

To understand Pinchot's concept, we must first examine the two European economists who influenced him.

Von Carlowitz and Sustainable Forestry Management

Seventeenth-century Germany greatly accelerated the destruction of its forests by nearly stripping them clean for the sake of building its navy and for copper and iron ore smelting during the seeds of European industrialism, which began during this era. They helped initiate industrialism by being among the first to venture into global markets fueling exploration and colonization.

Hans Carl von Carlowitz began his work in mining (1645–1714) when the degraded state of the forests in Germany started to present problems in economics, governance, and social equity. As the environmental realities of a small country with limited resources constricted the growth and development of the population, he led Germany in turning forestry management into a science. After touring Jean-Baptiste Colbert's (1619–1683) managed forests in France in 1713, von Carlowitz published *Sylvicultura oeconomica, oder Haußwirthliche Nachricht und Naturmäßige Anweisung zur wilden Baum-Zucht* (loosely translated as *Forestry Economics' Nature Decree: Moderate Instructions for Wild Tree Breeding*) in 1732.

This manuscript was a comprehensive treatise that tied the endurance of the mining industry directly to the development of German forestry. Von Carlowitz employed economics as the dominant metaphor and wrote primarily to keep the copper and iron mines, and the German colonial apparatus running (Grober, 2012). In it von Carlowitz coined the term *Nachhaltigkeit* or "lastingness" to describe forests that remain eternally productive and autonomously regenerative, while still producing enough harvest to profit economically. This firmly established economics as integral to sustainability. During the early enlightenment, new "forest managers" employed von Carlowitz's term in increasingly strict regulatory measures that geared forests toward the productivity of the nation.

Malthus and Population Ecology

The English cleric and scholar Thomas Malthus (1766–1834) may be most remembered for economics being named "the dismal science," but he is probably the most referenced historical figure among writers on sustainability. As one of our first population ecologists, Malthus would set the stage for the modern economy as well as a very early understanding of how ecological limits would affect the economy. Environmental historian Donald Worster (1994) claims Malthus introduced ecology to classical economics. However, Malthus' viewpoint was strictly instrumental, placing an economic value on ecosystems only in terms of supporting the maximum possible number of human beings for maximum

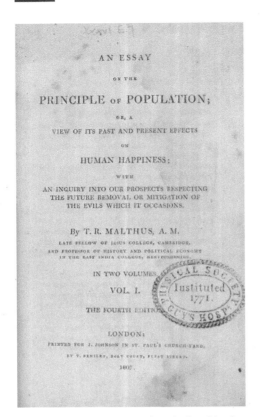

Thomas Malthus' 1798 book shockingly depicted a human race that would outgrow its environment

Source: King's College, London. (2014). Online Collections http://www.kingscollections.org/ exhibitions/specialcollections/charles-dickens-2/ victorian-tight-fi stedness/scrooge-and-malthus.

happiness, part of the utilitarian philosophy of Jeremy Bentham (1748–1842) and John Stewart Mill (1806–73) that dominated the thinking of this era of English history.

In his essay, "A Principle of Population," Malthus explores the production of food, population growth, and their effects on the future improvement of society. He states that food production grows *linearly* while population grows *exponentially*. This incongruity suggests that we will one day surpass the carrying capacity of the Earth, or the amount of life that can be sustained by available resources.

While Malthus is often referred to as an early sustainability thinker, environmental policy historian Lamont Hempel (2012) argues, "[t]he condition of overshoot described by Malthus was based on immutable mathematical logic, but offered very little of what we would call today insights on human behavior from social scientists or notions of resilience and sustainability" (p. 70). Presently, current population growth, unfulfilled food production, and unsustainable practices not being mitigated, "food security" is becoming one of the critical wicked problems of our time. This vindicated Malthus's earlier arguments.

Since the beginnings of the Industrial Revolution, Malthus' importance has waxed and waned in social sciences discussions. Climate change and other SES problems, however, have reopened sustainability conversations that are often characterized as "Malthusian" and "neo-Malthusian" because they involve the acceptance of the idea that the human race is outstripping its resources. Malthusian thought also feeds directly into the very controversial tradition of American population ecologists such as Henry Fairfield Osborn who wrote *Our Plundered Planet* in 1948, Paul and Anne Ehrlich in the 1960s who authored *The Population Bomb*, as well as environmentals Donella Meadows, Jørgen Randers, and William W. Behrens III who co-authored the highly influential document *Limits to Growth* in 1972, all of which suggest human economic development endangers the life-support systems of the Earth.

Gifford Pinchot and *Sustainable Yield*

The first prevalent use of the Greek root sustene in America can be traced to Gifford Pinchot (1865–1946), the first Chief of the US Forest Service. Pinchot was a friend of naturalist President Theodore Roosevelt (1858–1919) who was instrumental in creating America's first national park system. Pinchot, whose family made a small fortune in harvesting the Pennsylvania forests for profit, studied directly under Sir Dietrich Brandis, (1824–1909), a minor but textbook utilitarian forester. Pinchot's philosophy was geared toward Colbert's " bon usage" (Pinchot's "wise use" is a direct translation), and his 1905 manual "The Use of the

National Forest Reserves" is a document dedicated to the supply and demand of forests. These ideas would guide not only the US Forest Service, but also the first school of Yale School of Forestry in 1910 (founded by Pinchot's family). Pinchot was trained in, and heavily relied upon, the European von Carlowizian tradition of sustainable yield. His introduction of sustainable yield to the American dialogue had a longstanding impact and led to the wide acceptance of the maximum sustainable yield concept.

During his tenure, Pinchot's department began using the term *sustainable yield* as a guiding objective in the long-term commercial management of American forests—meaning that the largest harvests had to be such that they could be made without degrading long-term productivity. In sustainability, it has meant preserving natural capital. In practice, obtaining the *maximum sustainable yield* or the maximum level at which a natural resource can be routinely exploited without long-term depletion (*OED*, 2014) from an ecosystem without also damaging an ecosystem has been hard to achieve.

Controversies within Environmentalism and Sustainability

Several ideological polarizations have been observed throughout the course of environmental discourse. Most people refer to the *Brundtland Report's* (1987) definition of "future generations" and "**three pillars**" as the ultimate definition of *sustainability*. This focus has evolved partly out of a polarized political world of communist vs. democratic beliefs (Cronon, 2011), human-first vs. environment-first interpretations, and trade-offs between oppositions. Therefore, it is wise to be familiar with several key debates and the variety of lenses that are used to view sustainability problems. This section will examine *conservation* vs. *preservation* in the early twentieth century, issues of *developed* vs. *developing* in the post-WWII economic development discourse, and *anthropocentric* vs. *non-anthropocentric* views as they have taken shape over the last few decades.

Three pillars The three guiding areas of sustainability problems and solutions: *environment, economy,* and *social equity* (aka the *three e's*) as outlined in the *Brundtland Report* (1987) and the Rio "Earth Summit" Convention (1992).

Conservation versus Preservation

The controversy between Gifford Pinchot and John Muir first separated American environmentalists into conservationist/preservationist and anthropocentric/non-anthropocentric camps, one side stressing instrumental values (or values that represent human interests only), and the other side emphasizing intrinsic values (or values that represent ecosystem needs). While this controversy has often been misinterpreted through oversimplification, similar issues of ideological polarization among interpreters of sustainability and environmentalism persist today.

A devout utilitarian, Pinchot represented the *conservationist* worldview of his time—one that valued the environment but

Courtesy Library of Congress

John Muir was a friend of President Roosevelt's but when it came to the management of our national forests, Roosevelt eventually sided with Pinchot

primarily in terms of its value for economic development. Subsequently, Pinchot has been caricaturized in the literature as a villain. A more grounded characterization would be as an interdisciplinary scholar who identified the dwindling resources of the nation as an economic threat, and an ecologist and forester who sought to preserve American forests.

Pinchot's early sustainability worldview is one that was inextricably tied to the "progressive movement" of his era—utilitarianism and mass production. His terminology resonated into the next age of rapid growth in America with the "Multiple-Use Sustained Yield Act of 1960." The Act explicitly characterized forests for recreational use, timber, wildlife, and watershed management (Hempel, 2012, p. 96). Char Miller (2009), a contemporary Pinchot scholar, describes Pinchot as someone who saw the dwindling resources of the nation as a threat to national security and a pragmatic manager who sought to preserve American forests in the best way he could. But, Pinchot is probably most remembered as an early interdisciplinary conservationist outside of the naturalist vein. He is also viewed as one who helped unite the fields of ecology and economy and instigated environmental regulation in an era when none existed.

Pinchot, and his conservationist view, are often contrasted with the naturalist John Muir. Muir advanced a *preservationist* approach to conservation, lobbying his friend Roosevelt to accept his views. "When one tugs at a single thing in nature, he finds it attached to the rest of the world," said Muir. A naturalist, he advocated a biocentric, or **non-anthropocentric,** view of nature (Minteer, 2006). One of the first people to write about the Central Valley of California, Muir dedicated a large portion of his life to documenting its wildlife abundance. "He waded ankle-deep through the blooms, lay at night on them for a bed, shared their fragrance with the larks, antelopes, hares, and bees" (Worster, 1985, p. 9).

The construction of O'Shaughnessy Dam submerged the Hetch Hetchy Valley completely eradicating this biodiversity and seemingly along with it, Muir's preservationist vision for the national parks. The hydroelectric power plant project was one of America's first clear trade offs between the environment and economics. It was also the catalyst for a controversy between Muir and Pinchot. This separated American environmentalists into conservationist and preservationist camps, with the former stressing instrumental values of natural resources and the latter emphasizing the intrinsic values of Nature (Minteer, 2006).

Non-anthropocentric values Values that consider the environment for its own health and well-being, rather than for instrumental (*anthropocentric*) human uses (*biocentric, ecocentric, preservationist values* have a similar gist).

Issues of International Development

In ecology, *sustainability* refers to "a state that can be maintained over an indefinite period of time" (Du Pisani, 2006, p. 91). It also has a variety of meanings when referring to human activity, particularly in the context of legal arguments, land management, and other contemporary applications that include environmental considerations (*OED*, 2014). The majority of these definitions view nature and natural resources primarily for their economic value to human society. This is especially true within the context of economic development. This section will briefly look at this debate as it occurred between two American population ecologists.

As early as 1947, ecologist Barry Commoner (1917–2012) showed signs of deep concern for the planet's well-being. Commoner brought important high-risk scientific debates into the open as well as pressing the American Association for the Advancement of Science (AAAS) to play a more ethical role. In 1952, he was appointed to the AAAS' interim Committee on the Social Aspects of Science. Commoner was a harsh critic of technological optimism and conveyed this message in his seminal works, *Science and Survival* (1966) and *The Closing Circle* (1971). "In the eager search for the benefits of modern science and technology we have become enticed into a nearly fatal illusion: that through our machines we have at last escaped from dependence on the natural environment" (Commoner, 1971, p. 8).

Commoner argued that the "environmental crisis" (as it became to be known in the 1970s) resulted primarily from the technological advances designed to achieve a more sustainable future–an old idea, but one that seemed to be increasingly obfuscated when infused with capitalistic notions of progress (p. 17). Commoner also thought that most Americans had mistaken the "good life" as one primarily of affluence. He equated it to kitchen appliances, sporty automobiles, and national security in the form of the atomic bomb for Americans after WWII. Commoner defined a three-pronged platform to create a new scientific structure based upon 1) the necessity of dissent; 2) transparency of technical information; and 3) public discussions and decision-making (Egan, 2007).

In contrast, Paul R. Ehrlich believed that the environmental crisis was rooted in uncontrolled population growth and aggregate consumption. Ehrlich is an American Biologist and neo-Malthusian best known for these concerns. He is often perceived as both a pessimist and alarmist.

In 1968, he wrote *The Population Bomb*, a doomsday proclamation that by the "1970s and 1980s hundreds of millions of people will starve to death in spite of any crash programs embarked upon now. At this late date nothing can prevent a substantial increase in the world death rate, although many lives could be saved through dramatic programs to "stretch" the carrying capacity of the earth by increasing food production and providing for more equitable distribution of whatever food is available" (p. xi). He suggests that "We must have population control at home [US], hopefully through a system of incentives and penalties, but by compulsion if voluntary methods fail. We must use our political power to push other countries into programs which combine agricultural development and population control" (p. xii–xiii).

Debates between Ehrlich and Commoner in the late 1960s and early 1970s serve to illustrate the ongoing ideological debate and division within human development discourse. Commoner vehemently opposed Ehrlich's views on coercive population control. While Commoner agreed that there was a population issue, he believed that the crisis was actually rooted in poverty and not overpopulation. He believed that ending poverty and its underlying cause, the "grossly unequal distribution of the world's wealth" were the solution (p. 195).

Ehrlich publically attacked Commoner's views that social action constituted the best arena for solving population, consumption, and pollution problems, saying in a review of *The Closing Circle* (1971), "uncritical acceptance of Commoner's assertions will lead to public complacency regarding both the population and affluence" (Bulletin 1972, p. 55). Commoner fired back in the *Bulletin of the Atomic Scientists* a month later:

"Ehrlich took the position that ecological catastrophe is inevitable if the peoples of developing countries…are left to regulate population growth by their own actions . . . [which] include improvement of living conditions, urgent efforts to reduce infant mortality, social security measures, and the resultant effects on desired family size" (p. 55).

While the concept of sustainability has sought to integrate opposing views, it does not seem to have completely resolved ideological polarizations such as preservationist vs. conservationist and environmentalist vs. economist that manifest into instrumental and intrinsic value divisions among environmentalists. In fact, the construct of sustainability has often exacerbated ideological differences. Divergent social issues are the most hotly debated of the three sustainability pillars because they are the most vaguely defined and because frameworks often ignore issues such as environmental justice and poverty, despite the environmental movement's and sustainability's inherent concerns over these matters.

Deep Ecology and Environmental Justice

The environmental movement was marked by polar opposite stances on the state of the world and how to address its major problems. Despite substantial gains in pollution legislation that occurred in the 1970s, people began to view the world as a place that would one day become "uninhabitable." Within a few years of the first Earth Day in 1970, many social and environmental advocates became polarized over the approaches needed to interpret and solve environmental issues (Sale, 1993). These approaches transformed into issues that favored an environmental justice worldview and stressed social concerns vs. a much deeper ecologist point of view that favored the intrinsic characteristics of Nature as valuable, and its utmost consideration was perceived as the only way to preserve the planet.

Norwegian Arne Naess, who introduced the term "deep ecology" in *The Shallow and the Deep* (1973), revived thinkers from the distant past such as Baruch Spinoza (1632–77), who saw all "particularity" within Nature as equally representative of God and linking the human and natural world (Hay, 2000, p. 26). Naess described the differences between deep and "shallow ecology" in detail. Deep Ecology, or Ecocentrism, is a concept that applies to "the ecosphere as a whole" (Naess & Sessions, 1984: p. 5). With regard to pollution, a shallow approach seeks technology to help purify emissions that cause acid rain, while a deep approach attacks the causal economic and technological mechanisms responsible for the diminishing of biospheric integrity. This theory would facilitate the application of ethics to ecosystems. Deep ecology, however, was viewed by some as producing legislation that favored ecological values over human values, in practical situations such as national park products. The concepts of deep ecology have been applied, both theoretically and literally, in situations such as the re-introduction of predators in some American national parks. Non-anthropocentric values have also manifested throughout the world such as in Nepal's Chitwan National Forest, where extreme poverty exists right outside the forest, and pillaging rights have been eliminated due to its management.

The concept of environmental justice emerged as a social movement that was partly in reaction to this competition with the wilderness for its share of the benefits. But more importantly, it was a problem-driven response that grew out of the Civil Rights movement. Spurred by American housewife Lois Gibbs's campaign against the producers of a toxic dump in Niagara Falls, thousands of studies demonstrated that environmental problems of pollution and other environmental hazards unfairly affected not only poorer communities, but targeted minorities.

Leaders have relegated unsightly facilities, landfills, production, waste disposal and toxic waste disposal to parts of a city—and in some cases, entire cities—occupied by predominately minority populations. Center stage in many cases of this type of environmental injustice are issues of human health. The unfair distribution of environmental impacts has led to higher levels of asthma, cancer, lead-poisoning, mercury-poisoning, mental illness and death, especially among children and the elderly. Thus, environmental justice has better informed sustainability theory more than environmental studies alone ever could, by changing its orientation toward more pragmatic beliefs that more fully acknowledge both the benefits and impacts of environmental planning.

The idea of environmental justice has quickly spread around the world, where issues of pollution and human abuse often far exceed our problems at home. Problems of climate change, biodiversity loss, and toxic pollution are often produced by what is now referred to as the Global North, North America, Europe, and Australia (also referred to as the "Western" or the developed world). These problems are disproportionately distributed to the Global South—Central and South America, Africa and most of Asia—and are often framed in environmental-justice terms such as "water justice" and "climate justice."

As the concept of environmental justice captured momentum during the 1980s, the forefront of the public debate became skeptical of what was thought to be the biocentric or ecocentric views that were perceived as only a luxury of the educated elite in developed countries. Environmental justice placed precedence on the distribution of the environmental costs and benefits unequally between the Global North and Global South as well as within the Global North itself. Like Commoner's social platform, environmental justice advocates saw the capitalistic economy as the source of many of the problems, and social programs designed to increase transparency and encourage pluralism in decision-making as the cure.

The American Environmental Movement

During the 1960s and 1970s, many American ecologists were also writers and activists like Barry Commoner. Among the most well-known and influential is Rachel Carson, whose book *Silent Spring* (1962) helped to inspire an "environmental movement" among various interest groups in America. This section will focus on the contributions of intellectual thinkers like Commoner and Carson who inspired the environmental movement in America.

Rachel Carson (1907–1964) brought concerns about the pervasiveness of pesticides in our daily life into public discourse, and ushered in a new perspective on production and consumption. In *Silent Spring* (1962), she addressed pivotal issues like industrial pollution, persistent pesticides, and the bioaccumulation of toxins as they traveled up the food chain to human beings. Carson contributed to the understanding of coupled human-natural systems through her interdisciplinary work at Wood's Hole Naval Research Laboratory prior to and during World War II, when disciplines from all areas of epistemology came together to respond to the threat of global domination.

Before 1962, Carson had written many books about the interconnectedness of nature. Her greater body of work, including her letters and books on the natural history of the sea, focused on multi-generational environmental problems and risks. This was a novel approach that presented a new, holistic ecological vision. It also marked a break from some of the earlier conservationist writers and hailed a transition to "environmentalism." Her poetry was so compelling that she won the 1951 National Book Award with *The Sea Around Us*.

After realizing that much of the biodiversity in her local natural haunts along the East Coast was disappearing, Carson felt the need to switch gears. Specifically, she noted that in springtime fewer birds could be heard singing. She soon discovered that the use of DDT (*dichlorodiphenyltrichloroethane*) was responsible for thinning bird eggs so badly they could not hatch. She was able to do this by using her connections as a government scientist. She acquired what were then confidential reports of the deadly effects of DDT, allowing her to become an expert on the effects of herbicides and pesticides that could be found in everything we ate and drank, even the air we breathed.

Carson wrote, "The history of life on earth has been a history between living things and their surroundings" (1962, p. 297). Her holistic understanding of the intricate link between humankind and the world around us would become her legacy. Her primary contributions to sustainability included developing an understanding of *human ecology* as the link between science and ethics, and advancing a vision for a new and vital role for science and scientists alike in creating transparency between government and knowledge systems.

During the era of environmentalism (sometimes referred to as the "age of ecology" in Carson and Commoner's time), Commoner expanded Carson's thoughts on science and technology, injecting them into the political arena for over three decades. In *Science and Survival* (1963), he concluded that "[t]he age of innocent faith in science and technology may be over" (p. 14).

Commoner saw the gap between science and social science widening and took it upon himself to fill that gap. As early as 1966, he warned that we were "mortgaging for future generations not just their lumber or their coal, but the basic necessities of industry, agriculture, and life itself: air, water, and soil." with our current practices (Egan, 2007: p. 83). Like Carson, he conducted research during WWII, especially on the use of DDT to kill mosquitoes at Normandy prior to invasion, which influenced his beliefs. During the war and for some time afterward, social issues of science were put on hold and R&D gained primacy and escalated rapidly. Commoner identified another problem as well—science was declining in prestige as well as being the cause of new social concerns.

Because of writers like Carson and Commoner, the environment became a pivotal political theme. Their writings also shifted the dialog toward quality of life issues like the human health effects of pollution. The result of their work and other thought leaders of the era was a social-environmental movement that coalesced across a broad spectrum of society and mobilized citizens and governments to remedy environmental degradation and its effects. In the US, a platform of environmental laws were enacted, the likes of which have not been seen since.

A number of the most well-known pieces of legislation included the Clean Air Amendments of 1970, Federal Water Pollution Control Amendments of 1972, the Endangered Species Act of 1973, the Safe Drinking Water Act of 1974, and the Resource Conservation and Recovery Act of 1976. National environmental policy was promulgated though the 1969 National Environment Policy Act, which provided a process for decisions regarding environmental impacts. In addition, the Environmental Protection Agency was established in 1970 to monitor and enforce these acts.

During this period, researchers from the Massachusetts Institute of Technology (MIT) wrote and published *Limits to Growth* (1972). This document initiated sustainability discourse in its modern form, forecasting humankind's new global problems with the first computers and equations derived by Jay Forrester from MIT. *Limits to Growth* (Meadows et al., 1972) constituted the beginning of sustainability's first global toehold.

Sustainability in International Discourse

Limits to Growth together with the United Nations (UN) Conference on the Human Environment (UNCHE) held in Stockholm (Balboa, 1973) introduced the term *sustainable* to the development discourse focusing on the economic value of resources. The *Brundtland Report* (which reinforced the outcomes of UNCHE) further propagated and entrenched the human- and economic-centered focus of sustainable development. Preliminary and follow-up discussions such as the 1972 *Earth Summit* of Rio also embraced this definition, firmly establishing its framework for global discussions. *Sustainable development* had become the catchall phrase for environment and development (i.e., sustainability) discourses.

This section explores three important UN documents alongside three other revolutionary documents that all used sustainability terminology: *Limits to Growth* (1972), *The World Conservation Strategy* (1980), and *Our Common Journey: a Transition for Sustainability* (1999).

The *Stockholm Declaration* and *Limits to Growth*

The 1972 United Nations Conference on the Human Environment (UNCHE) held in Stockholm considered "the need for a common outlook and for common principles to inspire and guide the peoples of the world in the preservation and enhancement of the human environment" (Balboa, 1973). The *Stockholm Declaration* was praised by subsequent bodies (WCED, 1987; Summit, 1992), despite its predominant interest in the growth of economic markets for the developed and developing worlds alike. This document also influenced later UN documents such as the *Brundtland Report* and the *Earth Summit*.

The declaration lays out 26 principles that define common convictions for international cooperation. See the full document at http://www.unep. org/documents.multilingual/default.asp?documentid=97&articleid=1503. In the first principles, one can see ideals of protecting the ocean, wildlife, and preserving non-renewable resources; looking further, we begin to see many of sustainability's inherent contradictions. For instance, Principle 8 states, "[e]conomic and social development is essential for ensuring a favorable living and working environment for man and for creating conditions on earth that are necessary for the improvement of the quality of life" (Balboa, 1973). This places the two broad fields of inquiry, *development* and *environment,* at odds as well as operationalizing the benefits of the environment for development. Principle 11 goes even further by placing development first,

> "[t]he environmental policies of all States should enhance and not adversely affect the present or future development potential of developing countries, nor should they hamper the attainment of better living conditions for all, and appropriate steps should be taken by States and international organizations with a view to reaching agreement on meeting the possible national and international economic consequences resulting from the application of environmental measures."

As we travel further down the list of principles, the terms we see most are those framing sustainable development in terms of human-centric benefits. For example, Principle 14 states that conflicts (what we call tradeoffs today) "between the needs of development and the need to protect and improve the environment" will be solved by "rational planning" (Balboa, 1973). Principle 18 focuses on science and technology and "their contribution to economic and social development." It further states that they "must be applied to the identification, avoidance and control of environmental risks and the solution of environmental problems and for the common good of mankind." Finally, Principle 21 boldly claims that "[states] have . . . the sovereign right to exploit their own resources pursuant to their own environmental policies, and the responsibility to ensure that activities within their jurisdiction or control do not cause damage to the environment of other States or of areas beyond the limits of national jurisdiction" (Balboa, 1973). This last statement is considered impossible from most contemporary sustainability theorists' understanding of global interconnectedness and is illogical from a naturalist perspective.

Limits to Growth (Meadows, et al., 1972) framed sustainability quite differently. It predicted that population would overshoot the earth's **carrying capacity**, impairing the ability of the planet to support human and other biological life. The

Carrying Capacity The maximum sustainable population an area will support before undergoing environmental deterioration.

authors stated that global transformations were needed in five categories: *population, industrialization, pollution, food production,* and *resource depletion.* These were clearly anthropocentric in scope; however, the document linked humanistic and scientific horizons of interpretation. In this way, it paved a course for a new rapport between the previously bifurcated studies of social and hard sciences and named this new marriage *sustainable development.* Supporters of these concepts realized that both developed and developing countries needed to change their patterns of economic and population growth.

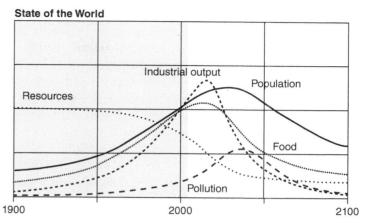

State of the World

Limits to Growth (Meadows, et al., 1972) illustrated that ecological limits dictated human limits of economic growth and expansion.

Source: Limits to Growth (1972) p. 40. Publisher: Chelsea Green Publishing Company (August 1993)

The findings in *Limits to Growth* (1972) were highly controversial then and continue to be today. The three prime findings were 1) the capital plant and the population are constant in size; 2) All input and output rates–births, deaths, invesment, and depreciation–are kept to a minimum; and 3) The levels of capital and population and the ratio of the two are in accordance with the values of the society (pp. 173–4). These limitations implied that developed countries must show constraint in production and consumption. For developing countries, it was interpreted as reinforcing the gross inequities between their countries and developed countries. As a result, it immediately received backlash from the business world and conservative think tanks. It was criticized from the left as well for its neo-Malthusian thinking. Yet, *Limits to Growth* (1972) demonstrated that the term *sustainable development* had began to take hold in the global discourse.

The *Brundtland Report* and the *World Conservation Strategy*

Commonly known as the *Brundtland Report* (WCED, 1987), "Our Common Future: Report of the World Commission on Environment and Development" was the first publication to use the term *sustainability* in its current usage (i.e., noun form). Its definition, "development that meets the needs of the present without compromising the ability of future generations to meet their own needs" (WCED, 1987: 1:1), may be the most often quoted definition for sustainability. Certainly future generations are crucial to sustainability, anything otherwise would defy its definition. In addition, its "three-pillar" framework has been an important foundation for sustainability discussions and can be found in many sustainability education textbooks and sustainability coursework at the university level.

The *Brundtland Report* (1987) The source of the modern paradigm of sustainability. It is significant for demanding ethical issues such as gender and economic equity be part of sustainability's definition.

Brundtland's central ethical and material issue concerned the rights of developing and developed countries to expand economically, while also maintaining ecosystems for future human and economic use. The document directs developing countries to develop and grow sustainably, yet it sets no clear boundaries for similar sustainable development of developed countries. While opening up important normative discussions, Brundtland only *began* to include the subject of environmental resources in development discussions. Until its publication, strictly economic bodies like the International Monetary Fund and the World Bank were the hegemons of all global development issues. Brundtland's intended scope was merely to add environmental resources as yet another dimension. It did not recognize the ecological phenomena that contribute to SES-problems, such as feedback loops, cascading socio-ecological effects, *ecological limits*, and *thresholds*. Nor did its authors intend to address or remedy the complex web of ethical issues surrounding development (Minteer, 2011; Orr, 2002).

The Brundtland Report's (1987) strength lay in its addressing of social inequities through democratic pluralism as a necessity for ecological sustainability (Boone & Modarres, 2006; Minteer, 2011). Nowhere before had universal social justice as an immediate goal been indissolubly linked to issues of conservation. This is the basis for contemporary sustainability theory, which posits that a sustainable society rests on three interlinked foundations–social equity, efficient economic allocation, and respect for ecological limits.

Also, often overlooked is the report's dedication to *intra*-generational equity. It inextricably linked the planet's well being to people of all generations and races, and solidified the link between the eradication of poverty and sustainability theory. For example, not only did the report secure sustainability's place at the center of human development discourse, it also provided women with an equitable share of the discussion. It recognized "the changing role of women" and "the right to self-determination" (WCED, 1987: 2.51, 4.6). Traditionally, child rearing responsibilities in developing countries—and often in developed countries—rest primarily on women. The report addressed women's shifting roles, including the observation that women were producing 60 percent to 90 percent of the food in Africa (WCED, 1987: 5.86).

It may have been in part the acceptance and endorsement of the term *sustainable development* by a group of ecologists in the *World Conservation Strategy* (1980) that led to Brundtland's success. The World Conversation Strategy was penned by the International Union for the Conservation of Nature (IUCN), one of the most prestigious non-governmental agencies for the protection of the environment today. This group's mission targeted "living resource conservation for sustainable development" (p. 18), opening the door for the term's universal use in dialogues about the environment.

Established in 1956, the IUCN (aka the World Conservation Union) is the preeminent body of global ecological preservation and conservation. It tackles the most difficult environmental issues like diminishing biodiversity, climate change, natural resource depletion, and endangered species. This non-UN conference has been applauded for being able to link inextricably the interest of human rights in developing countries to conservation efforts from the point of view of biologists and ecologists.

The *World Conservation Strategy* (1980) defined humans in ecological terms as a significant "evolutionary force" (IUCN, 1980: 3.1). In fact, the document's specific wording makes clear the use of sustainability as a socio-ecological interface. "The separation of conservation from development together with narrow sectorial approaches to living resource management are at the root of living resource problems. Many of the priority requirements demand a cross-sectorial interdisciplinary approach" (IUCN, 1980: 8.6). This admission by biologists, chemists, and ecologists—scientists dedicated to the study of natural systems—helped underscore sustainability's basic transdisciplinary premises, but with the understanding that environmentalists and developers should work together.

The *Earth Summit* and *Our Common Journey*

The UN's Rio Convention of 1992, also known as the *Earth Summit*, was a large, enthusiastic, and optimistic gathering of businesses and non-governmental organizations (NGOs) around environmental and social issues of combating poverty, deforestation, and the transfer of technology. National leaders from 180 countries gathered at the conference along with about 30,000 attendees. Coincidentally, this was 20 years following The Stockholm Declaration and publication of *Limits to Growth* (1972). Furthermore, it coincided with publication of Meadows' follow-up book, *Beyond the Limits* (1992), which reported that although the world had not collapsed as predicted, the carrying capacity of the Earth had been exceeded.

The *Earth Summit* (1992) concentrated on the environmental problems of new sources of pollution, the depletion of tropical rainforests, and the ozone layer. It proposed new approaches for contending with complex SES problems like climate change and biodiversity loss, including important new concepts like Agenda 21. Agenda 21 was a non-binding but important pledge that encompassed four areas: social and economic dimensions, conservation of resources in development, the strengthening of minority groups, and probably most important, the means of implementation often lacking in other treaties.

However, because the *Earth Summit* was not a formal treaty, the ideas in it lost momentum over time. At the 2002 Rio + 10 Earth Summit in Johannesburg, some of these ideas were reconsidered but became watereddown as discussions were "almost paralyzed by a variety of ideological and economic disputes, by the efforts of those pursuing their narrow national, corporate, or individual interests" (Meadows et al., 2004). Despite these obstacles, Agenda 21 initiated a shift from global ideals and principles to city initiatives, policies, and plans. The rise of myriad Municipal Climate Action plans illustrate this shift.

These municipalities and other organizations have embraced sustainability and helped promote Rio's key principles. For example, Local Governments for Sustainability, originally founded as the International Council for Local Environmental Initiatives (ICLEI), leverages these principles to support local governments in implementing sustainability plans. It has been instrumental in fostering urban sustainability efforts and voluntary greenhouse gas emission cuts in over a thousand cities in America and many more around the world. These and myriad other efforts have culminated in the science of sustainability. According to Robert W. Kates, a leading sustainability scholar, **Sustainability Science** "has attracted tens of thousands of research authors, practitioners, and knowledge users, . . . with a geographical, institutional, and disciplinary footprint very different from most science" (Kates, 2011).

Sustainability Science According to the Proceedings of the National of Academy of Science, Sustainability Science is ". . .an emerging field of research dealing with the interactions between natural and social systems and with how those interactions affect the challenge of sustainability: meeting the needs of present and future generations while substantially reducing poverty and conserving the planet's life support systems"

Sustainability became a formalized science when the National Academy of Science (NAS) published *Our Common Journey: a Transition for Sustainability* (Kates & Clark, 1999). It echoed the language in "Our Common Future," stressing sustainable transitions with a "normative vision." *Our Common Journey* (1999) links the ideals of Brundtland to the real world. Especially important is the second chapter that outlines the historical trends of population, economy, resource use, and pollution as humankind becomes an ecological force and begins to shape the planet in terms of changes to the life support system. In examining these trends and mapping them out to the year 2050, NAS researchers and many other experts anticipated that there will be over 9 billion people on the planet. In *Our Common Future*, they outline this and a number of other SES issues that they consider to be the largest threats to humans and the environment. Finally, this report is one of the first to establish reporting methods using "indicators" of human and land health, further establishing sustainability as a science.

Our examination of the UN documents, *Stockholm Declaration, the Brundtland Report,* and *Our Common Future*, trace the term sustainability from its first use in the early 1970s to its centrality in environmental discourse in the late 1980s. Together these distinct yet complementary documents tell the story of sustainability's rise from a virtual unknown to the leading term in development discourse. Each written by an international group of scientists and academics, they hoped to lead a logical and ethical discourse to solve global problems. These three documents helped bring sustainability into governmental, business, academic and environmental discussions and began to tease out the gamut of values that have made sustainability irreplaceable in contemporary environmental and policy discourse. The next section of this chapter posits that the values put forth to date are not all-inclusive. We look to American Naturalists Henry David Thoreau, Aldo Leopold, and E. O. Wilson to gain a greater understanding of a more encompassing, holistic approach to sustainability.

The American Naturalists: An Alternative Tradition

Naturalists The first scientists; field biologists and students of natural history.

As we saw earlier, John Muir, a life-long naturalist, represented a different worldview than the predominant instrumental view of nature. The use of the term **naturalist** has changed over time, but during the height of its use in the eighteenth and nineteenth centuries, it denoted, first, a natural philosopher or student of natural history, and second, a field biologist engaged in direct observation and experimentation. Naturalists at this time were synonymous with "SCIENTIST (q.v.): in practice those whom we would now call *physicists* or *biologists*" (Williams, 1976, p. 216).

The term naturalist also had a very important third aspect. The best naturalists systematically took their findings of direct observation and deductions, and applied them to the larger body of knowledge as a whole. This *holism* provided a more creative, frontier aspect and broader focus, which is often overlooked within today's intensely focused disciplinary tracts (Wilson, 1998). Was Muir part of a dying breed or can naturalist principles such as his help guide and enhance sustainability thinking today?

When we consider some of the most iconic and influential naturalists—Aristotle, Linnaeus, Darwin, E.O. Wilson—we think of generalists who are claimed by a multitude of disciplines. In fact, most people know them, not by their discipline, but by the influence their literature had on the environmental movement. This section examines three American naturalists , Henry David Thoreau, Aldo Leopold, and E. O. Wilson, who had an affinity for coupled human-nature systems and SES to show how their worldviews can broaden the field of sustainability science and inform sustainability efforts around the world.

Henry David Thoreau (1817–1862) was a naturalist during the rise of industrialism in America. His seminal 1854 work *Walden: or Life in the Woods* has inspired those looking for an antidote or alternative to the alienating effects of industrial development. Despite this, little attention has been devoted to his critique of society, and especially economics, in America. Thoreau's contributions to sustainability come *not* from his allegedly impracticable ideas about ethics or spirituality alone, but from his ability to integrate a widely varied spectrum of thought and values toward a wide range of practical applications and SES problem solving. His broader worldview is an ongoing challenge to sustainability thinking but there are three aspects that can be readily applied to sustainability today.

First, Thoreau created a roadmap for not only efficiency but also consuming less. *Walden* (1854) describes in detail what we call today living "off the grid." In describing his daily routine when building his own home, obtaining fuel and clothing, and living a contemplative life, he is in essense explaining how to have a minimal impact on the environment. Second, Thoreau's activism as described in "Civil Disobedience" (1949), a written protest against American capitalism and expansionism, provides a template of how to stage a non-violent protest. For instance, leaders such as Mohandas Gandhi (1869–1948) and Martin Luther King Jr. (1929–1968) as well as many leaders of the environmental movement have embraced his methods. Finally, his merging of philosophy and science through his own special breed of transcendentalism can guide us in linking the disparate fields of ecology economics and society with which sustainability thinking is most concerned.

A hundred years later, naturalist, ecologist, conservationist, land manager, and amateur philosopher Aldo Leopold (1887–1948) conducted his own Thoreauvian experiment with his family at a place in Wisconsin called "the shack." Unlike Thoreau, Leopold had devoted his entire life to conservation issues all over America, applying concepts devised by generalists like Thoreau who could link science and philosophy. Leopold (1949) was also able to consolidate intrinsic and instrumental values through his "land ethic" worldview. He demonstrates this integrated worldview when he explained ecosystem conservation in his manifesto, *A Sand County Almanac* (1949), "If the land mechanism as a whole is good, then every part is good, whether we understand it or not. . . . To keep every cog and wheel is the first precaution of intelligent tinkering."

Leopold is perhaps among the most widely recognized of pre-sustainability thinkers. Through practical applications in real-life scenarios in the US Forest Service, complemented by the university setting, Leopold's comprehensive land ethic advanced both the fields of environmental ethics and ecology. His extensive work with the Forest Service across the midwestern and southwestern states, together with his university teaching experiences, led him to propose expanded roles for both individuals and governments by instilling an environmentally based moral obligation. His "land ethic" presents a vision for inclusiveness, cooperation,

WALDEN;

LIFE IN THE WOODS.

By HENRY D. THOREAU,

BOSTON:
TICKNOR AND FIELDS.

Courtesy Library of Congress

Thoreau wrote Walden (1854) just a little over a mile outside his hometown of Concord, Massachusetts in the little shack depicted on the cover, providing a roadmap for an "alternative" lifestyle

and simplicity that people of all walks of life can support. Because Leopold was a hunter, outdoorsman, and land owner as well as a scientist, he often appealed to non-anthropocentric and anthropocentric views simultaneously and, for that reason, provides a template suitable for pluralistic discussions of sustainability.

Edward O. Wilson (1929-) is the only living and self-proclaimed naturalist mentioned in this section. Analysis of his work over the past five decades provides a contemporary, scientific, and comprehensive view of natural and social systems together. It covers an impressive breadth of topics, from evolutionary theory and genetics to his 300-year history of the sciences and humanities that examines the role of disciplines like economics, ecology, genealogy and many more as they arrive upon the academic scene. Interestingly, Wilson's own writing largely avoids the rhetoric of sustainability. In particular, his sophisticated understanding of sociobiology, biophilia, and conservation ecology contributes to an ethic of enlightened self-interest—an evolutionarily based argument for the human affiliation with nature—and a worldview driven by planetary survival that can help frame sustainability in such a way as to counteract the "natural economy crumbling beneath our busy feet" (Wilson, 2002, p. xxiv).

Wilson is an enormously prolific writer, having authored hundreds of articles and over 50 monographs. Nevertheless, his lifetime's work in the study of animal and human social behavior culminates in *Consilience* (1998) and *The Future of Life* (2002), and provides a broad platform for arguing for a changed economy, preserved biodiversity, and greatly improved social systems in order to survive. His work highlights the need for stability in bio- and socio-diversity, convergence of human and natural values, and the essential nature of multi-scalar coordination and public-private partnerships.

The Future of Sustainability

Today, the US—as well as most of the world—faces *socio-economic* problems such as widening income disparity, poor education status, high levels of poverty, and poor national health care. As significant as these problems are, there are parallel SES problems that increasingly threaten the integrity of our global life-support systems. Much of America's environmental footprint is externalized well beyond our borders, depleting arable resources and causing biodiversity loss around the world. As the US population nears 400 million, and the world 9 billion, the growth of middle-class consumers will further reduce our carrying capacity. This creates great uncertainty and complexity that can only be resolved by transdisciplinary, SES, and sustainability thinking.

Meadows, Randers and Meadows (2004) in their follow up publication, *Beyond the Limits: global collapse or a sustainable future* (1992), confirmed their prediction that the world's population would surpass its carrying capacity. Their chart depicting our growth begs the question, which of the following scenarios are more likely to follow: a slow easing down of resources or a sudden event that will shock socio-economic and ecological resources? These scenarios are: (a) where environmental resources are very plentiful, and population growth will not erode the carrying capacity, energy is abundant and/or efficiency will exponentially multiply the available energy; (b) where we have not reached carrying capacity,

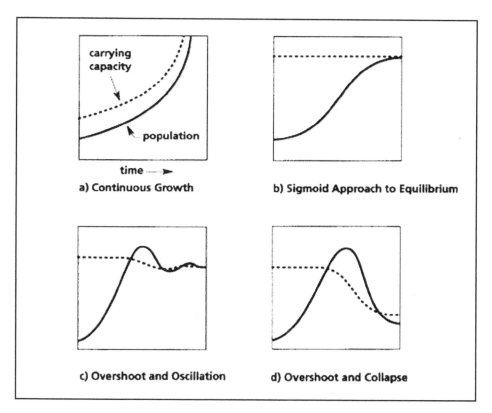

Scenarios of growth. In *Limits to Growth: The 30 Year Update*. Scenarios c. and d. are most likely

and signals from physical limits to the economy are instant or accurate; (c), where we have passed the carrying capacity, and signals and responses are delayed; or (d) where we have passed carrying capacity and signals are delayed and limits are irreversible (Meadows, Randers & Meadows, 2004, p. 158). The harsh reality is that of these scenarios, overshoot and oscillation (c) or overshoot and collapse (d) are the most likely given our slow response and continued overconsumption. Sustainability must be about cutting back net impact as an individual, as a community, as a nation, and as a global community. Unfortunately, most sustainability initiatives do not have such a transformational vision in mind, and, thus, will not address the scale and magnitude of the problem at hand.

Urban development has become the main forum for sustainability discourse, especially since William Rees and Mathis Wackernagel (1996) coined the term *ecological footprint*. In *Urban Ecological Footprints: Why Cities Cannot be Sustainable—and Why They are a Key to Sustainability* (1992), they describe how the ecological footprint of cities extends far beyond its borders and suburbs. In doing so, they address the foundational ecologist Eugene Odum's (1971) lament that "Great cities are planned and grow without any regard for the fact that they are parasites on the countryside which must somehow supply food, water, air and degrade huge quantities of wastes" (p. 371). Odum's, Rees' and Wackernagel's insights apply to countries as well.

Today, the environmental footprint of the world's most developed countries extend far beyond their borders. To make matters worse, developing countries are emulating developed countries' economies. If these countries proceed along the same trajectory of unchecked consumerism as most developed countries, particularly the US, consumption among a growing global middle class is likely to decimate the ecologies of many biodiversity hotspots and increase the rate of historical over-consumption patterns (Brown, 2011; Wilson 2002). This over-consumption together with continuing population growth is anticipated to have significant consequences. For example, the UN estimates that the world's population will level out at 11.5 billion people, far beyond the carrying capacity that most population ecologists estimate. This will be exacerbated even further by climate change (Rees & Wackernagel, 1992; Wilson, 2002). These and other wicked SES problems require a new way of thinking to prevent potential catastrophic and irreversible impacts.

The *Brundtland Report* (1987) can be thought of as the sustainability shot heard round the world, as it promoted a new "paradigm" for thinking about the environment in terms of human development (Cronon, 2011). Centered on efficiency issues, this thinking failed to address rampant production and consumption and the continued exploitation of the environment (Radkau, 2009). Thus, it fell far short of solving the wicked SES problems facing the world over the last few decades. The UN Earth Summits and other sustainability conferences since then have brought people from all around the world in an attempt to address this issue. However, the majority of agreements reached at these meetings have been non-binding. As a result, little progress has been made in protecting and preserving the Earth's carrying capacity (IUCN, 2004).

Legislators and leaders from around the world struggle now to give treaties and conferences some teeth to overcome these impediments. However, international agreement is proving difficult. This was evident in the Copenhagen and 2014 Warsaw conferences that witnessed ever-growing frustrations, especially by representatives from developing countries who often walked out of the proceedings in protest or to form their own sub-groups and decision-making bodies.

There may be light at the end of the tunnel, however. International bodies such as the International Panel on Climate Change (IPCC) and the Conference of Parties (COP) have begun to provide concrete analysis, assessments, and options for adaptation and mitigation of climate change. The IPCC specifically assesses the state of the climate and the science being undertaken to understand it. The COP meets every year to evaluate the progress made on implementing the 1994 UN Framework on Climate Change (UNFCCC), which seeks to stabilize and lower greenhouse gases (GHGs) that have been found to have a profound impact climate and the Earth's ecosystems.

Every five years since 1988, the IPCC has held a summit meeting to present the findings of its working groups and task force. It has engaged thousands of scientists and researcher from around the world as authors, contributors and reviewers. The primary activities of the IPCC is to produce reports on the "scientific, technical, and socio-economic knowledge on climate change, its causes, potential impacts and response strategies" as well as GHG inventory guidelines (IPCC, 2017).

The COP21 of Paris in November of 2015 was historical in that for the first time, the two largest emitters of CO2 on the planet—China and the US—made significant pledges in reduction. These two countries ratified this treaty in September. As of November of 2016, it had well over the mandatory 55 voters that represent 55% of the world's polluters. Its primary drawback, however, is that like many treaties since the 1997 Kyoto Protocol, it does not have binding agreements.

While the US pledged over 25% reductions by 2025 and China has pledged to reduce its current emissions by over 45%, it has yet to be seen how this will play out. Critics argue that while the COP21 represents progress, the proposed cuts will still allow an unacceptable 2.5 to 3 degree Celsius rise in temperature. Many scientists on the IPCC and elsewhere claim only net-zero global emissions sometime between 2030 and 2050 will prevent the complex and cascading effects of climate change. Unfortunately, the US has since pulled out of the Paris agreement making it harder for the world to meet the agreement's reduction targets.

Despite the difficulties of bringing together diverse ideas, institutions and knowledge systems, people who would have never before sat down together have reached agreement within the framework of *sustainability* and *sustainable development*. Keeping this in mind, we must continue to reinvent programs and policies that include the preservation of natural resources, biodiversity, and our life-support systems for many generations to come. The future success of the sustainability paradigm, or its failure as a constructive discourse, depends on its dedication to pluralistically driven debates and solutions that address the serious and global crises that face the world today.

Chapter 3

Earliest Urbanism

Charles Redman

What can History Tell us about Sustainability?

History holds many lessons for better understanding the operation of modern cities and possible directions for developing urban sustainability. Although more people reside in cities than in the past and the technologies we use are dramatically different, human interactions and the ways in which we organize ourselves are very much the same. If examined carefully, history can provide seemingly countless examples of how people around the world cooperated to live in ever larger settlements, how they solved problems, even how they sowed the seeds for their own collapse.

Sustainability is about solving problems as societies confront change, successful human interaction over long periods, and preventing imbalances between the extraction of resources from our natural environment and its ability to regenerate those resources. There are numerous advantages of using history in conjunction with contemporary studies. Among them is our ability to see how decisions and actions have played out over time, and whether they led to further growth or to decline and abandonment. Another is the large number of case studies, which include important processes such as experiencing global climate change or the introduction of fundamentally new technologies.

One particularly insightful message from the past is that the collapse of many cities and states occurred when they appeared to be at their peak of power and sophistication, and not after long periods of decline. This message has direct relevance for today's societies as it can lead to a better understanding of how and why the "great" societies of the past have repeatedly collapsed. This can help us redirect our trajectory toward sustainability instead of collapse.

One of the most significant and informative milestones in the history of human societies was the growth of the first cities in the Near East. This process, often called **The Urban Revolution**, involved much more than just an increase in the size of settlements. It included fundamental changes in the way people interacted, in their relationship to the environment, and in the very way they structured their

The Urban Revolution Processes that originated during the growth of the first cities in the Near East; encompasses the behaviors that simultaneously occurred as people began to reside in urban communities, such as the cultivation of crops, herding of animals, the mass production of goods, and the development of a writing system.

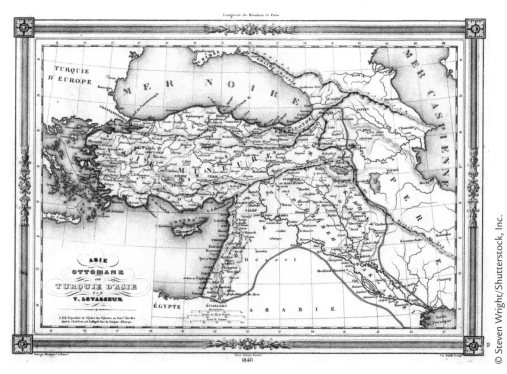

Historic map of Iraq dated 1872 showing "Mesopotamien," which is German for Mesopotamia

Human-environmental dynamics The study of complex interactions between natural systems and human activity that examines the causes and consequences of human impacts on the environment and human adaptations to environmental change.

communities. Processes and institutions that began at this time, some 5,000 years ago, have continued to evolve, forming the basic structure of urban society today.

Although ancient cities emerged in various regions of the world, this chapter focuses on the process as it occurred in the Near East, and in particular Mesopotamia (present day Iraq and southwestern Iran), for three important reasons. First, these cities appear in Mesopotamia as early as, or earlier than, anywhere else. Second, these cities and their societies have direct links with developments in neighboring regions and can be traced as ancestral to developments in many parts of the modern world. Third, we can identify early examples of **human-environmental dynamics** that are still operating today.

Pre-urban, Neolithic settlement (8000–5000 BC) of Mesopotamia concentrated in the uplands, with only a few settlements around the margins of the lowlands where the upper reaches of the rivers and their major tributaries intersected in what are now southeastern Turkey and the Iran-Iraq borderlands. The earliest that researchers have found evidence of any substantial settlement in southern Mesopotamia is approximately 5500 BC. Between 6000 and 4000 BC, however, the area of settlement expanded from the uplands to include more and more of the Mesopotamian plain. This was not a rapid migration in terms of a single lifetime. It took many generations to learn how to manage crops and animals in the heat and aridity of the lowlands. At first, expansion was limited to areas of possible, although unreliable, rainfall. Subsequently, with the aid of primitive irrigation systems, settlers moved into areas of the plain that previously could not be cultivated by rainfall alone.

The physical environment of Mesopotamia and the surrounding regions provided promising ecological conditions for the early introduction of agriculture and subsequent growth of the first urban society. Writing, a system of laws, the wheel, the plow, metallurgy, mathematics and engineering principles—all commonplace in our modern world—were first developed in the cities of Mesopotamia. Despite the vast scope of these technical innovations, the most significant developments were those of **social organization**. A quantum leap occurred in the number of inhabitants living in the largest settlements, and with this growth, transformation occurred in the way people dealt with each other and with their environment. Understanding these dynamics and how they supported the human-nature relationships that were to emerge is essential to a thorough understanding of **coupled human-nature interactions** today.

Social organization How people interact with each other, including the kinship systems they use and how they assign communal tasks, decide who has access to goods and knowledge, and make communal decisions.

Coupled human-nature interactions The connections or links between humans and their environment; expressed as a feedback loop, many of humans' actions affect the environment and in turn, changes in the environment affect the human population.

The Environmental Stage

Mesopotamia is a large, arid alluvial plain created by two major rivers—the Tigris and the Euphrates. It is surrounded on two sides by better-watered mountainous zones. The region's climatic pattern is one of summer drought and winter rainfall, although the lowland plain receives minimal rainfall and the people there derive most of their water from the rivers.

In the south, where the Tigris and Euphrates rivers join and eventually empty into the Persian Gulf, the land is almost flat and there are many marshy areas. Moving upstream, to the northwest, the slope is small, but it increases perceptibly, giving rise to more clearly defined watercourses surrounded by arid plains. Furthest from the rivers were lowland areas that were marshes during the flood season and supported natural grasses the rest of the year, making them useful for grazing animals.

Flat Alluvial Plain

Nearer the rivers, farmlands developed as they were within reach of irrigation water from the river, and were better drained due to the natural levees along the riverbank, making them more productive. These natural levees were the best locations for intensive cultivation and settlement as they offered several natural advantages: they were fertile, quickly drained after floods, and were least vulnerable to winter frosts. Equally important, as the density of farmers increased, the levees gave access to river water during years in which the river level was low and not all fields could be adequately irrigated.

Moving further upstream, to the northerly areas of the Mesopotamian plain, the gradient increases, the landscape becomes rolling, and the rivers cut deeper into the landscape, making irrigation more difficult. In the most northern regions, the major rivers and their tributaries cut across a series of increasingly high ridges, ultimately reaching the Taurus Mountains to the north and the Zagros Mountains to the

Irrigation Canal

Upstream Euphrates River in SE Turkey

Ecosystem management The use, protection, and conservation of our environmental resources in a way that seeks to ensure their long-term sustainability. This concept considers humans and the environment as a single system rather than as individual parts.

northeast. Proximity to these uplands and the large tributary rivers provided natural advantages to settlement in the north, such as access to stone and timber for building, while the vast stretches of relatively easily irrigated land in the south facilitated population growth there.

Processes of Change

History offers many lessons relevant to sustainability about how humans and their societies have recognized and responded to challenges and opportunities of their human-natural environment. Three of the basic approaches to problem solving in antiquity were: (a) mobility of people to available resources, (b) **ecosystem management** to ensure and enhance local resources, and (c) increasing social complexity encoded in formal institutions that guided an ever-expanding range of activities. These solutions were fundamental to the rise of early civilizations. More important, they are instrumental in the design of sustainable cities and continue to evolve.

The mobility of people to available resources has dominated the human approach to securing adequate subsistence for the vast majority of human existence. Until approximately 10,000 years ago (and more recently in many regions), virtually all people had to move between several locations each year to take advantage of the seasonal resources. This movement pattern was disrupted by the introduction of agriculture, which allowed the establishment of year-round settlements in most regions of the world. Agriculture is an example of ecosystem management for *enhanced productivity.* In fact, it has proven to be an astonishing successful solution to feeding an ever-increasing global population, thereby allowing virtually all people to live in permanent settlements.

The advancement of agriculture and the infrastructural improvements made to enhance productivity were strong incentives for the spread and growth of sedentary communities. Scholars believe that a highly effective socio-natural

Families on Annual Migration near Border of Northern Iran

pattern—village farming communities—emerged from millennia of experimentation and became the dominant settlement form across the globe. These communities were made up of between 1 and 500 people. They were the most enduring and widespread type of community because they had flexibility in their sources of subsistence, and balanced the extraction of natural resources and regeneration of the local ecosystem. This early form of human settlement spread to most continents, housing over half the world's population as recently as the middle of the twentieth century!

Traditional Plow near Village Farming Community in SE Turkey

It endured as the dominant community form for millennia because it proved to be a highly **resilient socioeconomic unit**. This resiliency allowed it to adapt and survive in the face of many challenges both natural and human induced. The modest number of people in a village meant that traditional face-to-face social organization was effective, the economy was flexible as it was based on a wide range of resources, and for the most part villages did not over-exploit the productive potential of the region.

Over time, some communities expanded their approach to ecosystem management enabling larger aggregations of people. The larger population required a division of labor beyond subsistence to support emerging institutions that resulted in a transformation in the social order. This change was largely achieved through innovations in social complexity and is at the heart of the Urban Revolution.

Resilient socioeconomic unit Characterizes the interconnectedness between the human inhabitants and the economy of a place in a way that promotes the ability to bounce back and recover after a disturbance or crisis.

The First Cities

The formation of the first cities and their linking together as one civilization on the Mesopotamian plain was relatively rapid. Approximately 2,000 years after the earliest known occupation of the region, about 5,500 BC, cities emerged along with attendant traits of urbanism such as writing, the construction of monumental buildings, and craft specialization. The rise of these early cities was not simply the growth of large collections of people—rather, it involved communities that were far more diverse and interdependent than their predecessors.

Relative independence and self-sufficiency characterized village farming communities, but they also limited the growth of these communities. Specialization in the production of various goods and complex exchange networks were necessary for the growth of urban societies. Cities were also interdependent with their surrounding towns and villages, developing mechanisms for obtaining goods and services from them. The development of effective irrigation agriculture, the manufacture and widespread exchange of goods, and the advance of science and mathematics were fundamental to the growth of cities. Additionally, changes in the social realm, such as a class-structured society, formalized systems of laws, a monopoly on the use of coercive force, and a hierarchical, territorially based government made cities possible. These traits continue to characterize the successful operation of cities today.

The landscape-productivity-human relationship that evolved in villages and towns enabled the growth of large, diverse populations that ultimately aggregated

Early Farming Communities in the Middle East

into cities. As positive as many of these "advances" were in terms of productivity and competitiveness, they also led to ambiguous relationships and increased long-term risks. For example, the concept of private property emerged to delineate individual ownership and community ownership. These new relationships evolved as farmers began producing more food than their families required and found ways to store this surplus for later use or trade. This led some farmers to produce a surplus to guard against future bad harvests. However, a variety of factors limited the amount of food that could be effectively stored. Thus, the stimulus to produce a surplus remained limited in most farming villages.

What may have changed this relationship, and was key to the growth of urban society, was the ability to transform locally produced surplus food and goods into enduring prestige items associated with elevated status. This only existed where a new social order occurred that acknowledged classes with differential wealth (greater access to productive resources, specialized labor, and exotic goods), power, and status. An ideology (through religion, myth, constructed history, and law) legitimized the existence of elite classes and the goods that helped identify them.

Another dynamic that coevolved with private property, surplus production, elite goods, and hierarchical class society was the reliance on inheritance for membership in these classes. Merit, strength, agility, and intelligence certainly were important, but the family, clan, or class a person was born into set limits on their future potential in the age of early cities. To some extent, this class structure continues to operate today both blatantly and subtly, presenting numerous social inequity and injustice challenges.

Cities Going Global

People have always sought to make sense of the world around them, developing myriad mechanisms for relating to their environment as well as other people. Organizing society into hierarchically stratified classes coincided with urbanism, and has come to characterize most of society today. This social framework, along with the widely accepted ideologies legitimizing it, was seen as an effective

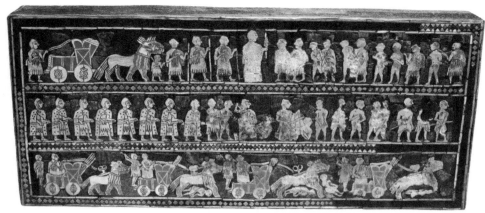

Legitimizing the Social Order

means of organizing large groups of people and large-scale productive activities. Myths, histories, and ethics often were organized into formal religions and social constructs that helped people understand their place in the world while simultaneously justifying the unequal allocation of rights and prescribing acceptable social norms.

In time, territorially based authority also emerged, largely through successful military action and monopolies on the use of coercive force. This secular authority also needed a source of legitimization, which emerged as constructed histories, legal codes, and institutions of management and enforcement. Although they often cited different rationales, governing authorities served similar purposes in creating a framework for the effective functioning of large populations and controlling vast tracts of land. Not surprisingly, the two interacted closely, and they have often been unified into a single entity or a closely cooperating team.

In Mesopotamia, sacred orders were established and widely accepted that legitimized the social order and prescribed appropriate behavior. The "ruling class"

Dating back to 3500 BC, Mesopotamian art intended to serve as a way to glorify powerful rulers and their connection to divinity

provided security through a monopoly on the use of force and formal systems of laws. Respect for this new social and governing order was reaffirmed through the construction of massive monuments, the performance of complex rituals, and large-scale art that represented their ordained rights. The concentration of people, stored supplies, and elite goods led to these early cities being targets for raiding and organized military activity, prompting further investment in defensive walls and armies to defend them. The relationship of the concentration of wealth leading to military aggression leading to the investment in armies is a cycle that dominates all of human history. This disruptive cycle is one that continues today and presents a significant sustainability challenge for many urban areas.

Are Traditional Cities Maladaptive?

Unfortunately, military and other responses that originated to solve problems have also established detrimental cycles that threaten the very existence of the societies that developed them. For example, large numbers of people aggregating into cities allowed for specialization of labor and other efficiencies, but it also meant that most people did not grow their own food. Hence, people in surrounding rural areas were responsible for growing enough food for themselves and city dwellers. This arrangement put a tremendous burden on farming communities to produce much more than they typically would.

As the population grew more divergent in their societal roles, so did their objectives and understanding of the world around them. For instance, farmers in the earlier village farming era more intimately understood the landscape and productive systems, and were inclined toward conservation practices that balanced extractive activities with the **regenerative capabilities** of the land. The urban elite, on the other hand, were more focused on the net produce they could extract from the countryside and insisted on maximum production, with little knowledge of or concern for the potential deleterious effects on the rural landscape.

In an ideal hierarchical society, knowledge travels up the hierarchy and informed decisions are made. It is somewhat surprising that this is seldom the case and that the dominant pattern has been **maximizing short-term returns with little concern for long-term consequences**. Archaeological evidence attests to repeated instances of intense environmental degradation in the region around cities, and the impact of urban demand on the rural countryside is still an issue today.

In addition to making ill-advised decisions leading to excess extraction of resources, individuals and societies are regularly forced to make decisions in response to changing conditions. These conditions might include changes in climate, such as less rainfall; in the local environment, such as soil erosion; or in cultural or technological contexts, such as new neighbors or the introduction of new subsistence strategies. Jared Diamond, in his book *Collapse: How Societies Choose to Fail or Succeed* (2006) points to errors in decision making as a primary cause of social collapse. Such errors occurred in a variety of contexts, such as not recognizing the significance of an important change in conditions; recognizing the change, but misdiagnosing the appropriate response; or waiting too long before responding. Today, many leaders are repeating some of these mistakes and exhibiting behavior that is strikingly similar to that of the collapsed civilizations.

Fortunately, Diamond also provides examples of leaders and citizenry that have acted in perceptive, positive ways to maintain and even enhance human-environmental relations. For example, he recounts how in the early 17th Century, Japan was at a crossroads where war and population growth could have led to serious environmental degradation and potential collapse. Insightfully, the new dynasty of Tokugawa shoguns (emperors) made the decision to protect and restore large areas of forests both as a potential natural resource and to protect against major soil erosion. This provided the foundation for centuries of high population density coupled with environmental stability.

Regenerative capabilities The processes of renewal, restoration, and growth that make ecosystems resilient to natural events in order to prevent irreversible damage.

Maximizing short-term returns with little concern for long-term consequences This concept encompasses the idea that many decisions made by those individuals with authority and power often focus solely on increasing short-term benefits, such as the accumulation of money, without considering long-term consequences of those decisions, such as degradation of the environment.

Other "urban efficiencies" have created their own challenges. While cities were the centers of economic and cultural activities, they were also the centers of disease. Many of the world's devastating contagious diseases were virtually nonexistent until the growth of dense urban populations. The spread of the plague, small pox, measles, cholera, and many other diseases can be traced to a combination of close association between humans and domestic animals, and living in large, dense populations. Large urban populations also created challenges that were unknown when the largest communities consisted of several hundred people or less.

For example, knowing who everyone is in the community and how to act towards them is not feasible when the community's population is in the thousands or hundreds of thousands. Similarly, security risks escalate as the population grows larger and less connected, requiring the introduction of formal, impersonal governance of human interactions. Certain other challenges become more complex as the population enlarges, such as the transport of people and goods, sanitation, and an adequate supply of potable water and food. Looking to the past to reflect on what has worked, and what has not, when responding to these challenges is imperative for not repeating the mistakes that led to the collapse of many of humankind's greatest civilizations.

Speculations for the future based on the past

Several lessons to be learned are apparent from this review of earliest urbanism. First, humans are amazingly successful at self-organization to promote their survival in the face of virtually any environmental challenge, but many of these solutions have unanticipated costs with continuing impacts on future societies. People manage their socio-ecosystems according to how they perceive the opportunities and risks and how they value the alternatives. Yet, this valuation process is very different among people in different social positions. Furthermore, the true "costs" of some alternatives are not easily recognizable and may even threaten the society's very survival.

Second, challenges may take unrecognized forms as they occur in a new context. Moreover, people may ignore or minimize the expected impacts due to a sense of powerlessness or lack of experience and/or urgency. In agrarian communities of the past, transforming biota (the flora and fauna of the region) and landscapes was done with the understanding that the community's livelihood, even its existence, depended upon a balance between human and natural systems. Today's emphasis on net yield for human systems without regard for the impacts on native biota and earth systems is already showing signs of irreversible, irreparable ecological changes. While we do not yet know the full extent of the impacts this paradigm holds, history provides us with ample examples that in all likelihood it will eventually undermine the sustainability of today's cities. To prevent this, we must devote ourselves to forming a better understanding of the impacts of our decisions and the pathway toward a more sustainable existence. Following are three suggestions for moving toward sustainable cities.

First, we must understand that environmental, social, and technological change is the norm and not the exception. Therefore, individuals and societies must prepare to respond to those changes in sustainable ways. Second, any actions taken to respond to these changes will involve trade offs. That is, some parts of the system will benefit while others will likely be diminished. To ensure that the impacts are both minimal and equitable, we must continually strive to 1) gain a thorough understanding of the interconnectivity of all social-ecological systems and 2) recognize that cascading and unintended consequences will result even from the best-intentioned actions. Third, the world and its challenges are perceived and valued differently by people in different positions within society. What may help one segment of society may hurt another. All too often, decision makers act in ways that benefit a specific segment of society at the expense of other segments of society. This inequality in power, wealth, and access to resources has very old origins. In designing a more sustainable world, we would be well advised to carefully study how the formation and growth of early cities resulted in societal inequality, and strive to discern and implement solutions to reverse this disparity.

Chapter 4

What Should Sustainable Cities Look Like? Programs, Policies, and Initiatives

Bjoern Hagen and Ariane Middel

Introduction

Today, more people live in city environments than in rural areas. Projections suggest that by 2050, approximately 75 percent of the world's population will live in cities. Ongoing rapid **urbanization** is transforming natural landscapes and ecosystems, and creating complex social, economic, and environmental challenges that require sustainable development solutions. Population growth in cities is coupled with transformative changes including the growth of megacities, expansion of urban poverty, especially in developing countries, and growing concern over food security, among other issues.

> **Urbanization** Urbanization describes the process of more and more people moving from rural to urban areas. As a result cities are growing in size and population.

As shown in Table 1, there are already 12 cities with a population of over 20 million (Demographia, 2016); the largest of which is the Tokyo metro area with over 37 million inhabitants. It is expected that numerous **megacities** will merge in this century with many transforming into megaregions with potentially over 100 million inhabitants (UN Habitat, 2012). Some of the existing megaregions include the Hong Kong-Shenhzen-Guangzhou region in China with about 120 million people and the Rio de Janeiro-São Paulo region in Brazil with more than 43 million people. The increasing number of cities and people living in urban areas offers an opportunity to tackle environmental and social issues through thoughtful sustainable policies and development strategies.

> **Megacities** Megacities are cities or metropolitan areas with more than 10 million inhabitants

There is a long history of sustainable urban planning and combining natural and man-made landscapes. In the last century, Ebenezer Howard's Garden City can be considered a sustainable urban development concept (Parsons & Schuyler, 2002). In his book "Garden Cities of To-Morrow" (Howard, 1902), he outlines a settlement structure consisting of self-contained communities with mixed residential, commercial, and recreational zones surrounded by greenbelts. However, only two garden cities were built. Letchworth and Welwyn were both located in England and loosely based on Howard's concept. Although the Garden City concept was mostly utopian, it incorporated many principles that underline today's sustainable urban development efforts (Caine, 2011). The concept aimed to combine the best of nature and urban environments in a planned and controlled way while providing high-quality urban fabric, economic development, community well-being, quality of life, and

Table 1 World's Largest Urban Areas in 2016

Rank	Urban Area	Country	Estimated Population 2016	Land Area (mi²)
1	Tokyo-Yokohama	Japan	37,750,000	3,300
2	Jakarta	Indonesia	31,320,000	1,245
3	Delhi	India	25,735,000	835
4	Seoul-Incheon	South Korea	23,575,000	1000
5	Manila	Philippines	22,930,000	630
6	Mumbai	India	22,885,000	340
7	Karachi	Pakistan	22,825,000	365
8	Shanghai	China	22,685,000	1,500
9	New York	Unites States	20,685,000	4,495
10	Sao Paulo	Brazil	20,650,000	1,045
11	Beijing	China	20,390,000	1,520
12	Mexico City	Mexico	20,230,000	800
13	Guangzhou-Foshan	China	18,760,000	1,475
14	Osaka-Kobe-Kyoto	Japan	16,985,000	1,240
15	Moscow	Russia	16,750,000	2,050
16	Dhaka	Bangladesh	16,235,000	142
17	Cairo	Egypt	15,910,000	680
18	Bangkok	Thailand	15,315,000	1,000
19	Los Angeles	United States	15,135,000	2,432
20	Kolkata	India	14,810,000	465
21	Buenos Aires	Argentina	14,280,000	1,035
22	Tehran	Iran	13,670,000	630
23	Istanbul	Turkey	13,520,000	525
24	Lagos	Nigeria	12,830,000	550
25	Shenzhen	China	12,240,000	675

Source: (Demographia, 2016)

social equity. Transportation by rail between urban villages was another idea that is showing promise today in the promulgation of transit-oriented developments.

While early sustainable development concepts focused primarily on creating new cities and communities, present day planners and urban decision-makers face the task of retrofitting existing cities with expanding populations (Vince, 2012) to be more resilient to climate change. Rapid urbanization presents dynamic sustainability challenges for the city's infrastructure, e.g., with regard to water and waste management, power usage, transportation, green building design, as well as GHG emissions and air quality issues. To achieve sustainability in urban communities, existing literature points to five dimensions of sustainable development: 1) economic sustainability, 2) social sustainability, 3) ecological sustainability, 4) sustainable spatial development, and 5) cultural continuity (Ahmedi & Toghyani, 2011).

A city can only evolve sustainably if all five dimensions are considered. City governments have started to recognize the necessity to plan for climate change by mitigating GHG emissions through a number of strategies and increasing urban resiliency to adverse climate change related impacts (C40 Cities: Climate Leadership Group, 2014). A recent study (Zottis, 2014) shows that 75 percent of 350 cities surveyed perceive climate change as a pressing issue for consideration in their overall urban planning initiatives and sustainable development strategies.

To evaluate city efforts to become more sustainable, recognize successful cases, and identify areas for improvement, cities are rated and ranked according to their sustainability-related performance. Different rating systems, such as the "The Green City Index" (The Economist Intelligence Unit, 2012), the "Resilient Cities Ranking (Grosvenor, 2014), the "Livability Ranking" (The Economist Intelligence Unit, 2016), and the "Sustainable Cities Index" (Arcadis, 2015), measure important sustainable development aspects. Table 2 shows the top and bottom ten rankings for these four rating systems.

The Green City Index (GCI) assesses the environmental performance of over 120 cities worldwide based on up to 30 indicators depending on the region and data availability. The indicators are grouped into categories, such as CO_2 emissions, energy, efficient buildings, land use, transport, water and sanitation, waste management, air quality, and environmental governance. The categories are then measured using comparative indicators (The Economist Intelligence Unit, 2012a). In Europe, the top three green cities are all from Scandinavia, whereas the bottom three are from Eastern Europe. In the United States and Canada, the top three cities are San Francisco, Vancouver, and New York, with Cleveland, St. Louis, and Detroit comprising the bottom three.

The Resilient Cities Ranking system examines the capabilities of cities to contend with negative impacts from events stemming from climate change, rural to urban migration, and globalization while simultaneously functioning as a hub for human, economic, and cultural development (Grosvenor, 2014). Assessing a city's resiliency is a two-step process. First, indicators are used to evaluate the vulnerability and its infrastructure to occurring or projected environmental stresses or disasters. The indicators determine the vulnerability to negative changes in climate, environment, availability in resources, infrastructure, and within the social fabric. A second set of indicators is used to determine the adaptive capacity of the city in terms of how well prepared the government and other vital institutions, the planning system, funding structures, and other key areas are. In terms of urban resilience, the Canadian cities of Toronto, Vancouver, and Calgary occupy the top three spots in a ranking of 50 cities worldwide.

The Livability Index ranks cities according to the quality of life they offer. Cities are scored based on over 30 indicators organized into five categories: a) stability,

Table 2 City Ranking Comparisons

The Green City Index (GCI) Europe		The Green City Index (GCI) US and Canada		Resilient Cities Ranking		Livability Index		Sustainable Cities Index	
Rank	City	Rank	City	Rank	City	Rank	City	Rank	City
1	Copenhagen Denmark	1	San Francisco United States	1	Toronto Canada	1	Melbourne Australia	1	Frankfurt Germany
2	Stockholm Sweden	2	Vancouver Canada	2	Vancouver Canada	2	Vienna Austria	2	London England
3	Oslo Norway	3	New York City United States	3	Calgary Canada	3	Vancouver Canada	3	Copenhagen Denmark
4	Vienna Austria	4	Seattle United States	4	Chicago United States	4	Toronto Canada	4	Amsterdam Netherlands
5	Amsterdam Netherlands	5	Denver United States	5	Pittsburgh United States	5	Calgary Canada	5	Rotterdam Netherlands
6	Zurich Switzerland	6	Boston United States	6	Stockholm Sweden	5	Adelaide Australia	6	Berlin Germany
7	Helsinki Finland	7	Los Angeles United States	7	Boston United States	7	Perth Australia	7	Seoul South Korea
8	Berlin Germany	8	Washington DC United States	8	Zurich Switzerland	8	Auckland New Zealand	8	Hong Kong China
9	Brussels Belgium	9	Toronto Canada	9	Washington DC United States	9	Helsinki Finland	9	Madrid Spain
10	Paris France	10	Minneapolis United States	10	Atlanta United States	10	Hamburg Germany	10	Singapore Singapore
...		...		...		...		...	
21	Dublin Ireland	18	Orlando United States	41	Sao Paulo Brazil	131	Kiev Ukraine	41	Doha Qatar

22	Athens, Greece	19	Montreal, Canada	42	Delhi, India	132	Douala, Cameroon	42	Moscow, Russia
23	Tallinn, Estonia	20	Charlotte, United States	43	Guangzhou, China	133	Harare, Zimbabwe	43	Jeddah, Saudi Arabia
24	Prague, Czech Republic	21	Atlanta, United States	44	Mexico City, Mexico	134	Karachi, Pakistan	44	Riyadh, Saudi Arabia
25	Istanbul, Turkey	22	Miami, United States	45	Rio de Janeiro, Brazil	134	Algiers, Algeria	45	Jakarta, Indonesia
26	Zagreb, Croatia	23	Pittsburgh, United States	46	Mumbai, India	136	Port Moresby, Papua New Guinea	46	Manila, Philippines
27	Belgrade, Serbia	24	Phoenix, United States	47	Manila, Philippines	137	Dhaka, Bangladesh	47	Mumbai, India
28	Bucharest, Romania	25	Cleveland, United States	48	Cairo, Egypt	138	Lagos, Nigeria	48	Wuhan, China
29	Sofia, Bulgaria	26	St Louis, United States	49	Jakarta, Indonesia	139	Tripoli, Libya	49	New Delhi, India
30	Kiev, Ukraine	27	Detroit, United States	50	Dhaka, Bangladesh	140	Damascus, Syria	50	Nairobi, Kenya

Adapted from Green City Index, 2016; Resilient Cities Ranking, 2016; Livability Ranking, 2016; and Sustainable Cities Index, 2016.

b) healthcare, c) culture and environment, d) education, and e) infrastructure. After the scores are collected, weighted, and combined, each participating city receives a score of 1–100. The higher the score, the better the living conditions. Out of the 140 cities surveyed in 2016, Canada and Australia have the most cities in the top ten of the livability ranking. The top three were Melbourne, Vienna, and Vancouver. The bottom three were Lagos, Tripoli, and Damascus.

The fourth and final ranking is the "Sustainable Cities Index" (Arcadis, 2016) focused on 50 cites and 31 different countries. Cities were chosen to provide a global representative sample of urban environments with different geographical, economic, and social characteristics as well as various sustainability challenges. They were ranked based on twenty indicators grouped into three sub-indexes. The first group of indicators assessed the social performance of the city, acknowledging aspects such as literacy, education, green spaces, health, property prices, or transportation infrastructure. The second group evolved around environmental factors including air pollution, GHG emissions, solid waste management, or energy use and renewables mix. The third and final group covered the business environment and economic performance of the selected cities. These indicators focused on the GDP per capita, the ease and costs of doing business, the city's importance to the global economy, and the total energy consumption per dollar of GDP. Overall, well established European cities performed best. Cities like Frankfurt, London, Copenhagen, and Amsterdam performed very well across all three sub-categories, whereas cities from North America are missing in the top 10 all together. The highest ranked North American city is Toronto in 12th position followed by Boston in the 15th spot and Chicago at 19.

Although these indicator systems (among others) measure different dimensions of sustainability, the same cities tend to rank fairly well, demonstrating common characteristics, policies, and initiatives. North American cities like Toronto, Calgary, Vancouver, Washington DC, and Boston are often found among the highest ranked cities. In Europe, Vienna, Copenhagen, Berlin, Zurich, and Amsterdam always perform well. The following sections of this chapter examine three of the most sustainable cities in the world more closely. Vancouver, Freiburg, and Stockholm are all well known for proactive policies and comprehensive planning, strategies emphasizing a commitment to becoming more sustainable, green, and resilient. They offer experiences from which other city governments can learn as well as best practices that can be transferred and implemented elsewhere. Stockholm and Vancouver appear on a variety of rating systems. Freiburg, however, is usually too small in terms of its population to be part of most indicator rating systems that only focus on major metropolitan areas. Yet, it is internationally recognized for its sustainable principles in its plans and buildings (Beatley, 2010).

Vancouver, Canada

Background

Vancouver is a coastal seaport city in British Columbia, Western Canada. With a population of just over 600,000 people and a land area of 115 square kilometers (Statistics Canada, 2011), the City of Vancouver is the eighth most populous

Canadian municipality and the most densely populated city in Canada. The Greater Vancouver area, with 2.3 million residents, is the third most populous metropolitan area in Canada. Archaeological finds suggest that the first settlements in Vancouver date back more than 3,000 years (City of Vancouver, 2014) with several Coast Salish First Nations, among them the Musqueam and Squamish indigenous peoples, having villages near the mouth of the Fraser River.

Today, shipping and trade are key to Vancouver's economy. The Port of Vancouver is Canada's largest and busiest port and a major gateway for Pacific trade. In 2013, the port traded goods worth $200 billion with more than 170 trading economies (Port Metro Vancouver, 2016). Other key economic sectors for the City of Vancouver include forestry, mining, film, and tourism.

The Economist Intelligence Unit (EIU) has consistently ranked Vancouver as one of the top five most livable cities in the world. Although the city is one of the most densely populated urban areas in Canada, it features abundant green spaces and parks, waterfront activities, and nearby nature parks and mountains. Vancouver is also one of the most ethnically and linguistically mixed cities in Canada and home to worldwide cultural groups. Fifty-two percent of the population speaks a first language other than English (City of Vancouver, 2014). Vancouver's diversity and multiculturalism are seen as a major source of the city's vitality and prosperity, attracting visitors and investment.

In the US and Canada Green City Index, Vancouver is ranked second overall, with top scores in the CO_2 and air quality categories (EIU, 2012a). The concepts of sustainability and livability are centrally placed on the city's development and policy agenda and constitute the basis for several initiatives and campaigns to address environmental challenges. The following section touches on the "Greenest City 2020 Action Plan," an initiative to make Vancouver the greenest, most livable city in the world by 2020 as well as presenting key aspects of Vancouver's sustainability.

Greenest City Plan

Although Vancouver currently has the smallest per capita **carbon footprint** of all US and Canadian cities as a result of policies promoting green energy and the dominance of **hydropower**, the city is not immune to environmental challenges. These include continued population growth, uncertainties of climatic change, and rising fossil fuel prices. To address these challenges proactively and remain one of the most livable and sustainable places, Vancouver started an initiative in 2009 to become the greenest city in the world by 2020.

The City Council developed a comprehensive plan, the "Greenest City 2020 Action Plan" (City of Vancouver, 2012), which builds on research findings from a team of local experts on best practices from sustainable cities around the world. Over 35,000 people helped develop the plan, providing input through social media, workshops, and events, setting an example for best practices in government-citizen collaboration. The action plan establishes ten measurable and attainable targets, each with a long-term (year 2050) goal and medium-term (year 2020) goal, baseline numbers to describe the current status of each target, highest priority actions for the next

Carbon Footprint The amount of greenhouse gases (GHG) emitted in a given time frame to directly and indirectly support human activities. The carbon footprint is usually expressed in "equivalent tons of carbon dioxide" (tCO_2e) and calculated for the time period of a year.

Hydropower Electricity that is generated using the energy of moving water.

three years, and detailed steps and actions required to achieve the goals. The ten targets focus on the overarching areas of carbon, waste, and ecosystems.

Climate Leadership

Renewable
Energy Energy sources
which are natural, such
as sunlight, wind, and
geothermal heat. Due
to peak oil and growing
environmental concerns,
more governments
support the use of
renewable energy sources

Vancouver's electricity mainly stems from **renewable energy** sources. The city benefits from its location close to mountains and draws most of its power from hydroelectric generating stations. Hydropower plants (usually situated at dams) use falling water to propel turbines to produce electricity. Renewable energy from water is highly efficient, with more than 95 percent of energy converted into electricity and exceptionally low GHG emissions. Because of Vancouver's green energy portfolio, per capita GHG emissions are as low as 4.2 metric tons of CO_2, considerably less than the Green City Index average of 14.5 metric tons (EIU, 2012).

Vancouver plans to further reduce its community-based GHG emissions 33% by 2020 from its 2007 levels (Table 3), despite expected economic growth and increasing population. A key strategy for realizing this goal is promoting local energy solutions and developing neighborhood energy systems, such as the Neighborhood Energy Utility (NEU). Currently, NEU provides a high-density residential area near the Olympic village with locally generated heat and hot water using thermal energy from sewage, a green technology that reduces GHG emissions from heating by more than 60 percent. Dense neighborhoods are particularly suited for this kind of renewable energy, as multiple buildings can be serviced simultaneously, making the system more cost effective.

Green Transportation

Due to Vancouver's geography and several municipal bylaws, Vancouver is one of the few Canadian metropolises without an extensive freeway network. Highway 1, the Trans-Canada Highway, is the only freeway within city limits, and it does not pass through the downtown area. A freeway through the center of the city was proposed in the late 1960's. However, public protest by residents, activists, and community leaders prevented its implementation.

Today, Vancouver has a long and dense public transit system (5.4 miles per square mile), a constantly growing network of bike routes, and the longest automated light rail system in the world (EIU, 2012). In 2012, the city adopted a transportation plan that sets targets for all transportation modes for the next 30 years, aiming for over 50% of all trips by foot, bike, or transit by 2020 and at least two-thirds by 2040. The City of Vancouver's Healthy City Strategy estimates that the city met it's 2020 goal of 50% trips made by walking, biking, or transit in 2014 (City of Vancouver, 2016 http://vancouver.ca/people-programs/getting-around.aspx).

Biking is the fastest growing transportation mode. The number of bike trips has steadily increased, with an overall trip number of 131,025 in 2015, up from 99,100 in 2014 and 83,300 in 2013 (CH2M, 2016, pg. v). Vancouver's extensive bike network consists of over 300 painted-lane kilometers. Most of the bike routes are bike boulevards, low-speed streets that have been optimized for biking and where traffic is calmed through circles and signals.

Table 3 Greenest City 2020 Action Plan mid-term targets for reductions in carbon emissions and waste, and improvements to the City's ecosystems

Goals	Targets	Indicator	Baseline	2020 Target
Green Economy	Double the number of green jobs over 2010 levels by 2020	Total number of local food and green jobs	16,700 jobs	33,400 jobs
	Double the number of companies that are actively engaged in greening their operations over 2011 levels, by 2020	Percent of businesses engaged in greening their operations	5% of businesses engaged	10% of businesses engaged
Climate Leadership	Reduce community-based greenhouse gas emissions by 33% from 2007 levels	Total tons of community CO_2e emissions from Vancouver	2,755,000 tCO_2e	1,846,000 tCO_2e
Green Buildings	Require all buildings constructed from 2020 onward to be carbon neutral in operations	Total tons of CO_2e from residential and commercial buildings	1,145,000 tCO_2e	920,000 tCO_2e
	Reduce energy use and GHG emissions in existing buildings by 20% over 2007 levels	Total tons of CO_2e from residential and commercial buildings	1,145,000 tCO_2e	920,000 tCO_2e
Green Transportation	Make the majority of trips (over 50%) by foot, bicycle, and public transit	Per cent of trips by foot, bicycle, and transit	40% of trips	50% sustainable mode share
	Reduce the average distance driven per resident by 20% from 2007 levels	Total vehicle km driven per person	N/A	20% below 2007 levels
Zero Waste	Reduce total solid waste going to the landfill or incinerator by 50% from 2008 levels	Annual solid waste disposed to landfill or incinerator from Vancouver	480,000 tons	240,000 tons
Access to Nature	Ensure that every person lives within a five-minute walk of a park, greenway, or other green space by 2020	Per cent of city's land base within a five-minute walk to a green space	92.6%	95%
	Plant 150,000 additional trees in the city between 2010 and 2020	Total number of additional trees planted		150,000
Lighter Footprint	Reduce Vancouver's ecological footprint by 33% over 2006 levels	Number of people empowered by a City-led or City-supported project to take personal action in support of a Greenest City goal and/or to reduce levels of consumption	600 people	to be determined

(Continued)

Table 3 Greenest City 2020 Action Plan mid-term targets for reductions in carbon emissions and waste, and improvements to the City's ecosystems

Goals	Targets	Indicator	Baseline	2020 Target
Clean Water	Meet or beat the most stringent of BC, Canadian, and appropriate international drinking water quality standards and guidelines	Total number of instances of not meeting drinking water quality standards	0 instances	0 instances
	Reduce per capita water consumption by 33% from 2006 levels	Total water consumption per capita	583 L per person per day	390 L per person per day
Clean Air	Meet or beat the most stringent air quality guidelines from Metro Vancouver, BC, Canada, and the WHO	Number of instances where air quality standards were not met	27 instances	0 instances
Local Food	Increase city-wide and neighborhood food assets by a minimum of 50% over 2010 levels	Number of neighborhood food assets in Vancouver	3,340 food assets	5158 food assets

Source: City of vancouver, 2014.

Separated bike lane

Olympic Line transit station

In 2009, Vancouver added separated bike lanes to key city streets in the cycling network. These dedicated lanes feature two-way travel for bikes on one side of the road. The bike lanes are separated from traffic using physical barriers such as medians, planters, bike racks, or car parking lanes. This makes biking safer and more comfortable, attracting more participants. A research team at the University of British Columbia conducted a study on safe biking (Winters et al. 2012) and found safety issues to be the biggest concern amongst active and potential cyclists. Pedestrians also benefit from the separated bike lanes, as fewer cyclists ride on the sidewalks.

The City of Vancouver plans to upgrade and expand the existing bike network to connect key destinations, such as schools, community centers, transit stations, and shopping areas. All city buses now have bike racks, and more bike corrals are being installed in and around Vancouver. In 2016, the City implemented a Public Bike Share (PBS) system that allows residents and visitors to rent bikes and helmets from automated docking stations distributed across the city center. The PBS system adds another green transportation option to Vancouver's transportation portfolio and helps reduce personal vehicle trips.

In addition, Vancouver has the world's longest automated rapid metro system, the SkyTrain. Built for the 1986 World Exposition (Expo '86), it has been expanded to serve most of Greater Vancouver. As of 2014, the SkyTrain has a network of three lines, comprising 68.7 km of track and 47 stations. Fully automated trains run underground and on elevated guideways to keep the SkyTrain cars on schedule. A fourth line, the Evergreen line, will be completed in 2016 and add 11 km of tracks and seven stations. The SkyTrain has significantly shaped the urban areas close to the stations by sparking new development. From 1991 to 2001, the population living in walking distance to a Skytrain station has increased 37 percent.

Vancouver plans to improve its transportation system further by re-establishing a streetcar network. During the 2010 Winter Olympics, a Downtown Streetcar called "Olympic Line" ran on the historic railway tracks in downtown as a showcase project and provided free public transportation. The City intends to continue operating this line, and three future line extensions are planned.

Access to Nature

The Vancouver region has historically been a temperate rainforest and the home of predominantly coniferous trees, such as the Western Redcedar and Coast Douglas-fir. Although urbanization has changed the landscape of the city, Vancouver's mild climate and rainy weather have contributed to a green and lush environment. Today, around 12 percent of the city area is green space and 92 percent of Vancouver's residents live within a five-minute walk from a park, greenway, or other green space (EIU, 2012).

Vancouver's Stanley Park, located in the heart of the city and designated a national historic site, is one of the largest urban parks in North America, covering approximately 400 hectares. In contrast to most large public parks, landscape architects did not design Stanley Park. Evolving over many years, parts of it are still densely forested with majestic trees that are hundreds of years old (City of Vancouver, 2014). It is almost completely surrounded by the sea and famous for its Seawall path along Vancouver's waterfront that is a recreational hotspot for walking, jogging, cycling, and inline skating. The park attracts about eight million visitors every year and greatly contributes both to Vancouver's identity as a green city and its quality of life.

Green spaces and trees are important contributors to the livability of a place and the health of a city. They create a sense of community by providing space for recreational activities as well as gathering and socializing (Beatley, 2010). Research has shown that urban green spaces improve the physical health and emotional well-being of urban dwellers, which is manifested in reduced blood pressure, cholesterol, and lower stress levels. In addition, green spaces provide various **urban ecosystem** services and socio-economic benefits. They absorb rainfall,

Urban Ecosystem Ecological system within a city or urban area, i.e. the community of humans, animals, and plants in conjunction with the urban environment.

Stanley park

Street trees in vancouver are part of the city's urban forest

reduce storm water runoff, filter out toxins, prevent flooding, reduce noise, create wildlife habitats, regulate air temperature through shading and evapotranspiration, and improve air quality by reducing pollutants. Vancouver ranks first in the Green City Index air quality category with the lowest rate of particulate matter emissions compared to other index cities.

Despite the large number of parks and other green spaces in the city, Vancouver's tree canopy cover has declined 20 percent over the past two decades. This is largely due to new developments, disease, and residents removing trees from their property. In response, the City is developing an urban forest strategy to provide policy direction and an action plan for growing and maintaining a healthy, resilient urban forest. An **urban forest** is not restricted to trees in parks and public spaces, it also includes street trees and trees in residential yards.

Urban Forest All trees in an urban area, including trees along streets, in parks and green spaces, forests, and yards of private citizens.

While one of the goals outlined in the urban forest strategy is to protect and preserve existing trees, it also calls for an expansion of the current tree canopy cover. This goal is highlighted further in the Greenest City 2020 Action Plan, which calls for 150,000 additional trees in Vancouver between 2010 and 2020. Since 67 percent of the existing urban forest is located on private property and 54,000 of the additional trees will have to be planted in people's yards, educating Vancouver residents about the benefits of trees and engaging the broader public to support new trees in the city is key to achieving the urban forestry goals.

To accomplish this, the City of Vancouver and the non-profit organization Tree City, in conjunction with the Environmental Youth Alliance, launched "Treekeepers" in late 2012 (TreeKeepers, 2012). Treekeepers is a $215,000 city-funded three-year initiative aimed at encouraging residents and businesses in Vancouver to plant more trees on their property. Various tree species suited to Vancouver's climate are available for purchase at subsidized rates between $10 and $20. The tree program provides a unique opportunity for private citizens to support the City's goal to become the greenest city in the world by 2020 and to continue to be named one of the most livable cities on Earth.

Box 1 Imagining Biophilic Cities—(Tim Beatley)

That we need daily contact with nature to be healthy, productive individuals, and indeed have co-evolved with nature, is a critical insight of Harvard myrmecologist and conservationist E.O. Wilson. Wilson popularised the term *biophilia* two decades ago to describe the extent to which humans need connection with nature and other forms of life. More specifically, Wilson describes it this way: "Biophilia . . . is the innately emotional affiliation of human beings to other living organisms. Innate means hereditary and hence part of ultimate human nature."[1] To Wilson, biophilia is really a "complex of learning rules" developed over thousands of years of evolution and human-environment interaction.

"For more than 99 percent of human history people have lived in hunter-gatherer bands totally and intimately involved with other organisms. During this period of deep history, and still further back they depended on an exact learned knowledge of crucial aspects of natural history . . . In short, the brain evolved in a biocentric world, not a machine-regulated world. It would be therefore quite extraordinary to find that all learning rules related to that world have been erased in a few thousand years, even in the tiny minority of peoples who have existed for more than one or two generations in wholly urban environments."[2]

Stephen Kellert of Yale University reminds us that this natural inclination to affiliate with nature and the biological world constitutes a "'weak' genetic tendency whose full and functional development depends on sufficient experience, learning, and cultural support".[3] Biophilic sensibilities can atrophy and society plays an important role in recognising and nurturing them.

The Nature of Cities

While we are already designing biophilic *buildings* and the immediate spaces around them, we must increasingly imagine biophilic *cities* and support a new kind of biophilic *urbanism*. As the planet barrels rapidly down the path of urbanisation the need for green and *nature-ful* cities is an ever more urgent need.

There is already much nature in cities, of course, more than we realise. It is both big and small, visible and hidden. It is intricate, yet sweeping. It is amazing in its biological functioning, ever-present yet highly dynamic, and vastly underappreciated for its ubiquity in cities. In understanding the nature of cities, it is necessary to think beyond our usual approach to visualising or imaging space and place, and to understand that nature is everywhere in cities if we look. It is above us, flying or floating by, it is below our feet in cracks in the pavement, or in the diverse microorganic life of soil and leaf litter. Nature reaches our senses, well beyond sight, in the sounds, smells, textures, and feelings of wind and sun. Understanding the natural history of a city helps us to see cities as ever-changing, ever-evolving palettes of life.

In the higher reaches of our cities, the rooftops and façades also harbour nature, sometimes by design, and sometimes by accident and natural volunteerism. New forms of nature are being created in cities all over the nation in the form of ecological rooftops and rooftop gardens. These miniature ecosystems host grasses and sedum, and are increasingly found (over time and with the right design elements) to harbour great diversity in terms of invertebrates, bird and plant life. We know, for instance, that butterfly species will visit rooftop

(Continued)

Box 1 Imagining Biophilic Cities—(Tim Beatley) (*Continued*)

plantings on high-rise structures. Thus, food, for humans and nature alike, can be grown here as well.

This nature in cities is the raw ingredient for a new global urban society organised around wonder. Few have made a more compelling and eloquent plea for the importance of wonder in the natural world than Rachel Carson more than half a century ago. In a 1956 essay entitled "Help Your Child to Wonder", she describes the value and pleasures of exposing her young nephew to the nature found along the Maine coast:

"If I had influence with the good fairy who is supposed to preside over the christening of all children I should ask that her gift to each child in the world be a sense of wonder so indestructible that it would last throughout life, as an unfailing antidote against the boredom and disenchantments of later years, the sterile preoccupation with things that are artificial, the alienation from the sources of our strength."[4]

Carson counsels looking at the sky, taking walks, uncovering and experiencing nature, even if (as parents) we are not able ourselves to identify a species or a constellation. It is about cultivating an awareness of the sights, sounds, and natural rhythms around us, paying attention and learning to see the mystery and beauty in everything.

We need wonder and awe in our lives, and nature has the potential to amaze, stimulate, and propel us forward, to ignite the desire to learn more about our world. The qualities of wonder and fascination, the ability to nurture deep personal connection and involvement, and visceral engagement in something larger than and outside ourselves offers the potential for meaning in life few other things can provide.

My landscape architecture colleague Beth Meyer argues that with matters of environment and sustainability we need also to emphasise the beauty, pleasure and enjoyment we derive. We often forget about the aesthetics, or try to reduce them to monetary values. At the end of the day, watching that circling hawk or turkey vulture, walking or bicycling through an urban woods, harvesting and eating produce from one's garden, and listening to the sounds of kadydids and tree frogs on a humid August evening are deeply pleasurable. They are the building blocks of a life enjoyed. We climb trees as kids because this is a fun and enjoyable thing to do. As adults, unfortunately, we often forget these pleasures (and of course rarely climb trees!).

In our recent documentary film *Nature of Cities,* we spent a stimulating several days in Austin, Texas filming the 1.5 million Mexican free-tailed bats that have inhabited the underside of the city's Congress Avenue bridge during the summer months. People lineup hours before nightfall to get a good look at the wondrous columns of bats emerging from their roost. Merlin Tuttle, founder of Bat Conservation International (BCI), dutifully recites the many environmental (and economic) benefits provided the city by these bats. And they are considerable, including the millions of mosquitoes they eat each day. But ultimately, the sight of thousands of bats flying off in distinct columns that can be seen for several miles is an immense and beautiful thing. It is the raw emotion and beauty of the natural word, a primordial spectacle unfolding against a backdrop of high-rise buildings and a human-dominated (at least we think) urban environment.

In many American cities, the biodiversity is aquatic and sometimes off-shore, as in Seattle with its abundant and wondrous life in the not-far depths of the Bay and Sound. Much of the biodiversity of King Country, in which the City of Seattle lies, is found in the "deep subtidal habitat" of Puget Sound. In some places, it is almost 900 feet below the surface and includes "over 500 benthic and 50 pelagic invertebrates."[5] And while some are known and recognisable to residents, such as the king crab, many are not. That the Seattle metro region is also home to such unique marine critters as the giant Pacific octopus and giant Acorn Barnacle suggests a wildness and mystery very close at hand.

New forms of nature can be fostered in the many "leftover" spaces of a city as well. A visit to the Green Roofs Research Center in Malmo, Sweden shows the extent of possibilities. Here they have planted and monitor hundreds of green roof test plots, testing different plant and soil combinations. Some plots are for so-called brown rooftops—places in the urban environments (there are many) were plants can be used to restore end even take up pollutants in highly contaminated and degraded settings (Phytoremediation). The Center's immense rooftop also shows the potential of different, sometimes surprising delivery methods. For example, their standard green roof, as Trevor Graham who runs many of the Center's green city efforts explains, is made from recycled polyurethane car seats and in several places. There are small mounted frames with sedum growing vertically, showing the potential for a kind of natural artwork suitable for hanging in one's living room!

These new forms of nature are catching on, and are now encouraged and in some places mandated by codes, and we will see more of this happening in every city around the world. And new creative developments in cities, such as *Via Verde* (the green way), a 200-unit complex of affordable housing planned for a 1.5-acre site in the South Bronx of New York, will find ways to insert and grow nature. In this case, the nature takes the form of a connected multi-functional garden "that begins at street-level as a courtyard and plaza, and spirals upward through a series of programmed, south-facing roof gardens that end in a sky terrace".[6] Increasingly biophilic cities will understand rooftops, courtyards, and façades as places to cultivate nature.

What is a Biophilic City?

Exactly what is a biophilic city, and what are its key features and qualities? Perhaps the simplest answer is that it is a city that puts nature first in its design, planning and management. It recognises the essential need for daily human contact with nature as well as the many environmental and economic values provided by nature and natural systems.

A biophilic city is at its heart a *biodiverse* city, a city full of nature, a place where in the normal course of work and play and life residents feel, see, and experience rich nature. This nature is both large and small—from tree-top lichens, invertebrates, even microorganisms to larger natural features and ecosystems that define a city and give it its character and feel. Biophilic cities cherish what already exists in and near cities (and there is much as we have already seen). They also work hard to restore and repair what has been lost or degraded, and integrate new forms of nature into the design of every new structure or built

(Continued)

Box 1   Imagining Biophilic Cities—(Tim Beatley) (*Continued*)

project. We need contact with nature, and that nature can also take the form of shapes and images integrated into building designs, as we will see.

A biophilic city ought to be judged by the existence of nature and natural features, but also in some way its biophilic sensibilities or *spirit*. For instance, how important is nature and how central is it to the lives and *modus operandi* of the city, its leaders and its populace? A bit harder to quantify, this *biophilic spirit* or sensibility, suggests a value dimension that residents and public officials alike recognise the importance and centrality of nature to a rich and sustainable urban life. This quality could easily fit as both an activity and an approach to governance.

Every city will have its natural spectacles–some large, others more nuanced. But a biophilic city is one that pays attention to its natural attributes, a city that sees and projects a sense of beauty and wonder and caring. It may be the running of the Steelhead trout in the Niagara River, or the appearance of Orcas in Price William Sound, or the migratory return of robins along the east coast of the US. A biophilic city celebrates this wonder and sees in these events the opportunity to connect, to strengthen bonds, to mark the cycles of life and seasonality. This celebrating often involves the direct experience of that biodiversity and nature, such as watching migratory birds, or visiting a park or green area. Or it might be a more referential form of biophilic expression.

As Table 1 suggests, how actively citizens enjoy the nature around them and participate in this nature is also an important measure of a biophilic city. Participation is an interesting word to use here because it implies a level of active engagement beyond passive observation. It suggests a keen and active interest in the subject. Citizens of a biophilic city and their leaders are not removed from the nature around them, but are highly aware of it and present in its midst. A biophilic city is a city in which a large percentage of its population actively enjoyes nature. This enjoyment and engagement can take many different forms. It can range from walking and hiking in natural areas, to bird-watching and plant and tree identification, to organised nature events and activities such as fungi forays to nature festivals.

Biophilic cities make it easier to enjoy nature and reflect an understanding that exposure to and enjoyment of nature are key aspects of a pleasurable and meaningful life. There are many potential outlets and venues for fullfilling our need to connect with nature. Most are intensely social. Facilitating contact with nature has the great potential to help create new friendships and build social networks, in turn, helping to make urbanites healthier and happier. In San Diego, the activities of a number of "friends" of the canyon organizations help to conserve and protect the region's canyons as a neighbourhood and community resource, but also provide opportunities for neighbours to interact and socialise. In Rose Canyon, for instance, residents from different sides of the canyon have places and opportunities to converse and come together, something that would have been difficult without the pull of nearby nature.

Cities must also begin to see the value and importance of facilitating such connections with nature, perhaps offering help and support in the Australian Bushcare model. Here local groups of citizens and community volunteers organise around a specific urban ecosystem–a patch of green space, a stream,

Table 1 Some Important Dimensions of Biophilic Cities (and Some Possible Indicators)

Biophilic Conditions and Infrastructure • Percentage of population within a few hundred feet or metres of a park or green space • Percentage of city land area covered by trees or other vegetation • Number of green design features (e.g. green rooftops, green walls, rain gardens) • Extent of natural images, shapes, forms employed in architecture, and seen in the city • Extent of flora and fauna (e.g. species) found within the city
Biophilic Behaviours, Patterns, Practices, Lifestyles • Average portion of the day spent outside • Visitation rates for city parks • Percent of trips made by walking • Extent of membership and participation in local nature clubs and organisations
Biophilic Attitudes and knowledge • Percent of residents who express care and concern for nature • Percent of residents who can identify common species of flora and fauna
Biophilic Institutions and Governance • Priority given to nature conservation by local government; percent of municipal budget dedicated to biophilic or nature-based programmes • Existence of design and planning regulations that promote biophilic conditions (e.g. mandatory green rooftop requirement, bird-friendly building design guidelines) • Presence and importance of institutions, from aquaria to natural history museums, that promote education and awareness of nature • Number/extent of educational programmes in local schools aimed at teaching about nature • Number of nature organisations and clubs of various sorts in the city, from advocacy to social groups

Source: Beatley, 2010

a park–and with the help of a municipal staff person ("bushcare officer" in Australia), spend weekends and spare hours cleaning up, repairing, and tending over these spaces. The result is not only ecological repair, but also making friends and rebuilding communities, as well as becoming more embedded in place and environment.

Creatively involving citizens in conducting science is another way to intimately engage people with the nature around them. In San Diego, citizens have been trained to become "parabotanists", helping collect plant specimens in this highly biodiverse area. There are now 200 citizens serving as parabotanists, collecting plant data for the San Diego Country Plant Atlas Project (begun in 2002). The project records plants on a three-square-mile grid.

(Continued)

Box 1 Imagining Biophilic Cities—(Tim Beatley) (*Continued*)

San Diego is considered a . . . biodiversity "hotspot" and one of the most floristically biodiverse county in the US. Thus, recording and protecting this biodiversity takes on special importance. The Plant Atlas will eventually result in an "internet-accessible plant atlas based upon vouchered specimens". There are more than 1,500 native species of plants in San Diego County. So there is much to document and record, and citizens here play an important role. Volunteers go through training by San Diego Natural History Museum, and once trained, collect and press the plants and record date about the plant's location. A museum botanist verifies the plant's identification.

A biophilic city then is a city with an extensive and robust *social capital*, to extend Robert Putnam's concept.[7] Evidence is compelling that we need extensive friendships and social contact to be healthy and happy, as well as our contact with nature. So finding creative ways to combine these needs becomes an important goal in the biophilic city. I have been calling this *natural* social capital, acknowledging that there are many ways that learning about and experiencing nature can also help to nurture friendships and help to overcome the increasing levels of social isolation felt by many Americans. How many social organisations or clubs, or community events or activities, explicitly focus around the unique nature of cities? The extent of creative social possibilities is almost limitless: weekend fungi forays, wildlife tracking clubs, *bioblitzes* and nature festivals, wildflower and birding clubs, among many others.

Judging what happens is often a function of the range of the organisations, some public, some private, that exist in a city and that can help in supporting the educating and engagement of citizens. One measure of a biophilic city is the extent of the organisational support, the quality and reach of the biophilic organisations that exist in a city that can actively work towards nature. Bird watching and nature hikes through the city might be one option, but there should be many: swimming, canoeing, and kayaking in urban waters, visiting parks near and far, experiencing nature on a sidewalk or rooftop or building façade as one walks to work or to the subway, among many others.

Many cities around the world are located on or near water bodies and a measure of their biophilic tendencies is how easy it is for residents to enjoy these aquatic environments. In some cities such as Boston, non-profit organisations have worked to make it economic and easy to learn how to sail. In that city, a junior sailing programme run by the non-profit Community Boating, Inc. offers kids the chance to learn how to sail for only $1, during the entire June-August season. Many American cities, moreover, have worked hard to reestablish direct physical contact and connection with rivers, creeks, and harbours through waterfront parks and trails and opportunities to get out on a kayak or canoe.

Biophilic cities are cities that work to expand the opportunities to spend time outside and in close proximity to nature. Partly this means rethinking the ways parks and green spaces are used. New York City has been a leader in creating opportunities for urbanites to camp on weekends in city parks. The programme occurs in the summer months and is quite popular. In 2009, family camping took place in every borough of the city. These camping evenings are, especially from the perspective of kids, quite enjoyable and exciting. The City's

Parks and Recreation Department provides the tents and sleeping bags, and there are typically barbeques, night hikes, skywatching and even S'mores!

Biophilic cities should be identified not just by the presence or absence of nature, green spaces, and green infrastructure, but other forms of investment that also facilitates a biophilic life. A biophilic city invests in a robust network of public (and private) institutions that will educate about, restore and protect, and nudge residents toward enjoying nature. These include traditional environmental education and natural science institutions such as local botanical gardens, zoological parks, and natural history museums, among others. Environmental education centers have been very effective in some cities, and in some cases based in urban neighborhoods.

And biophilic cities are also concerned about and work to protect nature beyond their borders. Each city has opportunities to express care about the environment and other life in the world. Large cities exert a tremendous pressure on global biodiversity through their material flows and consumption patterns, and one measure of a biophilic city is the extent to which it seeks to moderate or reduce those impacts.

New York City, for instance, has recently acknowledged that it purchases a large amount of tropical hardwoods, an estimated $1 million worth each year. The city uses this wood–South American species such as *Ipe* and *Garapa*–for such things as benches, boardwalks, and ferry landings. The ten-mile long Brooklyn Bridge Promenade is constructed of *Greenheart*, another South American hardwood. In recognition of the destructive impact of such purchases Mayor Bloomberg announced a plan in 2008 to significantly reduce the city's purchasing of such wood–a 20 percent reduction immediately and larger reductions later as the city researches and pilots alternative wood sources and alternative materials that could be used.[8] Describing tropical deforestation as an "ecological calamity", and noting that it may be responsible for as much as 20 percent greenhouse gas emissions. Mayor Bloomberg has made an eloquent plea for cities to become better stewards of the global environment. "New Yorkers don't live in the rain forest. But we do live in a world that we share. And we're committed to doing everything we can to protect it for all of our children."[9] City purchasing policies and decisions is an important opportunity for biophilic values to gain expression.

Biophilic Cities In Our Future?

What constitutes a biophilic city is still very much a matter of discussion and debate. Less a definitive list or set of principles, the categories described above are meant to identify at least some of the potential building blocks of a biophilic city. It is unlikely that a singular coherent vision of a biophilic city will emerge. Rather, perhaps there are many different *kinds* of biophilic cities, and many different expressions of urban biophilia. And they might be expressed by different combinations and emphases of the qualities and conditions described here. At the simplest level, though, a biophilic city is a city that seeks to foster a closeness to nature–it protects and nurtures what it has (understands that wild nature is usually abundant), actively restores and repairs the nature that exists, while at the same time finding new and creative ways to insert and inject nature into its streets, buildings, and urban living environments. And a biophilic city is an outdoor city,

(Continued)

Box 1 Imagining Biophilic Cities—(Tim Beatley) (*Continued*)

a city that makes walking and strolling and daily exposure to the outside elements and weather not only possible but a priority.

As the above discussion indicates, a biophilic city is not just about its physical conditions or natural setting, nor is it just about green design and ecological interventions. It is just as much about a city's underlying biophilic spirit and sensibilities, about its funding priorities, and about the importance placed on support for programs that entice urbanites to learn more about the nature around them. A biophilic city might be measured and assessed more by how curious its citizens are about the nature around them, and the extent to which they are engaged in daily activities to enjoy and care for nature, than the physical qualities or conditions, or for instance the number or acres of parks and green spaces per capita that exists in a city.

There are a variety of important research questions about designing and planning biophilic cities in the future. We still have, for instance, relatively little knowledge of the *cumulative* recuperative and healing powers of urban nature. How do the many smaller green features in a city or urban neighbourhood contribute to our closeness with nature and what are the interactive effects? Is access to a large forest more effective than a neighbourhood full of smaller green features, such as street trees and green rooftops? And what is the actual daily minimum level of nature needed by urbanites, and in what form, to live a healthy life?

There are also a host of research questions that relate to how effective our biophilic strategies in fact are. For instance, what are the most effective planning and policies for getting people outside? What will it take to nudge urban populations to adopt a more outdoor nature-oriented lifestyle? Furthermore, our very understanding of the science and ecology of cities remains quite limited and there is much work to be done here as well. New research is needed to better understand the biology and lifecycles of fauna found in cities and how it changes or is modified in an urban setting (think of coyotes!) as well as the management implications therein. There are many, almost countless, research questions and opportunities that arise from the agenda of biophilic cities.

A major task in the future, certainly for those in city planning and urban design, will be in offering an alternative future vision of cities and urban neighbourhoods. As Stephen Kellert of Yale University has said: "We need to do more than just avoid all the bad things that we have done in terms of our adverse effects on natural systems. We also have to create the context for thriving, for development, for meaningful exchange with the world around us, and the people around us, And for that we need to restore that sense of relationship with the natural world which has always been the cradle of our creativity."[10] That context will be one of dense, sustainable, walkable cities and places that are full of nature and are profoundly restorative, magical, and wondrous.

References

1. Wilson, E.O. 1993. "Biophilia and the Conservation Ethic," in Kellert and Wilson, *The Biophilia Hypothesis*, Washington, DC: Island Press.
2. Wilson, op cit, p. 32.

3. Stephen Kellert, *Building for Life: Designing and Understanding the Human-Nature Connection*, Washington, DC: Island Press, 2006, p. 4.
4. Rachel Carson. "Help Your Child to Wonder," *Woman's Home Companion,* July, 1956, p. 46.
5. King County, Washington, King County Biodiversity Report, 2008, p. 57.
6. New Housing, New York Legacy Project, "Phipps-Rose-Dattner-Grimshaw Selected to Develop City-Owned Site in South Bronx," press release, January 17, 2007.
7. Robert Putnam, *Bowling Alone: The Collapse and Revival of American Community,* Simon and Shuster, 2001.
8. See "Mayor Announce Plan to Reduce the Use of Tropical Hardwoods." February 11, 2008. found at www.NYC.gov, accessed on February 17, 2009.
9. ibid.
10. Stephen Kellert interview, in *The Nature of Cities,* documentary film, 2009.

Freiburg, Germany

Background

The city of Freiburg is located in the southwest corner of Germany, at the edge of the Black Forest, and very close to the borders of Switzerland and France. Founded in the year 1122, the city was strategically located at a junction of trade routes between the Mediterranean Sea and the North Sea. Freiburg was heavily bombed during World War II and carefully reconstructed based on the city's medieval plan that includes a unique system of small waterways throughout its historic center.

Today, the purpose of the small canals/waterways is no longer to channel fresh water into the city, rather they add to the pleasant and comforting atmosphere of the city's pedestrian environment. The city is home to approximately 220,000 people and is a hub for regional eco-tourism. It is also a center for academia and research as it is home to the Albert Ludwig University, one of the oldest universities in Germany. The University has been the city's largest employer since the end of World War II.

© katatonia82/Shutterstock, Inc.

City center, Freiburg, Germany

© g215/Shutterstock, Inc.

Pedestrian zone with historic waterway canal (left)

Green Movement Green movements are often grass-roots movement advocating social reforms and the protection of the natural environment and resources

Eco-City Eco-cities, or sustainable cities, consider environmental impacts in their design and policy decisions. In addition, eco-cities strongly support the use or renewable energy sources, public transit, and compact and walkable neighborhood design.

Global Transport Concept (GTC) The Global Transport Concept is a traffic management plan that is updated every 10 years. The main goals are to reduce traffic in the city and support public transit, cyclists, and pedestrians. Furthermore, the plan focuses on creating a rational balance between all modes of transportation.

Green Movement

Freiburg became a center of the country's **"green" movement**, which started in the 1970s, because of its large academic community. The movement led to the founding of the Green Party in 1979, with a political agenda containing numerous environmental goals such as the nuclear energy phase-out and stricter environmental protection laws (Bündnis 90/Die Grünen, 2014). An important event that led to the Freiburg of today stemmed from the successful 1975 citizen protest against the plans to build a nearby nuclear power plant. Many protest leaders and other people involved in Freiburg's green movement remained in the area and became involved in local and regional politics. They were often involved in the city administration and found employment in educational or research activities, or founded environmentally based companies. As a result, the mayor and more than 25 percent of the council are currently members of Germany's Green Party (City of Freiburg, 2014).

Over the last few decades, Freiburg has focused heavily on becoming a recognized green and sustainable city. The city has won various national and international environmental awards for their policies and developments. Freiburg is especially well known as an **eco-city** for its efforts in transportation, alternative energy systems, and sustainable place-making (Newman et al, 2009). City administration emphasizes other sectors as well, including land conservation and a green economy, to increase sustainability (Green City Freiburg, 2014).

Transportation

Transportation plays a pivotal role in Freiburg's urban development policy. In 1969, the city established its **Global Transport Concept (GTC)**, a traffic management plan that is updated every decade. The main goals are to reduce total automobile traffic volume in the city and support public transit, cyclists, and pedestrians. The GTC also focuses on creating a rational balance between all modes of transportation. The first GTC policy created a bicycle path network. Today, more than 300 miles of bike paths and bike-friendly streets exist throughout the city as well as 8,000 public bike parking spaces. This network was quickly followed by considerable expansion of the light rail network and the transformation of the city center into a pedestrian zone in 1972.

The light rail network is now the most used transit system of the city. Almost 75 percent of all public transit users rely on the light rail for their daily travel. Furthermore, 70 percent of the population lives within 550 yards of a tram stop. This short distance promotes walking to a transit station and Transit-Oriented Development (higher density development) around stations. During peak hours, the tram runs every 7.5 minutes. All stations have park-and-ride as well as bike-and-ride facilities. What makes Freiburg's public transit system truly sustainable, however, is that it is powered completely by renewable energy sources. Approximately 80 percent of the energy used is generated by hydropower; the remaining ~20 percent is provided by solar and wind energy.

Besides aiming to make public transit convenient, fast, reliable, and comfortable, city administration strives for affordability as well. A monthly transit pass costs about 50 Euros ($54.50 USD) and provides unlimited access to more than 1,800 miles of routes from numerous transportation companies throughout the entire region. This is much cheaper than the monthly cost of owning and operating a private vehicle. Between 1982 and 1999, the number of people using public transit increased from 11 to 18 percent. Bicycling increased from 15 to 28 percent in this same period. Among university students nearly 90 percent use public transit or bike. It is estimated that, only 30 percent of the entire traffic volume is generated by private automobiles. Because of Freiburg's transportation policies, vehicle miles traveled by car have decreased, and as a result, GHG emission have declined as well.

Green Energy

In addition to the GTC, Freiburg is characterized by its progressive energy policy. The city's energy policy is based on three pillars: energy savings, efficient technologies, and renewable energy sources. Since 1992, the city has enforced strict building design standards for new houses. These standards have reduced heating oil consumption from about 12–15 liters to 6.5 liters per square meter. To improve energy efficiency in existing buildings, Freiburg instituted a comprehensive support program for home insulation and energy retrofits. This program, which cost around 1.2 million Euros in subsidies and 14 million Euros of investments, decreased energy consumption by around 38 percent per building.

In terms of efficient energy technology, Freiburg relies heavily on more than 100 **combined heat and power plants (CHP)**. These power plants reuse the waste heat from electricity production to generate more electricity and useful heat for buildings. They are powered by natural gas, biogas, landfill gas, geothermal energy, wood chips, and/or heating oil. The majority of CHPs are small scale and located directly in the neighborhoods they serve. In addition, 14 large CHP plants are located at the city's outskirts. These efforts align with Germany's mandate to reduce dependence on nuclear power for creating electricity.

The steady increase in the number of CHP plants has allowed Freiburg to decrease its reliance on nuclear power from 60 to 30 percent (The EcoTipping Points Project, 2011). Today, slightly more than half of the city's electricity is produced by CHP plants or other renewable energy sources, including solar, wind, hydropower, and biomass. The remaining energy is imported, about a third of which comes from external power plants. For the near future, Freiburg's goal is it to further increase the amount of renewable energy and to reduce reliance on nuclear power and fossil fuels.

Although four different renewable sources are utilized, the city is most famous for its support of **solar energy**. The use of solar energy is widespread with more than 400 photovoltaic installations occurring on both public and private buildings. The 1.6 million square feet of photovoltaic cells produce more than 10 million **kWh/year** of energy. Yet, solar still only provides a small fraction of the city's electricity needs as the city's total electricity demand is well over 1,000 million kWh/year. Nevertheless, solar energy is an important element of Freiburg's sustainable economy.

Combined Heat and Power Plants (CHP) These power plants reuse the waste heat from electricity production to generate more electricity and useful heat for buildings. CHP plants can be powered by natural gas, biogas, landfill gas, geothermal, wooden chips, and heating oil.

Solar Energy This technology allows generating energy by converting sunlight into electricity using solar panels. Solar panels use solar cells made up of photovoltaic material.

kWh/year kWh stands for kilowatt hour and is used as a unit of energy

Solar settlement in Freiburg

© Gyuszko-Photo/Shutterstock, Inc.

The highly visible solar installations attract eco-tourism as well as numerous scientific and educational organizations to the city (Newman et al., 2009). Overall, 1,500 companies from the green economy sector employ nearly 10,000 people. This includes research facilities and environmental education programs for university students as well as solar technicians, and installers, manufacturing. Other companies produce solar cells and machinery to create the cells. These companies have also attracted numerous suppliers and service providers in the green-supply-chain system. In the solar industry alone, more than 80 businesses employ 1,000 people or more.

Sustainable Living and Place-making

Vauban is one of the most sustainable neighborhoods in Freiburg, relying heavily on solar energy. This neighborhood is 38 hectares in area (Vauban, 2014) and located on a former French military base close to the city center. The final development plan was approved in 1997. Vauban is now an attractive, family-friendly community of about 5,000 people. Due to carefully designed zoning regulations and policies, low-energy buildings are obligatory in this district and **zero-energy** or **energy-plus buildings** with solar technology are prominent. The 60 energy-plus homes in this neighborhood create more energy than they consume. Their residents earn 6,000 Euros per year on average by selling their surplus energy back to the grid.

Based on the Vauban district master plan, the neighborhood is characterized primarily by a dense pattern of attached housing and multifamily housing. Green spaces between the housing clusters ensure good climatic conditions and provide play areas for children. The green corridors channel fresh air into the neighborhood while vegetation filters the air and reduces the air temperature during the summer. A set of U-shaped access roads limits automotive access to the neighborhood, allowing community life to take place in the interior pedestrian spaces. An extensive set of walkways and paths connects the different housing areas and makes the entire neighborhood both bicycle and pedestrian friendly. Furthermore, Vauban has its own CHP plant that uses wood chips from local forestry and provides all residents and local businesses with power.

In addition to these efforts the neighborhood has received significant international attention for its efforts to promote car-free living. Before the construction of Vauban, the state zoning law required builders to provide parking space for every housing unit. The law was changed due to organized lobbying by Forum Vauban, a non-profit group formed in 1993 that has been a driving force for making Freiburg a sustainable community. Today, Freiburg can waive any parking requirements if the developer can prove that future residents will not own a car and that extra land is available to create parking spaces if residents subsequently choose to own a car. In the case of Vauban, costs of housing and parking are separate. If residents choose to own a car, they must cover the costs of a garage parking space at the fringe of the neighborhood. The one-time charge is about $14,400 per car. This

Zero-Energy Buildings Zero-energy buildings use different technologies such as solar and wind to harvest energy on site. These type of buildings are very energy efficient and do not have to rely on the city's energy grid. Furthermore, zero-energy buildings and energy-plus buildings do not produce any carbon emissions.

Energy-Plus Buildings Due to good insulation, special design guidelines, and the use of renewable energy sources, energy-plus buildings create more energy than they consume. Residents can make extra money by selling the extra energy back to the grid.

is a strong incentive for residents to find alternatives for getting around, such as using the tram system or a bicycle. New residents are offered a special mobility package as an alternative. The package includes membership in a car-sharing company, a one-year local transit pass, and a 50 percent reduction on train tickets. Official numbers report about 250 motor vehicles per 1,000 Vauban residents, which is much lower than the national average in Germany of 500 automobiles per 1,000 residents.

Zero-Energy and energy plus buildings in Vauban

Stockholm, Sweden

Background

Stockholm, the capital of Sweden, has a population of close to 900,000 and is located on the country's south-central east coast (City of Stockholm, 2014). Often referred to as the "Venice of the North," Stockholm consists of 14 islands that are connected by 57 bridges. The history of the city dates back to the thirteenth century. It became the formal capital-in 1634. Following the Great Northern War early in the eighteenth century, Stockholm became an economic hub and cultural center experiencing strong population growth. Since then Stockholm has fully evolved into a modern and cosmopolitan city.

Presently, just over 2 million people live in the greater Stockholm metropolitan area which accounts for almost one quarter of the country's total population (Statistikomstockholm, 2013). The core of the city is densely populated with 31 percent of the population occupying only 8 percent of the total land area. Although

The city of Stockholm

the total area of the municipality is spread out over 73 square miles, only one third is urbanized. The remaining two-thirds are equally water and green space (Berggren, 2013).

In the past few decades, Stockholm has been focused on sustainable urban development and planning. As a result, Stockholm received the 2010 European Green Capital Award for its commitment to environmental protection, GHG emission reduction, and sustainable development policies and urban designs (www. europeangreencapital.eu). One of its key challenges is balancing the demands of an increasing population and the infrastructural needs of a modern city while aspiring to become 100 percent fossil fuel-free by 2050.

Transportation

Stockholm continues to experience rapid population growth. Forecasts suggest that by 2030, its population will increase by 25 percent (City of Stockholm, 2012). As a result, roads and public transit systems need to be updated and extended without compromising the overarching goals of becoming more sustainable. A specific area of concern is the transportation sector's ongoing reliance on fossil fuels and the relatively high degree of air pollution from carbon emissions (LSE, 2013). To address this, a number of performance targets have been set and policies implemented to make the transport system more environmentally friendly and reduce emissions. The goal is to encourage residents to switch to alternative modes of transportation such as bicycle, public transit, or walking for their daily trips.

Overall, the initiatives are not much different to the policies discussed in the previous cases of Freiburg and Vancouver. Measures include regulations and **zoning ordinances** that lead to compact and **mixed-use development** that support public transit, encourage walking and cycling through extensive path networks, and investment in a green transit fleet. Some transportation policies and strategies do stand out, however. These have contributed substantially to today's Stockholm's environmentally friendly image and sustainable development pattern. Notably, the bicycle is a major component of the city's efforts to make the transportation sector more sustainable, while accommodating an increasing population.

Today, more than 150,000 people commute daily to work by bike, taking advantage of the over 750 kilometers of bike paths within the city limits (Fourteenislands.com, 2014). To increase safety and encourage even more people to use bikes for daily travel, bike lanes are often situated between sidewalks and roadside parking, providing a buffer between moving cars and cyclists. In addition to the already existing extensive network of bike paths, the city's current mobility plan emphasizes bicycles as an important form of transportation. As a result, automobiles are no longer considered a priority in the city's transportation concept and overall development (Berggren, 2013).

Another aspect of the city's transportation policy that has reduced reliance on the private automobile is the high intermodal connectivity of the different public-transit modes. One can easily switch between public-transit services such as regional trains, trams, or buses using the same fare card. The different transit stations are in close proximity and the schedules are timed to accommodate each other. Since the public transit lines run every five to ten minutes, people can

Zoning Ordinances Consist of regulations and laws dictating how property in specific areas can be used. They determine whether particular zones within a municipality can be used for residential, commercial, or recreational purposes.

Mixed Use Development Are developments that are not limited to a single use. Instead mixed use developments often combine residential and commercial use in the same building.

commute to work or travel within the city with minimal walking involved and without needing a car.

The transportation policy that Stockholm is probably best known for is its congestion charge, which was first tested in 2006 (LSE, 2013). Stockholm was only the second European city, London was first, to introduce such a system. Cars that enter the downtown area pass unmanned electronic control points that recognize the license plate and the owner is charged a fee depending on the time of the day. The maximum amount that can be charged is about SEK 105 or $11.54 a day (Transport Styrelsen, 2016). Due to the congestion charge and other transportation policies, downtown traffic has decreased on average by 20 percent annually since 2007 (OECD, 2013). Simultaneously, public transit use has increased continuously by about 7 percent and now accounts for roughly 70 percent of all motorized trips within the city (Stockholm, 2012). This has led to 12 percent reduction in CO_2 emissions by motor vehicles (OECD, 2014). This reduction is quite remarkable, considering that the city's population is growing by around 40,000 people a year.

Land Use & Urban Form

Land use and urban form are other key areas of Stockholm's efforts to reduce energy consumption and become fossil fuel free in the foreseeable future. In order to reduce the use of fossil fuel, one of the main goals is to reduce the number and length of trips people take on a daily basis for work, shopping, or recreation as well as improve the cost effectiveness of public transit systems through a high-density, mixed-use settlement structure. By developing land-use policies that create an urban form characterized by shorter commuting distances, increased use of public transit, and a high degree of walkability, the city expects to further reduce energy consumption and GHG emissions. "The Walkable City" plan outlines four core strategies to achieve a more integrated, interconnected urban structure (City of Stockholm, 2010). The overarching objective is to create space for about 200,000 new residents by 2030 without compromising the city's attractiveness or its commitment to sustainable development.

The first strategy focuses on strengthening the downtown area by densifying the urban environment while simultaneously protecting the high quality of existing green infrastructure such as parks. The second strategy focuses on node development to establish strong links between the public-transit network and adjacent neighborhoods. These corridors not only extend public transit into the suburban regions of the city but also increase neighborhood density along public-transit lines (OECD, 2013). The third strategy aims to connect different parts of the city. One of the most important projects underway is the completion of a cross-town rail tunnel that will link southern and northern Stockholm and integrate suburban neighborhoods in these areas to the core of the city. The fourth strategy is to create a vibrant urban environment (City of Stockholm, 2010).

Hammarby sjöstad district

© Estea/Shutterstock, Inc.

Brownfield redevelopment plays an important role in creating this vibrant environment. Since the mid-1980s, densification and revitalization of old industrial complexes has been an integral part of the city's sustainable development strategy. In recent years, more than half of urban development taking place inside the city involves infill development (City of Stockholm, 2012a). Between 2000 and 2007, approximately 25,000 new housing units were built with more than one third built on large-scale brownfields. A very good example of a brownfield redevelopment is Hammarby Sjostad.

Started in 2000, this eco-neighborhood is located south of Stockholm's south island and, when fully completed, will have 11,000 residential units for about 25,000 inhabitants. Overall, the building environment is characterized by high density and modern architecture that utilizes primarily sustainable materials. Although Hammarby Sjostad will not be completed until 2025, it is already considered a success story attacting over 10,000 visitors a year (Ignatieva & Berg, 2014). The underlying masterplan focuses on new public transit routes, centralized heating and cooling as well as an underground waste collection system (Ignatieva & Berg, 2014). Furthermore, the city set highly ambitious targets for very low car ownership (0.5 cars per household) and integrating twice as much nature compared to other similar housing projects in Stockholm. What makes this neighborhood truly unique, however, is its "closed-loop" system for treating water, waste, and energy (Future Communities, 2014).

Often referred to as the "Hammarby Model" (City Climate Leadership Awards, 2013), the city's infrastructure system enable water, waste, and energy to support each other, thus reducing the energy and natural resources needed to operate the

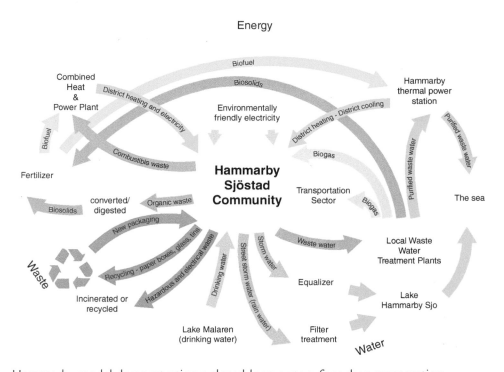

Hammarby model demonstrating a closed-loop system for urban regeneration

neighborhood. Other technologies such as fuel and solar cells, and solar panels are also integrated into the neighborhood, allowing residents to meet half of their energy needs once the area is fully developed. Moreover, the city imposed stringent environmental restrictions on buildings, technical installations, and traffic, significantly limiting their impact on the natural environment.

Integrating and Preserving Nature within the City

Ekoparken

The integration and preservation of nature is integral to Stockholm becoming more sustainable and resilient. To preserve important green areas and protect surface and ground water as well as integrate new parks, open spaces, and green corridors, Stockholm developed the "City Plan 99" in 1999. It updated this plan in 2001 with a regional development plant that was the first of its kind for the metropolitan area. These plans specifically identify sites for brownfield redevelopment to accommodate the increase in population while limiting the destruction of existing green spaces for urban expansion.

Today, natural infrastructure is clearly visible throughout the city. For example, connective green corridors spread out like wedges from the city center to the suburbs, providing recreational space as well as ecological services to maintain the local natural biodiversity. In addition, twelve large parks function as green anchors throughout the city. Each park is at least 200 acres. Together, they account for approximately one third of the city's open space. Among these parks is the "Ekoparken," which became the first urban national park in 1995. This park has a high cultural value for Stockholm residents and is home to rare insects and birds (Nelson, 2014). Smaller neighborhood parks and community gardens make up another third of the overall park network. In addition, the vast majority of Stockholm's 160 km of shoreline is accessible as well. Today, 13 percent of the city's land area is covered by water and 42 percent by parks and forest (OECD, 2013). To further increase the number of green spaces, the city supports and provides assistance to households and property owners in creating greener and more sustainable courtyards.

Characteristics & Principles of Sustainable Cities

If we examine cities such as Vancouver, Freiburg, and Stockholm, among others, various characteristics and principles emerge that make these cites more sustainable. These include:

1. Low Ecological-Environmental Impact
   Using solar structures on private and municipal buildings, passive energy designs, and carbon neutral buildings, the cities significantly reduce CO_2 emissions. Moreover, the associated reduction of fossil fuel use for energy production is a very effective strategy for reducing GHG emissions and thus mitigating climate change.

2. Cluster Housing and Densification
    New housing developments are zoned and designed as high-compact and high-density neighborhoods, while simultaneously preserving and integrating nature and walkways into the areas. Overall, many of these new settlements combine solar-based housing, cluster development, open space, and connections to public transit.

3. Green Employment
    Concomitant with implementing ecological friendly and sustainable designs, city policies also focus on boosting the green economy. For example, as Freiburg utilized its recognition as a solar energy community to generate green jobs in energy research at the university, employment through eco-tourism, and manufacturing of solar technologies.

4. Pedestrian-& Bicycle-Friendly Environment
    These cities implement policies to support and encourage walking and bicycling. They provide extensive bicycle pathways, monetary incentives for not owning a car, punitive action for automotive parking, well organized and widely available public-transit options, and a settlement structure that encourages walking short distances. These types of sustainable urban design and planning efforts also result in numerous social benefits, including the creation of viable areas for social space and interaction, community building, an enhanced quality of life, and a sense of wellbeing and community efficacy.

5. Emphasis on Resilience
    Another dimension that sets sustainable cities apart from other places is often their diverse, small energy systems located in neighborhoods. The "Hammarby Model" and Freiburg's energy mix consisting of wind energy systems, solar collectors, hydropower, biogas, and geothermal add to the community's resilience. They provide security because the small technologies are not dependent on fossil fuel and its attendant political and economic constraints and fluctuations.

6. Place Making
    Sustainable cities also recognize the importance of building or restoring special places of history, culture, and community. These elements are an important part of the social domain in sustainable development. In the case of Freiburg, the historic city destroyed in World War II was rebuilt with a similar spatial organization, using open public spaces and compact development without suburban sprawl. Later, the small canals that once brought water into the cities were reestablished as a means of visually reintroducing history to the community and enhancing aesthetics. Open space in the form of a farmer's market surrounding Freiburg's central church and largest structure provides a dominant place-making, pedestrian, and cultural experience.

7. Political and Policy Support for Sustainable Development
    The policies of "sustainable urbanism" are highly entrenched in cities focusing on becoming more sustainable. City government support for

solar use, research, and manufacturing; high-density living; habitat preservation and protection within the city; and increasing public-transit all work toward the greater goal of an overall sustainable development pattern. This particularly applies to urban planning that is central to governmental activity such as housing requirements and design, pedestrian environments, and bike trails.

8. Interconnectivity of Economic, Ecological, and Social Dimensions
Following the sustainable development framework developed by the Brundtland Commission, sustainable cities encourage and advance sound ecological and environmental goals (such as the reduction of GHG emissions or a significant increase in public-transit ridership) and support green employment agenda. Social- and community-based urban planning objectives are also advanced and supported.

Social Dimensions of Sustainable Neighborhood Development

Case Study Freiburg – Rieselfeld

When it comes to sustainable development projects and environmental friendly neighborhood design, the social dimension often receives less attention compared to the environmental and economic side of sustainability (Vallance et al., 2011; Murphy, 2012; Woodcraft, 2012). Too often decision-makers tend to focus only on technical aspects such as energy reduction, sustainable building materials, or compact settlement structures, without acknowledging the importance of building social capital or social networks. Although different definitions exist, the main purpose of social sustainability is to create strong, vibrant, and healthy communities, which

© karmizz/shutterstock, Inc.

Light Rail and Community Center in Freiburg – Rieselfeld.

(Continued)

Social Dimensions of Sustainable Neighborhood Development

Case Study Freiburg – Rieselfeld (*Continued*)

enhance the quality of life and the overall resiliency of the neighborhood and its population by establishing a built-environment of high quality that provides appropriate and accessible local services, contributing to the overall physical, social, and cultural well-being (Department of Communities and Local, 2012; Bacon et al., 2012).

A good example of a development where the social dimension of sustainable neighborhood design was considered through all stages of the planning process is the Rieselfeld neighborhood in the city of Freiburg. Very similar to the previously described Vauban district, Rieselfeld is characterized by simple building designs, mixed use, strong support for public transit, as well as many green spaces, playgrounds, and other public areas. Divided into four construction stages, the building of the neighborhood began in the year 1994 and was fully developed in 2010. The first residents moved to the district in 1996 and today about 12,000 people live in Rieselfeld's 4,500 housing units.

As mentioned above, the social aspects played an important role during the development of Rieselfeld. One of the major goals of development concept for Rieselfeld was to establish social and cultural life simultaneously with the construction of the physical build environment. As a result, the whole neighborhood development process was not only guided by design, transportation, and ecological principles, but also by a social concept acknowledging the need for a community based participatory foundation. In particular, the social concept addressed the following community – based infrastructure (Siegl & Kaiser, 2002):

Childcare, Schools, Youth Programs

One of the cornerstones of Rieselfeld's social concept was to provide facilities for childcare, education, and youth programs as quickly as possible and before the completion of all the residential units. From the start is was very important to provide kindergarten space for every child as well as specifically tailored childcare for children between 1–3 and 6–10 years old. Therefore, the first elementary school was ready only 2 months after the first people moved into the neighborhood and the first Kindergarten was completed in 1998. Just 8 month after the completion of the elementary school, the high-school was also completed in September of 1997. Today, the elementary school has the highest enrollment statewide. Because all these institutions were present from the beginning and children did not have to leave the neighborhood to attend kindergarten or school young families not only felt comfortable moving to Rieselfeld, but also built strong social networks among each other. Since 2002 a community center was also established with a children library and youth workers on site, offering different recreational activities in the afternoon ant creating another opportunity for social interaction.

Healthcare for Elderly People

Another important aspect of social sustainability is the provision of healthcare inside the neighborhood, especially for the elderly. Riselfeld does not only have medical practices, but also offers a private nursing home next to regular housing units. Although, only a minority of the people living in Rieselfeld belong to the elderly age group of 65 and higher, the early construction and central location of the nursing home allows its resident to be still an active part of the community. Moreover, people who grew up in the district do not have to leave their social network and support system behind, once they are in need of care.

Church, Community Center, Public Space

Both major congregations in Germany, Catholic and Protestant, were present in Rieselfeld from the beginning and play a central role in the district's social and cultural life. The joint store, offering faire-trade products, was one of the first public spaces available to the residents to establish contacts with others. Moreover, both congregations celebrate interfaith services on a regular basis which is greatly supported by the residents and offers another opportunity to strengthen social ties. Whereas, at the beginning of Rieselfeld church services took place at the local gymnasium, both congregations built an integrated religious center in the most central location of Rieselfeld offering not only religious services, but also common community services and public meeting spaces.

The Community Center "Glashaus" was finished in 2003 and is the main social meeting point of Rieselfeld and located next to the church in the center of the neighborhood and in very close proximity to the elementary and high school. In addition, to the already mentioned youth library, which is a very popular meeting point after school, the "Glashaus" also has office space for the local citizen's organization K.I.O.S.K, which stands in German for Contact, Information, Organization, Self-Help, and Culture. Part of K.I.O.S.K and also located inside the community center is a café that is run by volunteers from the neighborhood.

The well planned public spaces such as parks and playgrounds also contribute to today's strong social network in Rieselfeld. Green infrastructure and path networks were established early on in the development process, providing a high quality public realm before the settlement structure was completed and all units were occupied. Especially the concept of creating semi-public spaces inside the courtyards of the different building blocks, strengthened the social networks between neighbors and improved the overall quality of life. Besides the green infrastructure, streets were also

(*Continued*)

Social Dimensions of Sustainable Neighborhood Development

Case Study Freiburg – Rieselfeld (*Continued*)

designed to encourage social interaction and provide a safe environment for children. With the exception of the main road with its light rail line all roads a very narrow and the maximum speed limit in the entire neighborhood is less than 19 miles per hour. In many areas the allowed speed limit is only walking speed.

Quatiersarbeit and Public Participation

Public participation plays a major role in Rieselfeld and started even before the first residents moved into the neighborhood. The first public participation projects started in April 1996 with the goals to engage future residents in the early stages of the design and implementation of social and cultural infrastructure as well as to help create the district's culture of everyday life. Furthermore, the city founded project, provided an opportunity for the public to provide feedback on the proposed project during the planning stage and thus influence the final design of Rieselfeld. Involving citizens from the beginning of the process created a strong sense of identity and responsibility among today's residents. People who live in Riselfeld generally care about their environment and value the social infrastructure and networks they helped to put in place so many years ago.

References

Bacon, N., Cochrane, Douglas, Woodcraft, S., 2012. *Creating strong communities: how to measure the social sustainability of new housing developments,*
London: The Berkeley Group

Department for Communities and Local, 2012. National Planning Policy Framework. Available at: http://www.communities.gov.uk /planningandbuilding/planningsystem/planningpolicy/planning-policyframework/ (accessed 8.18.2014)

Murphy, K. (2012). The social pillar of sustainable development: a literature review and framework for policy analysis. *Sustainability: Science, Practice & Policy, 8*(1)

Siegl, K., Kaiser, P. 2002. *Rieselfeld: Wo Freiburg wächst.* Stadt Freiburg: Baudezernat Bauverwaltungsamt/Geschäftstelle Rieslefeld.

Vallances, S., Perkins, H.C., Dixon, J.E. 2011. What is social sustainability? A clarification of concepts. Geoforum, 42: pp. 342–248.

Woodcraft, S. 2012. Social Sustainability and New Communities: Moving from concept to practice in the UK. *Social and Behavioral Sciences, 68*: 29–42.

Chapter 5

Urban Sustainable Design

Lauren Allsop

We shape our buildings, and afterwards our buildings shape us.
—Winston Churchill

Introduction

In September 2016, the United Nations Habitat III conference outcome document identified urbanization as one of the 21st century's most transformative trends (UN Conference, 2016). The conference adopted a New Urban Agenda that envisions "cities and human settlements that:

    a. Fulfil their social function, including the social and ecological function of land, with a view to progressively achieving the full realization of the right to adequate housing as a component of the right to an adequate standard of living, without discrimination, universal access to safe and affordable drinking water and sanitation, as well as equal access for all to public goods and quality services in areas such as food security and nutrition, health, education, infrastructure, mobility and transportation, energy, air quality and livelihoods;

    b. Are participatory; promote civic engagement; engender a sense of belonging and ownership among all their inhabitants; prioritize safe, inclusive, accessible, green and quality public spaces friendly for families; enhance social and intergenerational interactions, cultural expressions and political participation, as appropriate; and foster social cohesion, inclusion and safety in peaceful and pluralistic societies, where the needs of all inhabitants are met, recognizing the specific needs of those in vulnerable situations;

c. Achieve gender equality and empower all women and girls by ensuring women's full and effective participation and equal rights in all fields and in leadership at all levels of decision-making; by ensuring decent work and equal pay for equal work, or work of equal value, for all women; and by preventing and eliminating all forms of discrimination, violence and harassment against women and girls in private and public spaces;

d. Meet the challenges and opportunities of present and future sustained, inclusive and sustainable economic growth, leveraging urbanization for structural transformation, high productivity, value-added activities and resource efficiency, harnessing local economies, and taking note of the contribution of the informal economy while supporting a sustainable transition to the formal economy;

e. Fulfil their territorial functions across administrative boundaries, and act as hubs and drivers for balanced, sustainable and integrated urban and territorial development at all levels;

f. Promote age- and gender-responsive planning and investment for sustainable, safe and accessible urban mobility for all, and resource-efficient transport systems for passengers and freight, effectively linking people, places, goods, services and economic opportunities;

g. Adopt and implement disaster risk reduction and management, reduce vulnerability, build resilience and responsiveness to natural and human-made hazards, and foster mitigation of and adaptation to climate change;

h. Protect, conserve, restore and promote their ecosystems, water, natural habitats and biodiversity, minimize their environmental impact, and change to sustainable consumption and production patterns" (p.4).

Smart Growth An urban planning term that focuses on compact growth in walkable urban areas. Sprawl is its antonym.

Good bones A phrase used as a synonym for a building (or collection of structures) that was constructed with solid, long lasting materials.

Green Building This term refers to an environmentally friendly and resource efficient building as well as a concept that the built environment impacts the natural environment. By practicing Green Building, the impacts of construction and the dwindling use of resources are minimized.

To meet this call, it is imperative that urban design—the process of shaping cities—play a key role in changing the human footprint towards sustainability. Sustainable urban design must examine the environment, local climate, existing regulations, **Smart Growth**, and—perhaps most importantly—the well-being of its inhabitants and the health ramifications of what is designed carefully. Existing built stock and the surrounding neighborhoods also are significant features to consider, and not just because tearing them down increases our landfills. The **good bones** of our existing, and in some cases historic, infrastructure meet many of the criteria for sustainable design and often bring character, place, and stability to a burgeoning city.

By carefully examining the many elements that make a place memorable, a place where people return to again and again or even move to, urban design professionals can and will create vibrant, resilient cities where **Green Building** certification points will not need to be sought. Rather, they will happen automatically. Through case studies old and new, cities and planners can identify key features of urban sustainable design and scientific research that will move them forward responsibly and enable them to create better habitats for the expected growth by 2050 (UN, 2014).

A Brief History of Urban Design

Successful and sustainable cities were developed long before zoning regulations existed. The past readily teaches the successes and failures of urban settlements, and identifies the factors that played into the sustainability of urban places. In the 1st century CE, Rome had a population of 350,000 whereas the majority of other cities on the continents now defined as Asia, Europe and Africa were 50,000 or less (Russell, 1958). It was a prospering city and the inhabitants enjoyed a good living. This happened because of forethought and ingenuity. The politics and prescriptive zoning regulations that often enter into today's city discussions were not present at the equivalent Roman table.

Emperor Augustus (27 BCE–14 CE) saw the need for housing and supporting infrastructure in Rome to accommodate the growing population. In a city considered by many to be pre-industrial, certainly pre-zoning, his engineers set about building massive new aqueducts to channel water into the city and an underground sewage system to protect the health of the residents and the purity of the Tiber River. They also invented pozzolana cement, known today as Roman cement and the forerunner to concrete. This material enabled lightweight, multistory insulae (apartment blocks) to be constructed.

Despite the ability to construct buildings higher than the six-story limit for wood, height restrictions remained to retain human scale. Augustus saw the need for **mixed use** housing to enable the wide variety of jobs to be filled without the need for commuting. Open spaces abounded throughout the city and were not linked to income levels. Agricultural lands were identified and marked as sources strictly for the Roman markets. When Emperor Vespasian came to power in 69 CE, he continued these themes.

> **Mixed use** The co-mingling of different uses and socio-economic groups within the same building and/or neighborhood.

This demonstrates that Rome's leaders and civil servants understood the needs of a city and its inhabitants for many decades. This pattern of urban design continued well into the 19th century, even in the US. For example, Collinsville, Connecticut was developed as a result of Samuel Collins building the Collins Company Axe Factory there in 1826. He was a key player in laying out green spaces, a mixed-use downtown and housing. Collins understood the need for his employees and their families to be able to easily get to work, shop for necessities, and live in a wholesome town that provided amenities so workers would stay. He even saw the importance of creating a village center, a green with the sides of the streets lined with elm saplings (Leff, 2004).

By the late 19th century, planners began sharing these successful elements of urban design. The Garden City movement initiated by Ebenezer Howard in 1898 in England was one of the first. Published as *Tomorrow: A Peaceful Path to Real Reform* (Howard, 1898), Howard created a plan that limited the size of a city before it was developed. Each city was separated by an agricultural **green belt**, yet linked by various forms of transportation. The plan retained human scale and easy access to parks and other amenities. The first Garden City was Letchworth, designed by Raymond Unwin and Barry Parker in 1904 north of London. It served as a prime example when the country created the New Towns Act of 1946 (Newton, 1971).

> **Green belt** Unbuilt land often designated as forest, park or farm land, surrounding an urban settlement of any size.

The 20th century saw these visions change with the introduction of the automobile and the often-prescriptive measures of zoning. Development, sprawl and the drive for

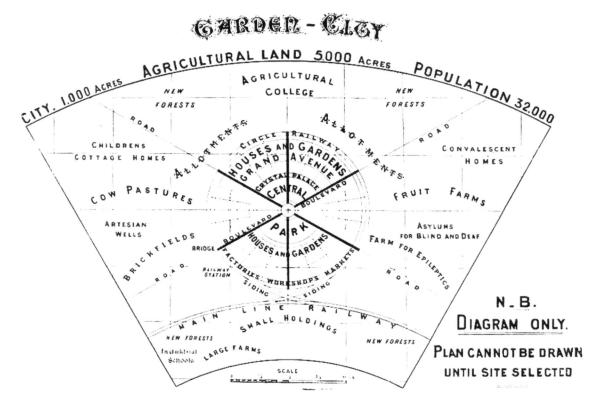

Figure 1 Ebenezer Howard's Garden City Plan of 1898.

Source: Howard, Ebenezer. 1902. Garden Cities of To-morrow: 22.https://archive.org/details/gardencitiesofto00howa (accessed 2.23.2017)

the almighty dollar quickly followed. Out-of-town developers, for example, largely created Phoenix; they built then left. By the last decades of the 1900s, bankers and developers dictated development to cities. Recently, a banker said that developers were not incentivized to change to sustainable design because the methods of building cities have shown to turn a profit and anything new is a risk (Allsopp, P. 2015).

This has led to low-density, auto-dependent cities that have negative impacts on the environment and human health. Planet Earth will survive without us, but there is an ever-increasing need for humans to adapt to our changing environment in order to survive. With attention given to sustainable design rather than one-size-fits-all zoning measures, sustainable and even off-grid cities can become the norm.

Policy Impacts on the Quality of Place

Sense of place The aesthetic feeling or perception people experience as they interact with a given space. The responses, whether good or bad, shape the impression of that place. In the last decades, the phrase has become a buzz word with only positive connotations.

The rate of population growth in the latter half of the 20th century accelerated well beyond the imaginations of earlier generations. In addition, the Industrial Revolution brought an ever-increasing number of people to urban areas to live, run businesses, and raise families. Increasing technology along with massive advances in transportation and communication created the urban areas of today.

Cities have benefitted human populations throughout history. They promote a **sense of place** and encourage the sharing of ideas and collaboration. But not all urbanization, particularly that based on low density and auto-dependency, has been positive. For instance, suburban residents endure long, increasingly dangerous commutes; inhospitable streets designed for traffic rather than people; and tract neighborhoods where there is nothing to do and nowhere to walk. These all impact human health as well as shape human emotions, outlooks, prejudices and ideas.

Amidst all of the climatic, financial, social and economic problems nations are experiencing, there is an increasing number of committed and competent people pushing for policies necessary to ensure better, more affordable living conditions in cities and rural areas (Vitalyst, 2016). Grassroots efforts to reshape policies and regulations have been underway for over 15 years across the globe. In addition to reducing **carbon emissions**, these efforts have sought to drive improvements in the design and performance of built environments targeted toward mitigating the environmental drivers of chronic disease and improving levels of well-being and health.

Carbon emissions The release of carbon into the atmosphere.

The rising social and economic costs of chronic diseases are gaining a great deal more attention as social justice and social responsibility are added to the equation of economic success. Today the costs of treating 130 million Americans with one or more chronic diseases is $2.5 trillion annually—about 75% of all health care costs in the US (Partnership to Fight Chronic Disease, 2015). As a comparison, this is over four times the annual US defense budget. It is a massive and unnecessary drain on productivity, well-being, and the ability of millions to pursue and enjoy fulfilled lives.

The **urban fabric** needs to perform better to enable, indeed encourage, people to live healthier, more active lives. Deterring professionals from creating these environments is the profit-centric business model of real estate speculation and land development, which exerts huge pressures to design buildings along formulaic lines. It continually produces the same types of places with the same built-in limitations and biases of recent decades. This is especially true for US urban areas. Today, financiers too often see design as nothing more than cosmetics for buildings. There is little to no regard for community and streetscape.

Urban fabric The physical elements of an urban environment; these include buildings, parks, streets, trees and sidewalks, to name a few.

Yet, a person's environment shapes perceptions, attitudes, and emotions as much as interpersonal relationships do (British Council, 2016). Codes come in three varieties, prescriptive, performance and outcome. Prescriptive codes provide for a fast way of complying with energy and other environmental standards. Performance-based codes are designed to achieve specific results, thus freeing professional designers (architects/engineers) to develop innovative solutions. Outcome-based standards are applied generally after a building or a development is completed. The objective is to discover whether and how well the building or development achieved the desired outcomes as well as how well the applied sustainability standards actually worked once the building or development has been in use. Inhospitable urban environments can lead to breakdowns in community relations, isolation, depression, and a host of other chronic diseases and conditions. Connectedness to neighbors and community, so critical for **social capital** and innovation, declines while depression and violence increase.

Social capital The connections and relationships between people that enable them to work together, generally for practical benefits.

In order for new policies to gain traction and encourage more hospitable, human-oriented cities, development has to promote much higher levels of livability and environmental health than many offer today. It is timely to ask the question of whether it is enough to leave the development of solutions for better performing human habitats to the '**market forces**' driving existing construction, real estate, and land development sectors. There has to be a better and more efficient way of creating and adaptively reusing man-made environments so that human health and well-being are at the core of solutions and not an afterthought.

Market forces The factors, often self-interested and profit-oriented, that determine the price levels in a specific economy.

Design Elements for Sustainable Cities

Urban design is the framework for public space. It ranges from the layout of an entire city to managing the details—such as pedestrian zones, incorporating nature, character and meaning, and even street furniture (Center for Design Excellence, 2016). Sustainable design contains a wide range of detailed elements, such as configuration, space, security, light, noise, temperature, and air quality. These details are worthy of discussion, particularly after the American Society of Civil Engineers released their Report Card for America's Infrastructure in 2013.

They evaluated infrastructure aspects such as energy, transit, bridges, public parks, wastewater, and drinking water—to name a few—and America's Grade Point Average (GPA) overall was a D+ (American Society of Civil Engineers, 2016). In the meantime, tax dollars are repeatedly syphoned to roads, encouraging more and more personal automobile use. Clearly, US infrastructure investment needs rethinking. Simultaneously, sustainable solutions must be created to alleviate this problem. The following are some of the critical elements that should be a part of any urban design discussions based on sustainability.

Human Scale

Eye candy An aesthetically pleasing environment, ranging from building details of character to full neighborhoods or cities, that engenders a strong connection with one's surroundings.

The invention of the elevator (1852) by Elisha Otis and the introduction of cement in the US (c. 1870) forever changed the scale of the built environment. Prior to this time, six stories was generally the limit of wood construction for habitation and the average height people could comfortably climb. These restrictions also lent a human scale to the environment in which people lived and worked. With the ability to construct higher, and later the advent of the automobile, people and foot traffic became a secondary consideration in urban design.

Johann Wolfgang Von Goethe said 'Architecture is frozen music.' In other words, the built world affects the emotions of those interacting with it. The more character and texture to a structure or diversity within a city, the more music or 'eye candy' is visible and the stronger the connection people will have with their surroundings. When the scale moves skyward, perceptions, including unconscious feelings of fear or lack of desire to stay, may change and encourage people to leave.

Symmetry and proportion are tools that have been used in the past to maintain scale while enabling cities to grow taller. Phi or the Golden Section is a ratio found everywhere in nature including the proportions of the human body. The closer a ratio between two parts of an object , i.e.,

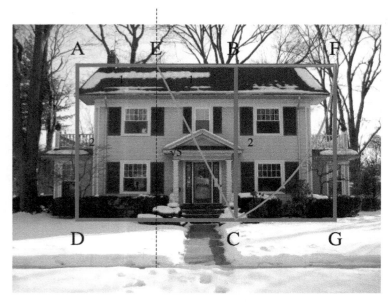

Figure 2 Phi or the Golden Section applied to an early 20th century House (illustration by Philip D. Allsopp)

Source: Republished with permission of Birkheauser, from The Modulor: A Harmonious Measure to the Human Scale, Universally Applicable to Architecture and Mechanics, Le Corbusier, © 2004; permission conveyed through Copyright Clearance Center, Inc.

features on a human face for example, is to Phi (1:1.618) the more appealing people find it. This ratio can be and indeed is used to create a sense of harmony, proportion and even beauty for large- and small-scale urban structures. Its discovery dates back to the ancient Greeks and Egyptians. For instance, Marcus Vitruvius Pollio, a Roman architect who practiced in the first century BCE, authored *The Ten Books on Architecture,* the oldest book that delves extensively into the principle of symmetry, harmony, and proportion in architecture (Pollio, 1914).

In the 20th century, Le Corbusier developed a Phi-based system of measurement and proportion called Modulor based on the human body for the design of buildings and large-scale urban developments. He employed it in many of his buildings, including the Indian Capitol complex at Chandigarh and his famous high rise apartments and streetscapes in the air: Unité d'Habitation in Marseilles. The practice of designing building, plaza, and street dimensions to the human scale is

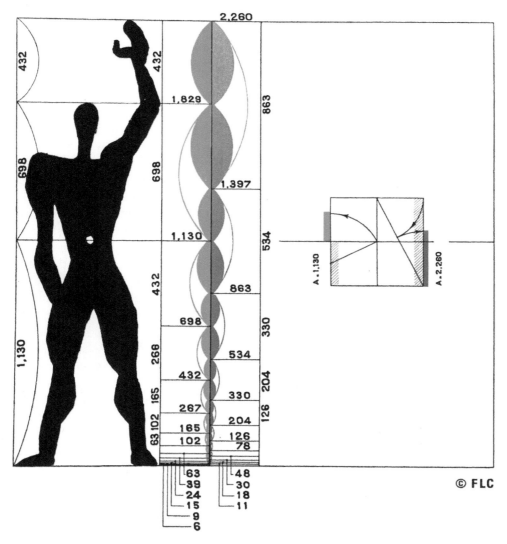

Figure 3 Le Corbusier's Modulor Man; the 20th century tool created using Phi.
Source: Le Corbusier. 2000. Modulo. Basel, Switzerland: Birkhauser: back cover of ISBN 3-7643-6188-3.

still used today by some leading planners and designers including Renzo Piano, who used Phi-based ratios in every aspect of his design for The Nasher Gallery in Dallas.

Harmony, proportion, and symmetry in urban environments are important *design metrics* for creating human space. These tools can serve as templates for making more sustainable places. Human scale encourages increased interactions with a city and sets the stage for healthier lives, a key sustainable factor in urban design.

Walkability and Sidewalks

Walking saves lives. It helps to maintain a healthy weight, which prevents obesity, strengthens bones and muscles, and improves balance and coordination; decrease the risk of and manage various conditions, including heart disease, high blood pressure and type 2 diabetes; and offers a promising low-cost treatment to improve neurocognitive function, by reducing stress, anxiety, even depression (Curtis, 2014). Walking clearly has more benefits than merely exercise, however, and it deserves equality during discussions on transportation improvements. For example, travel by automobile not only distances people from the empty lots and run-down, graffitied properties, it also separates us from each other.

Research has shown that we lose eye contact with others at about 20 miles per hours and crime increases with higher speeds (Curtis, 2014, p. 206). In a 2010 survey, the majority of people agreed that walking and a sense of community were linked. Yet since the middle of the 20th century, urban design has focused almost exclusively on the automobile as the preferred mode of transportation. Cities, ranging from downtowns to residential neighborhoods, have become inhospitable to human beings and the scale is geared more to traffic than people.

In 1969, roughly half of all students still walked to school, 41% of all students lived within one mile of their school, and 89% of these students walked. Today, only 13% of American students walk to school. Instead of exploring the world, burning calories, gaining independence and socializing with friends along the way, children are chauffeured in cars and buses (Curtis, 2014, p. 54). Furthermore, local businesses left downtowns for suburban malls, catering to the automobile and further discouraging walking.

Sidewalks, once a normal element of a street's development, are often being omitted during road construction or relegated to corners only, perhaps as a token gesture to walkers by the local transportation department. Paved corners were perhaps a solution, albeit a poor one, for addressing the fact that over 40% of pedestrian deaths in the US in 2007 and 2008 occurred where there was no crosswalk (Smart Growth America, 2016). This is beginning to change.

People are no longer willing to merely accept their surroundings. More and more voices,

Figure 4 Sidewalks for crossing purposes only (photo by Lauren Allsop)

particularly young ones, are demanding alternatives to sprawl and congestion. Founded in 2007, Walk Score has become a tool to determine which neighborhoods are safe and walkable to nearby amenities (Walk Score, 2016). A user types in an address to generate a map dotted with icons that identify parks, schools, groceries and other services. The tool then provides a walk, transit, and bike score using a point scale from 0–100. The higher the points, the closer to 'paradise' the address is. In addition, a growing number of Internet services, such as SpotCrime.com, are available that enable people to evaluate the amenities a particular subdivision or location offers and choose where to live and work. Walking is one of the strongest elements of a sustainable city as the benefits are healthy bodies, minds, and surroundings (Montgomery, 2014).

Jeff Speck, author of *Walkable City* and co-author of *The Smart Growth Manual*, cites ten key points of walkability, ranging from appropriate parking, mass transit and bikes to pedestrians and the diversity of the spaces they pass (Speck, 2013). He delves into the need for vehicles and the equal need for pedestrian-only zones. Neighborhoods with diversity, be it mixed use, design of space, or architectural details, increase walkability and its frequency. While no book can provide a city with step-by-step guidelines for improving walkability, there is sufficient evidence for the success of providing visually diverse walkable areas and various transportation options.

Trees and Their Canopy

Trees bring numerous social, environmental, and economic benefits to urban settings. It is common knowledge that trees produce oxygen and absorb carbon dioxide from the atmosphere. Greenery and shade tree canopy also encourage walking and reduce stress and noise levels (Figure 5). In addition, benefits include ecosystem services that range from reduced energy use—shaded buildings decrease the demand for air conditioning—to storm water management and improving water quality—vegetation reduces runoff and helps purify water (Speck, 2013). So it is perplexing as to why cities cut trees down, especially in climates where their shade offers documented benefits.

Shade in the desert or the hot summer of northern climates, for instance, is critical. Research has shown that the shade under the canopy of an average tree can reduce radiant heat by up to 8%. Yet, arid urban centers such as Phoenix, Arizona have replaced natural habitats with buildings, streets, and parking lots. Many in the greater Metro Phoenix area argue that the cost of watering trees is too high or budget cuts no longer include payments to tree services for periodic pruning.

A University of California study of urban forestry in five US cities used the "i-Tree" software tool developed

Figure 5 Urban landscape and trees

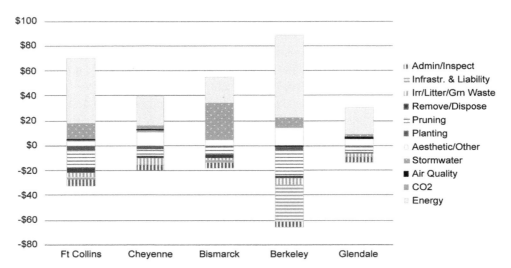

Figure 6 The benefits and costs of trees in five US cities.
Source: McPherson, G. et al. (2005) p. 4.

by the US Department of Agriculture's (USDA) Forest Service to assess the true cost of urban trees. This software allows cities to calculate both tree benefits and management costs and determine the full return on investment (ROI) (USDA Forest Service, 2017). The study found that the cities spent between $13 and $65 annually per tree. They also showed the ROI for every dollar invested in trees and management ranged from $1.37 to $3.09 (McPherson, G. et al., 2005). This data has prompted organizations such as the US Environmental Protection Agency (US EPA) to argue for tree retention, citing their long-term human and environmental values and the significant financial returns that accrue by keeping them (US EPA, 2016).

Extensive urban tree removal began with the automobile. As streets were widened to accommodate increasing volumes of motorized traffic, trees in the **road verge or planter zone** were eliminated. Many were also cut down as a pre-emptive measure to minimize the risks of automobiles hitting them. Research in 2002 showed that less than 1% of vehicle accidents involved a tree on an urban street, however (Wolf, 2017).

Road verge/Planter zone
A narrow strip of land before the curb and road surface, usually between a sidewalk and a curb. Typically, the space is filled with grass, but can include plantings and trees. The name varies from state to state and country to country.

To help cities improve their neighborhoods, some utility companies have established tree-planting programs. In Phoenix, the Salt River Project (SRP) offers two free native trees to residents they serve if the people come to a class to learn more about trees, their benefits, and care. Until recently, Arizona Public Service (APS) did as well. The evidence demonstrates that planting and maintaining trees offers far greater financial, environmental, and human well-being benefits than eliminating them.

Complete Streets

Complete Streets, as defined by Smart Growth America's website, are streets for everyone (Smart Growth America, 2016). All stakeholders—pedestrians, people with disabilities, bicyclists, public transportation users, and automobile drivers—are important and need to be included in street design plans. Thus, the Complete Streets Policy is to routinely design and operate the entire right of way to enable safe access for all users, regardless of age, ability, or mode of transportation (Smart Growth America, 2016). This is the future of urban design in cities.

Before the rise of the automobile, and the urban design and zoning policies that catered to it, cities and towns had Main Streets that were the hub or center of activity surrounded by residential neighborhoods. These hubs included shops, restaurants, other amenities and social spaces that residents could walk to from their houses on the surrounding streets. They were complete public spaces. The move to wider streets, void of trees and other amenities like sidewalks, and sprawling suburbs has created incomplete public spaces in towns and cities. They do not accommodate—nor encourage—**place making** or a sense of community.

Grand Avenue, part of US Route 60 in Phoenix, is an example of how a Complete Street can impact an area. Originally a major thoroughfare from downtown Phoenix to Los Angeles, it supported stores, motels, and numerous other businesses in what is known as Lower Grand Avenue (7th Avenue to 19th Avenue) and industrial areas in the northern corridor. With the completion of the US-10 Interstate, Grand Avenue was bypassed and the once busy highway abandoned by many tourists. This led to the area's decline, resulting in scores of abandoned buildings and lower-income neighborhoods.

Over the last twenty years, the demographics have begun to change again and the area has become a mecca for artists. In 2013, the Greening of Grand, a Complete Streets initiative, began with a grant from the Environmental Protection Agency (EPA). The community worked closely with the City of Phoenix to create the new streetscape and improve livability for the residents. Much of Grand Avenue between 7th Avenue and 15th Avenue was converted to two lanes and bike lanes installed. Planters were set in what used to be the outer lanes, and artists decorated them with paint, tiles and other artwork.

The project began with a three-day community design charrette, during which concerns were identified. These included high traffic speeds, and the lack of on-street parking and bike lanes (City of Phoenix, 2012). Upon project completion, all participants felt that the design made the urban area sustainable by:

1. Engaging the public;
2. Encouraging community events;
3. Enhancing economic development opportunities;
4. Supporting existing businesses (visibility, parking, etc.);
5. Providing a variety of mobility choices (bicycle, pedestrian, etc.);

Figure 7 Grand Avenue, Phoenix, AZ post-Complete Streets initiative with bike lanes and art (photo by Lauren Allsop)

Place making Encourages communities to re-envision their public spaces and strengthen the connection between people and the places they share. It not only promotes more favorable urban design, but also nurtures the physical, cultural, and social identities of a place.

6. Reducing storm water runoff and increasing recharge areas (future);
7. Contributing to a reduction of the "Urban Heat Island"; and
8. Developing a model for streetscape design and community outreach (City of Phoenix, 2012).

The success of the Grand River Complete Street project demonstrates the far-reaching benefits of such programs. They help city planners and engineers learn how to incorporate new elements into street design. Successful projects also enable city officials to understand the importance of community involvement, see first hand the resulting economic benefits and vitality of a community, and gain experience in changing government policy for the sustainable future of their city.

Community residents gain a strong sense of pride for being a part of change. They come to understand that they can help implement better pedestrian safety for children and adults alike, and make a difference in their own environment—by creating healthier alternatives to the automobile or improving the local **microclimate** by lowering GHG emissions. These are strong elements for improving future sustainable urban design.

Microclimate Localized atmospheric conditions.

Transportation Choices

Transportation choices must go beyond merely reducing automotive use to create safe pedestrian environments or reduce pollution emissions in cities, however. The automobile has been predominantly responsible for many transportation problems, ranging from traffic congestion and accidents to significant cost of maintaining roads, bridges, and parking structures, as noted in the American Society of Civil Engineers' 2013 Report Card. These costs have very little to do with the automobile itself and everything to do with the volume of motor vehicles in use. Other problems include fracturing or elimination of once vibrant communities and deserted downtowns.

For instance, developers in the latter half of the 20th century carved roads through once cohesive communities or razed downtown businesses for parking lots to service people driving to work. This coupled with the exodus of both people and businesses to the suburbs turned cities into ghost towns after working hours. The Warehouse District in Phoenix is a perfect example (Figures 8 and 9).

Figure 8 An intersection of remaining warehouse buildings gives an overall feel for the character of this once thriving district in Phoenix, Arizona (photo by Lauren Allsop)

Figure 9 Vast acreage of parking lots in the Warehouse District, where buildings servicing Phoenix, Arizona once stood (photo by Lauren Allsop)

Once full of businesses such as meat packing plants, ice factories, or laundries to name a few, the district served other downtown businesses and residents. Today, one corner of the district is nothing but parking lots allocated for the city's sports venues, which lie vacant on non-game days. Other cities are now faced with similar situations. Cities such as Detroit, Michigan and Cleveland, Ohio are struggling with significant abandonment, urban blight, and increasing crime.

Providing alternate transportation offers some answers, particularly when facts such as 'increased bus frequency reduces auto trips by 15–30%' are added to the equation (Coyle, 2011). Per capita motor vehicle travel peaked around the year 2000 in most Western countries. Between 2004 and 2012, transit ridership increased in the US by 14% while motor vehicle travel declined by 1% (Litman, 2017). Clearly, the world is moving in the right direction, but we have a long way to go.

Increasing mobility choices enables everyone to participate in an urban setting. An incremental approach is one solution that begins with Park and Ride locations where mass transit lines end. The closer one gets to downtown the more options, such as bike lanes, sidewalks, and more buses, become available. London, England has imposed a congestion tax for anyone driving a car into downtown inside the designated zone rather than using public transportation options. Bike and car sharing have also become popular in many cities. Portland, Oregon is just one city offering smart cars placed strategically around the downtown and available whenever needed. Phoenix, Arizona has installed bike-share locations throughout the downtown (See Chapter 12 on Sustainable Transportation). These programs are mobile app-based and enable the user to return the vehicle at any designated location around town.

Such options are enabling cities to strengthen the sustainability of their urban fabric both in human health and congestion. Blighted or vacant lands, many resulting from fewer automobiles within urban boundaries, are being redeveloped with positive outcomes. The City of Burlingame, California, for example, is constructing affordable housing on several defunct parking lots (City of Burlingame, 2015). The proximity to downtown enables residents to walk there. While transportation will remain a critical element of urban design, selecting the right amount of the various choices is what will make for a sustainable and integrated future in urban settings (Frederick, 2017).

Smaller Scaled Design Elements

National Aeronautics and Space Administration (NASA) just reported that Earth's surface temperature in 2016 was the warmest since recordkeeping began in 1880 (NASA, 2017). Whenever these temperatures—be they seasonal or year-round—are high enough to have inhabitants thinking of turning on air conditioners, alternatives are needed to cooling those microclimates. Buildings need to perform appropriately to the climates they are in, especially before dwindling resources are used. Thus, it is vital that urban design discussions not only include the larger community picture as examined above, but also consider what individuals can do to actively participate in making the urban envelope sustainable. To that end, the following are localized solutions that can be implemented on a case-by-case basis.

Green, Cool, and Solar Roofs

A green roof is the generic name given to creating a sustainable microclimate on a rooftop. The term is typically applied to systems involving plants, also called living, eco, or vegetated roofs. The design phase is essential to determine the load on the roof members as well as the type of waterproofing needed. Once in place, green roofs can help reduce rainwater runoff, reduce the heat island effect, and improve the quality of air. However, a green roof is not the perfect solution for every building.

Shutterstock Stock photo ID: 186834584

Figure 10 Green roof in Manhattan, NY

Under desert conditions in the southwestern US, planted green roofs are not as feasible as they are in more temperature climates. The lack of precipitation and extended summer heat make the watering and maintenance cost to cooling benefit ratio less than desirable, especially if non-native plants are used. In these areas, cool roofs or solar installations are a better alternative.

Cool roofs refer to those that reflect more sunlight and absorb less heat than a standard roof. They can be made of highly reflective paints, a sheet covering, or highly reflective tiles or shingles (US DOE, 2017). These materials reduce the heat retention of a roof, which reduces the transfer of heat to the interior of the building. This decreases the need for air conditioning. Other benefits include reduced atmospheric air temperatures, lower electricity demand during peak times, and lower GHG emissions. See Chapter 14 for a discussion on **urban heat island**.

Eco-friendly paints provide an inexpensive and sustainable solution to the radiant heat emanating from any exterior surface, particularly concrete and asphalt. They contain bio-based, renewable, non-hazardous ingredients, while offering a solar reflective surface that reduces heat absorption with a two-coat process. The first layer—a base coat—is a white insulation layer. The second coat, not limited to a single color, is the reflective coating comprised of microscopic ceramic particles that capture and reflect any solar rays back. A study in New York City showed that white roof coatings reduced rooftop temperatures in summer by an average of 39.7°F (Gaffin et al., 2012).

Another alternative, particularly on roofs that cannot support the weight of a living roof, is the installation of solar panels. The roofs of commercial buildings are often filled with mechanical equipment. As such, the roofs are flat and well suited for solar panels, unobstructed from sunlight yet hidden from street view so as not to visually mar the character of place. For homeowners, the initial costs of installing a solar system is often prohibitive. This is rapidly changing as the technology becomes more readily available and refined.

For example, Tesla Inc. announced the introduction of solar roof tiles that mimic traditional roofing materials in fall 2016. Each glass tile is a solar cell and eliminates the large panels mounted on roofs. This system is ideal for new construction or whole roof replacements. In April 2017, the company revealed its new Panasonic-made low-profile residential solar panel for those installing solar on an existing roof. These innovations are enabling building owners to define the type of green roof best suited for their needs.

Urban Heat Island (UHI)
The "island" of stored, radiative heat caused by urban buildings, roads, and other impermeable infrastructure that replace open land and vegetation.

Eco-friendly Not harmful to the environment.

Green Walls

Stanley Hart White first patented the concept for living green walls in 1938 (Hefferman, 2013). These walls are a vertical version of living roofs, though the design is simpler and the benefits are far more extensive. Like trees, green walls reduce the UHI effect, reduce noise, and improve air quality. They offer the flexibility of being freestanding or attached to a building. Freestanding green walls provide privacy as well as the aforementioned benefits. Walls directly against a structure serve as a form of insulation, stabilizing temperature fluctuations. This also reduces the costly need to energy retrofit building interiors.

Figure 11 Vertical Garden, Caixa Forum, Madrid, Spain (photo by Philip D. Allsopp)

Figure 12 Close-up of Vertical Garden (photo by Lauren Allsop)

The potential diversity of green walls makes them one of the best urban design features. One of the better-known walls is the Vertical Garden at Caixa Forum in Madrid, Spain. Designed by Patrick Blanc in 2007, the project began by covering an existing hotel's four-story wall with a sheet of polyurethane. This protected the building from moisture problems and provided space between the wall and the building for the irrigation and fertilization system, which is zoned at different heights. The system includes a network of pipes arranged in layers with emitters, fed by a pump to get the water to the top of the wall. The wall itself is plastic mesh covered with a non-biodegradable felt-like wool blanket with pockets. Gravity pulls the water through the different irrigation layers to wet the wool and feed the plants. Over 15,000 plants were initially set into the pockets, with an estimated weight of 30 kilograms per square meter (6.14 lbs./sq. ft.) (Hefferman, 2013).

Reflective Paving Surfaces

Dark surfaces absorb 80–95% of sunlight, so any effort to reduce this has enormous impacts on a building's performance and the overall sustainability of an urban environment. Cool smart pavements, developed by the University of California Davis and Lawrence Berkeley Lab scientists in 2012, are either a cool-colored coating with higher solar reflectance or a lighter colored traditional pavement material. Either way, the goal is to achieve a reflectance of 35%, 21°F cooler than traditional asphalt (Cool California, 2016).

Permeable pavements provide another tool for reducing the UHI effect during both day and night. Typical examples include pavers laid with gaps or concrete blocks embedded on their side and filled with micro-gravel for driveways and parking lots. Not only are temperatures lowered due to their color, porosity, and insulating qualities, but also the open mesh system absorbs runoff water, enabling local groundwater to be recharged into the watershed. Permeable paving also enhances neighborhoods by bringing character and texture to it.

While the design elements cited here are some of the most important for a city to implement for sustainability, they are by no means sole-end solutions. For example, a city can create one of the best economic and environmentally friendly

Figure 13 Permeable Paving in Phoenix, AZ (photo by Philip D. Allsopp)

transportation systems available. But without an engaging, human-centric urban core, the system is destined to fail. City planning and economic development departments need to examine and reevaluate their current protocols and implement new standards that better match sustainability measures as well as their residents' needs. It is also important that ways are found to fund and conduct assessments of how well urban development and settings perform.

Design metrics are valuable tools and can be used across the spectrum of design elements discussed in this chapter. Creating design metrics that enhance activity, engage the community in both the design and the physical environment, and promote safety and comfort together with periodic assessment also enables new ideas and sustainability measures to be introduced. New York City, for example, was the first in the United States to install a protected bike lane (New York City Department of Transportation, 2012). The area selected was 8th and 9th Avenues. The City's Department of Transportation (DOT) created the following design metrics:

1. Crashes and injuries;
2. Volume of vehicles and people;
3. Traffic speed;
4. Economic vitality, including growth in retail activity;
5. User satisfaction; and
6. Environment and public health benefits.

The first three were used in designing the bike lane. The others were devoted to street calming, improving bus services and creating public spaces. DOT chose to buffer the lane by installing parking between it and traffic. The metrics were periodically monitored and proved to be successful as evidenced by a 35% decrease in injuries to all street users on 8th Avenue, a 58% decrease in injuries on 9th Avenue, and a 49% increase in retail sales on 9th Avenue between 23rd and 31st Streets (NYC DOT, 2012).

The City of Mesa, Arizona is re-envisioning all of its economic development zones. They employed a community-design approach in rethinking its downtown

Figure 14 Downtown Mesa, Arizona (photo by Lauren Allsop)

area. The city ran a visioning competition that encouraged design teams to work with residents and the city to determine the needs of the community. The result is an active hub peppered with outdoor benches, bronze statues, and artwork that maintains the city's low-intensity historic district, enhances its cultural district, and encourages people to enjoy their town and inspire a return visit. These features also promote walking and help create an environment filled with the diversity humans need for well-being.

Post Development Assessment

The current mechanisms for developing and creating places for a wide range of human endeavors largely encompass a one-way journey—from policy and funding to a development solution (Partnership to Fight Chronic Disease, 2015). For all the benefits and great ideas illustrated and marketed by developers for a particular design scheme, there is little to no follow up—and minimal incentives—to measure whether or not the design actually delivered the benefits originally envisioned. This results in built environments being largely experimental prototypes rather than well-researched products that are put into production and comprise all of the attention to detail, engineering, and human-factors research necessary for high performance results.

Post development assessments can improve policies directed at human health, real estate financing, and zoning and planning laws and regulations (Allsopp, P., 2016) Many economic sectors, such aerospace, automotive, and electronic consumer products, employ evaluations to determine whether products or services actually work as planned. In the field of urban development, however, there are significant financial pressures for real estate to generate rapid profits so much so that there is little time or money to conduct research into what worked and what did not. Given that data from research clearly shows that well-being is shaped largely in part by the places people inhabit, it makes sense to find ways of rebalancing these incentives so that the act of discovering the real performance or outcomes of a particular development become routine and appropriately funded.

Sustainability Standards and Rating Systems

In addition to mandatory national and municipal building, planning, and zoning codes, there are at least 15 different sustainability-rating systems for buildings today (National Institute of Building Science, 2016). Each takes a different approach to defining what constitutes a sustainable building design. Some are focused on building systems and appliances, while others attempt to deal with a more holistic assessment of buildings, nature, people and community. One of the most comprehensive analyses of green building standards and certification systems can be found in the Whole Building Design Guide, a publication of the National Institute of Building Science. They identified three types of code—prescriptive, performance, and outcome. Prescriptive codes provide for a fast means of complying with energy and other environmental standards. Performance-based codes are designed to achieve specific results, thus freeing professional designers, architects, and engineers to develop innovative solutions. Outcome-based standards are generally applied after a building or development is completed. The objective is to discover whether and how well the building or development achieved the desired outcomes as well as how successful the applied sustainability standards worked once the building or development was in use (National Institute of Building Science, 2016). Why were all these standards developed and where did they come from?

Historic Drivers leading to Modern Sustainability Standards

Designing cities to account for local climatic conditions, rather than in spite of them, is not a new concept. Ancient civilizations were designed explicitly to maximize the sun's warmth in winter and minimize its heating effects in the summer. The city of Priene in Ancient Greece, for example, was situated on the east coast of the Aegean in ancient Ionia, now present-day Turkey. It is generally considered to be one of the first examples of city planning based on a grid as well as a prime example of passive solar orientation. After Rome fell in 476 CE, vast amounts knowledge about building and city planning were lost.

In the early years of the Industrial Revolution, concerns were growing over the headlong rush to profits at the expense of people's health and well-being. By 1770, Britain's Royal Society for the Arts, Manufactures, and Commerce, of which Benjamin Franklin was a Fellow, issued a competition open to all factory owners for a way to reduce air pollution from smoke and soot belching out of thousands of chimney (Royal Society for the Arts, 2017). Science was also on the move during this time.

Joseph Fourier, known more widely for his mathematical work, was the discoverer of the greenhouse effect in the 1820s. He concluded that the earth's atmosphere was warmed because it could absorb visible light by converting it into infrared energy. Since the atmosphere is not an efficient emitter of infrared energy, temperature rises. Further work by Eunice Foote in 1856 concluded that atmospheric warming was increased by the presence of carbon dioxide and that if the earth's atmosphere were rich in this gas, it would have a higher temperature. By 1896, Svante Arrhenius, a Swedish scientist, concluded that the human emission of carbon dioxide would eventually lead to global warming. However, because of the relatively low rate of carbon dioxide production at that time, Arrhenius thought the warming would take thousands of years and would be a benefit to humanity.

Rapid population growth in industrial cities during the 19th century meant that more coal and wood were being burned. Much of this occurred within poorly ventilated dwellings and cramped tenements or in manufacturing near these housing settlements, making air pollution a daily hazard for millions. This, coupled with the great Cholera Epidemic in London in the early 1850s, forced the eventual passage of new public health and city planning laws. Two examples were the separation of drinking water from sewage, and the establishment of green belts and public parks to give city dwellers respite from the choking fumes of thousands of workplaces. The field of city planning was jump started by this legislation.

Frederick Law Olmstead's travels to Britain in 1850 made an indelible impression on him and his future career. After visiting Birkenhead Park, one of the first public open spaces designated and funded by the British government, Olmstead submitted a paper to A. J. Downing's *Horticulturalist* from which Downing's support of Central Park began. These initiatives led to environmental protection and public health laws as well as numerous ethically based movements aimed at improving the lot of the average person. Many were designed to make city living and working healthier and provide growing populations with access to fresh air and nature. Businesses too benefitted greatly. Over time, they enjoyed the economic benefits of a fitter, healthier, more energetic and innovative workforce.

Reacting to Mass Production: The Green Building Movement

Continued industrial development and population growth through the 20th century saw corporations expand their influence over almost every aspect of daily life, despite two devastating world wars. People began to be referred to as 'consumers' whose main purpose, it seemed, was to work and buy things. The goal became focused on keeping the engines of industry moving forward, particularly the corporations building out suburbia. Gradually, well-being and the ability of people to live and work in urban settings that were healthy, efficient, and pleasant places to be were secondary considerations.

In 1990, the United Kingdom's Building Research Establishment's Energy and Environmental Assessment Method (BREEAM) was published to address the need to improve the quality, durability, and performance of built environments. First employed to rate office buildings, BREEAM is the world's longest established method of assessing, rating, and certifying the sustainability of buildings (Santa Monica Farmers Markets, 2017).More than 250,000 buildings have been BREEAM certified and over one million are registered for certification—most in the UK with others in more than 50 countries. BREEAM includes the following sustainability categories for assessment including:

1. Management
2. Energy
3. Health and wellbeing
4. Transportation
5. Water
6. Materials
7. Waste
8. Land use and ecology
9. Pollution

In the US, the US Green Building Council's LEED (Leadership in Energy and Environmental Design) is a commonly applied system for certifying buildings and the performance they can achieve. LEED is based on a subset of BREEAM, but over the years it has expanded its scope to include extant buildings, landscapes, and urban settings. Other, more holistic standards such as Green Globes and the Living Building Challenge offer alternative ways of rating built environments. These systems provide a moderating influence on the negative tendencies of real estate speculation and have been adopted by some banks to reduce their risks of funding poorly performing buildings. While all these actions are laudable, chasing sustainability points to garner a higher level of certification for real estate marketing purposes does not necessarily result in a more sustainable building or healthier urban environment.

At the core of these rating systems lie the same concerns that arose in the early years of the Industrial Revolution—clean air, separation of waste from clean drinking water, clean and healthy indoor environments, energy conservation, global warming, and community resilience. Of these, global warming presents us with a multi-generational problem to be solved. Human health and well-being depend on it as do secure and thriving nations and cultures. Even at the highest levels of policy making and finance, it is incumbent upon practicing professionals in the overlapping fields of public health, architecture, planning, engineering, and economics to engage knowledgably and with insights about the use and improvement of sustainability standards and rating systems. One of the paths available for developing solutions to population growth and global warming is to provide professionals responsible for shaping human habitats and settlements with sufficient room and time to innovate. Standards play a role in these endeavors, but they are far from guaranteeing success. When applied prescriptively, they can impede the innovation that is required to achieve sustainable and healthy environments for future generations.

Overcoming Ingrained Problems

The 20th century was one of the hardest on our built environments. It was the century that saw zoning first implemented as well as change over time. Some cities had prescriptive regulations, but relaxed them when developers came and offered to help them grow. Developers saw it in their best interest to influence building regulations (Smart Growth America, 2016). Today we look back and understand that profits, not healthy communities, were behind much of what was built. Financial institutions, concerned about mitigating their investment risks, continue to prioritize lending toward projects that maintain the status quo rather than spur evidence-based innovation. More and more people are realizing that this model does not work, nor is it sustainable. They see the need for proper **stewardship** over existing built environments, rather than continually razing structures and further impacting dwindling resources.

Many cities are moving into form-based codes, which address form, mass, character, scale, and other aspects of the built environment in relationship to the public realm (Form-based Codes Institute, 2017). Conventional, or Euclidean, zoning focuses on micromanaging and segregation of uses. The newer codes are better suited to saving and preserving entire extant—and even historic—neighborhoods that were often previously leveled in the face of progress. Now cities are working to preserve them, helping owners utilize tax credits and other initiatives. The ingrained problem, though, is that cities often do not look back and reevaluate past projects until there is a need.

For example, numerous shopping centers were created with vast parking lots under conventional zoning codes. Prescription stated that there must be a specific number of parking spots per the square footage of the store. Yet, many of these locations have never experienced a full lot and the spaces exist with no thought to reducing the landmass, installing permeable pavements, trees, or other sustainable measures that would greatly alter the character of the area. Even if a city did change this formula, some stores, particularly big box stores, demand that the formula be retained or they will leave. The vast open acreage tends to be void of shade and adds enormously to the heat island effect.

Cities are striving to create a sense of place. By employing urban sustainable design tools, they are making forward progress. They must, however, overcome the ingrained problems from the past to envision a city for everyone. More and more organizations are coming forward to help. One is Smart Growth America, whose mission is to work with officials, developers, planners and others to improve life for everyone (Smart Growth America, 2016). Through partnerships with local commissions and organizations across the US, they advise state leaders on sustainable growth, particularly on transportation.

The Texture of the Future

The design features and assessments cited in this chapter will, hopefully, serve to make readers more aware of the logic behind their observations when walking through their city. It is important to note that there is no single path to creating a sustainable city, no checkboxes to define success, and understanding the needs of all city inhabitants is a difficult task. However, designing an urban fabric that is scaled to the climate, location, and inhabitants and *managed* growth of a city is one step forward. It will take a broader, more holistic understanding of the true nature of cities and redefining their purpose to meet the needs of its inhabitants.

There is no such thing as a successful city without **human capital** (Green City Freiburg, 2016). Therefore, involving the public to gain valuable information enables

Stewardship The responsible care and preservation of something—in this context, the built environment.

Human capital The measurable economic value each person, individually or collectively, brings to a defined area or situation. Values can range from knowledge and experience to social attributes and wisdom.

planners to more fully understand what measures are best for creating connectivity among the inhabitants and an overall sustainable community. The city of Freiburg, Germany, cited by many as the greenest city in the world, believes active contributions from its people are crucial as local people are the ones with the necessary detailed knowledge of an area (Green City Freiburg, 2016; also see Chapter 4).

For decades, the bulk of studies leaned towards motorized forms of transportation. Now with health data becoming more readily available, such as that on obesity, the focus is moving towards human-scaled modes of transportation like walking and bicycling. The studies are not solely transportation-centered, but include human reactions while employing slower forms of movement. One of the early studies was by Robert Cervero in 1988. Analysis of his findings showed that mixed use was positively and significantly related to walking and bicycling (Cervero, 1988). In a later study, it was found that pedestrians responded differently to the different scales at which uses are mixed (Frank et al., 2003).

For planners, this means that measures should be created that lean away from the amount of square feet of different uses and towards the absolute number of different commercial land uses within a community (Frank et al., 2003, p. 180). It also enables planners to understand the communities and neighborhoods within their city and fit codes to local needs versus the generality of prescriptive codes and zoning. Remember, a different view is obtained when on foot versus speeding by in a car.

By including simple actions such as reflective paint or green walls, neighborhoods can more easily participate in improvements and know that their contribution is sustainable. It is important, though, to define improvements. City websites can help building owners decide when improvement means new or merely restored. They can link to the National Park Service, which publishes a series entitled, *Preservation Briefs* and provides guidance for preserving, rehabilitating, and restoring buildings (National Park Service, 2017). For example, *Preservation Brief #47* covers exterior building maintenance. *Preservation Brief #13* discusses windows. New windows—perhaps the current popular vinyl sliders—are not always the best and first course of action. Removing existing postwar steel casements is throwing out good windows that merely need double-glazing. The key is that a different window alters the visual character and texture of the neighborhood. Incentivizing the preservation and restoration of original features and providing easily obtained information on this process from the city is vital.

Many cities have taken measures to ensure that these considerations—large and small—are part of discussions and plans. Key questions, ranging from eliminating pollutants to social cohesion, are asked and responsible requirements put in place (Roseland, 2012). The City of Santa Monica, California, for example, has created a Sustainability Report Card for itself to monitor its continued success (Roseland, 2012).

It has defined eight goals that are annually reexamined:

1. Resource Conservation
2. Environmental and Public Health
3. Transportation
4. Economic Development

5. Open Space and Land Use
6. Housing
7. Community Education and Civic Participation
8. Human Dignity.

Under the second goal, one policy measure is to increase the consumption of fresh, locally produced, organic produce. To this end, on several days per week, several streets in or near the downtown are closed for a Farmer's Market (Vitalyst Health Foundation, 2016). A measure under Open Space and Land Use is trees. In 2012, 1,384 new trees were planted for a total of 35,884 (Sustainable Santa Monica, 2012).

Creativity and innovation will become a stronger design feature of sustainability in the future. Transition towns, a phrase coined in 2004, focus on energy independence through grassroots community projects. One feature pertains to the human emotional side of life in uncertain times—psychology. The Transition Network of towns proposes an alternative to the doom and gloom reaction to peak oil through low carbon living and acknowledging the emotional impact of change it will have (Connors and McDonald, 2010).

Urban sustainable design is indeed a transitioning element of planning. It encompasses logic from the past, such as human scale and phi, but looks to the future through features like the use of green walls as a form of building envelope insulation. Preserving our built environment with stronger tools against demolition in combination with implementing the results of post-development performance assessments will enable residents and visitors alike to take delight in the urban world around them. As mentioned above, research cites essential requirements to wellbeing that must include open spaces, light and order. Each of us as individuals is critical to the success of our cities, especially when nearly 70% of humans will be residing in them by 2050.

A man who builds a house owes a duty not alone to his family and himself, but to the neighborhood as well. He really has no moral right to construct a home that will be a blight on the landscape.
—Developer William Radford (1908)

Chapter 6

Examining Urban Sustainability through Urban Models

Subhrajit Guhathakurta

Sustainability and Urban Models

It is widely accepted that the sustainability of human existence requires the protection of earth's life-support systems as well as the efficient stewardship of natural resources so that they continue to be available for future generations. Sustainability thinking is about imagining a sustainable future. It is critically dependent upon understanding how the decisions we make today help to shape the future of our societies and settlements. Models are among the most important tools that help us anticipate that future. We use models to analyze how current trends play out over time, ask "What if?" questions about specific interventions in social and ecological processes, test our hypotheses about social trends, and provide guidance for planning. In other words, models are indispensable for sustainability thinking.

Urban areas are the primary hubs of human settlements and the most challenging places for advancing sustainability. Since 2007, the world population has become predominantly urban and the urbanization trend is continuing unabated. Cities are the centers of concentrated consumption of resources and the major sources of pollutants, which include climate changing greenhouse gases (GHGs). Therefore, the future sustainability of the planet will be determined by our ability to transform urban living along sustainable principles. Urban models help us find sustainable alternatives to our current (often unsustainable) urban lifestyles. This chapter shows how urban models have evolved from basic concerns about land use and transportation to providing important metrics for advancing sustainability in our cities.

Understanding Models and Modeling

Models are the microcosms of real-world phenomena that cannot be experienced or examined in their actual spatial and/or temporal contexts. We use different models for different purposes. Yes, some are indeed human models used by the fashion design and media industries to promote overpriced products. When Geiorgiou Armani or Calvin Klein want to convey how you might look and feel when you buy their newly designed outfits, they will find a model to demonstrate the virtues of these outfits. Of course, what is being modeled is far from our reality given that many of us will probably look uncomfortable or unseemly in the same outfit. In this case, the reality that is being modeled may only exist in peoples' imaginations.

Most models, however, are not meant to cater to the imaginary world but are designed to bring our imaginations in line with reality. In fact, they are quite distinct from the objectified humans described above. Scientists, innovators, and researchers often use physical models to test critical properties of a designed or engineered product before the product is distributed for broad use. For example, scaled-down airplane models are tested in a wind tunnel to examine how the shape and structure of the model perform under different wind conditions. Similarly, structural engineers use miniaturized models of buildings designed with different materials and technologies on a vibrating platform to test how these structures perform for different intensities and types of earthquakes. In both these cases, models allow us to test the robustness of our critical infrastructure under laboratory conditions. They also offer researchers the ability to improve on existing designs.

Models can also be concepts, ideas, or plans. These non-physical models are expressed through words or symbols—or sometimes not expressed at all. When I leave my home in the morning to go to work, I use a mental map—or a model—to choose my mode of travel and route to my destination. This model mostly remains in my unconscious (unless I am asked to describe the journey) since I have used it hundreds of times. We use such informal "**mental models**" to accomplish most of our tasks. Without mental models, we are unable to act deliberately.

These mental models are often inadequate when we face novel or complex situations. In such cases, we tend to formalize them so that they can be examined and perfected. Take, for example, travel to various destinations in a new country. Most travelers in new situations will study maps and make copious notes of places to stop for conveniences or sightseeing and alternative routes in the case of road closures. These maps and notes together constitute a formal model of travel in a relatively strange land.

Understanding complex processes and organizations requires **formalized models**. It would be difficult, for example, to have a clear understanding of how a city functions with only mental models. Our mental models may provide bits and pieces of the total picture. For instance, we may know that population growth and an auto-oriented culture will lead to high levels of traffic congestion. We could also intuit that high levels of congestion may induce some people to seek homes closer to work, other forms of transportation, or, in the long run, seek out options in other cities with better quality of life. Each of these processes triggers changes in individual and social behavior that can either reinforce congestion levels or counteract them.

Reinforcing processes are known as **positive feedback** loops while the counteracting processes offer **negative feedback**. When individuals try to leave the pollution and noise of congested roads by fleeing to the suburbs, they may actually reinforce congestion levels by increasing commute distances. On the other hand, when enough commuters decide to give up on driving (because it is no fun

Mental model An image we hold in our minds about how processes and events are related to each other.

Formalized model A specific set of procedure or mathematical formulae that show how different processes result in a particular outcome or set of outcomes.

Positive feedback A connected causal chain of events in a closed loop where the intensity or scale of the initial driving force increases with each cycle.

Negative feedback A connected causal chain of events in a closed loop where the intensity or scale of the initial driving force decreases with each cycle.

to drive at 5 miles an hour and be abused by irate drivers) and choose to travel by public transit, congestion levels may ease. This would be a negative feedback of congestion. Our mental models are inadequate to tease out the interactions among all the direct, indirect, and induced feedback loops of individual and social behavior in a city. In such complex situations, we use formal models with specific syntax and notations. These formalized models can both illustrate the interactions among the different parts of the system (such as individual and social decisions) and communicate how the system can change under different scenarios.

One such model for cities was developed by Jay Forrester who became interested in questions about urban policies and how they affect the quality of life of urban populations in the long run. He wrote a book called *Urban Dynamics* (1969) based on his findings from this modeling exercise. His model incorporated specific kinds of syntax, notations, and mathematics, known as **system dynamics**, which Forrester himself had developed to examine logistical aspects of growth and decline. Through his system-dynamics model, Forrester demonstrated that policies meant to alleviate urban problems like housing shortages, unemployment, or neighborhood decline in fact often exacerbate those problems. These models show that our intuitive "mental models" are often wrong when situated in a complex dynamic context.

As you can guess, there is a reason why I am using models of urban areas as prime examples of formalized models. Formalized models with complex mathematical expressions are used in numerous domains—from power systems to meteorology to economics. But urban areas pose a specific challenge given that the dynamics of urban processes include complex interactions among human, social, and natural systems. **Urban models** can never capture the richness of all interacting processes within an urban region. However, they can be useful in predicting future location and intensity of activities based on prior trends and show how different decisions we make today might play out in the future. Importantly, modeling can greatly inform urban sustainability. In the next few pages, I will demonstrate how to examine future sustainability impacts with the help of **urban-environmental models**. Before I delve into the questions of sustainability, a brief background on urban modeling is warranted.

> **Urban dynamics** Changes in the total connected urban system induced by changes in one component of the system. It is also the title of a book on examining a systems model of an urban region by Jay Forrester.

> **System dynamics** A methodological approach for characterizing a connected system of processes and events, and examining how changes in one component of the system affect all other parts of the system.

> **Urban odels** Mathematically derived expressions for connecting aspects of urban processes, most commonly the relationship between land use and transportation.

> **Urban-environmental models** Connecting urban models with environmental processes such as determining air quality from land use and transportation attributes.

The First Generation of Urban Models

While Forrester was developing his system dynamic framework, another form of urban model was already transforming land use and transportation planning in large urban regions across the US. This model was based on the concept of **"spatial interaction,"** which assumed that activities that were closer together had a higher intensity of interaction than those further away. This premise was an analogue to the Newtonian law of gravitation, which states that every mass in the universe attracts every other mass by a force equal to the product of their masses and inversely proportional to the square of the distance between them. In mathematical terms, this translates to:

> **Spatial interaction** A term used to characterize movement across spatial domains based on distance and other factors that impede such movement.

$$F \propto \frac{m_1 m_2}{d^2}$$

Where: F = Force of attraction between two masses; m_1 and m_2 = masses of the two bodies; and d = distance between m_1 and m_2

If you are wondering whether the inverse relationship with distance is always to the second power for all situations, your curiosity is not without merit. A more generalized version of the above equation is:

$$F = K \frac{m_1^\alpha m_1^\beta}{d^\lambda}$$

In this case, α, β, γ, and K are constants to be determined. Note that both the numerators and the denominator have power values that are not fixed but determined through empirical observations for each context of their use. If you now consider that m_1 and m_2 can also represent intensity of activities (e.g., number of people working, living, or shopping in an area), it is not difficult to see how the interactions among these activities can be modeled as an extension of the general spatial interaction model above.

Consider, for example, you want to allocate future population growth to the three existing neighborhoods in a small city rather than build new neighborhoods. Also, you expect that the attractiveness of these neighborhoods will be based on the amount of space available in them for accommodating new housing units (a proxy for price of entry) and how close the neighborhoods are to employment locations relative to the number of jobs in these locations. In other words, the *accessibility* (A_j) of a neighborhood j to employment locations is proportional to the sum of the number of jobs in these employment locations ($E_1, E_2,$ and E_3), each weighted by the inverse of distance to the neighborhood j (d_{ij}). That is:

Where K and λ are constants estimated empirically

$$A_j = K \sum_{i=1}^{3} \frac{E_i}{d_{ij}^\lambda}$$

Once the accessibility of each neighborhood is known, the allocation of growth to these neighborhoods can simply be apportioned according to the ratio of the product of land availability and accessibility of each neighborhood in relation to the total of similar products of all neighborhoods. Again, in mathematical terms this would be:

$$G_j = G_T \left(\frac{L_j A_j}{\sum_{i=1}^{n} L_j A_j} \right)$$

Where L_j is the indicator of amount of space available; G_j = Growth allocated to j; and G_T = Total growth in the city

The Lowry Model

In 1964, Ira Lowry of the Rand Corporation unveiled a land use and transportation model that became the precursor to the first generation of such models. These first-generation models allocated activities spatially using a form of the generic spatial interaction model described above. Similarly, the Lowry model included jobs and housing and their location relative to each other as well as several key innovations.

First, Lowry distinguished between two types of jobs—*basic* and *non-basic*. Basic jobs were those in export-oriented sectors. That is, the products of this sector were mostly consumed outside the region such as the output of manufacturing, mining, agriculture, tourism, and similar economic activities. All other jobs were considered non-basic, which indicated that they mostly served the local population. The number of basic jobs and their locations, current and future, are provided as *exogenous* inputs to the model (i.e., decided outside the model). The model then figures out where the non-basic jobs and households will be located *endogenously* (i.e., estimated by the model).

The Lowry model consists of two sub-models. The first, known as the *economic base sub-model*, jointly determines the additional service-related jobs and the households and population increments resulting from new basic employment. The second sub-model is the *spatial allocation sub-model* that allocates the new households, population, and service employment to zones using the familiar spatial interaction format. As you can tell, there is positive feedback between new households and new service-related jobs. As we add households of basic employees, more service-related jobs are needed, which in turn generates more households (families of those service employees), and subsequently more service jobs in a chain effect. This iterative process is stopped when additional increments are too small to account for. A simple schematic of this process is provided in Figure 1.

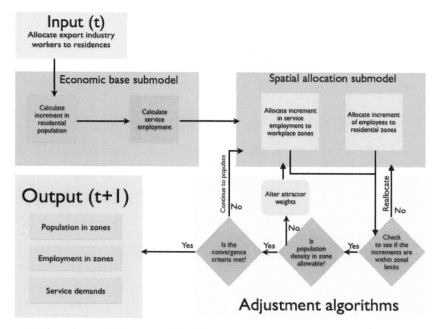

Figure 1 Flowchart of the Lowry Model

The Lowry model inspired a surge of successors that improved upon its basic framework. Within six years after Lowry's publication of *A Model of Metropolis* (1964), about a dozen such models were formulated both in the US and Europe. These included a Time Oriented Metropolitan Model (TOMM 1964); the Bay Area Simulation Study (BASS I 1965); the Cornell Land Use Game (CLUG 1966); a Dynamic Model for Urban Structure (TOMM II 1968); the Projective Land Use Model (PLUM 1968); and a portfolio of eight models developed by various individuals for different regions in the UK (see Goldner 1971 for a detailed review). Even today, over 50 years past the initial Lowry's formulation, aspects of his approach continue to be applied in urban land use and transportation models (i.e., DRAM, EMPAL, and METROPILUS models by Putnam 1983, and Putnam and Chen 2001).

Subsequent developments in urban land-use-change models included several innovations that extended the theoretical and empirical scope of the Lowry model. Most of these models included a more disaggregated set of households and jobs. Many broke away from the rigid square grid and adopted census tracts for zones. Several later models also experimented with different forms of land use constraints and limitations on household behavior and budgets. In addition, new techniques for calibration and evaluation were also introduced to make the models better "fit" the data available for specific places.

Despite the relatively rapid pace of development of the first generation of urban models, these models were perceived as being too large, too expensive, too reliant on extensive data, and therefore, inadequate in offering critical and reliable information planners needed in formulating long range plans. A seminal article written by Douglas Lee in 1973 called *Requiem for Large Scale Models* was significant in substantially slowing down research and application of urban models. Lee noted "seven sins of large–scale models" that together rendered the first generation of models irrelevant to the new directions in which planning was headed (Guhathakurta, 1999; Wegener, 1994). Regardless, urban models continued to progress and mature, while computers became more powerful and data management less onerous with the advent of Geographic Information System (GIS) technologies. A second generation of models began to take shape, albeit more self-consciously and with more humble objectives.

Second Generation of Urban Models

Second-generation urban models followed two general approaches. The first is rule-based **land classification** that assigns all land parcels a "suitability" score for various forms of development. The second approach follows from the well-known **discrete choice models** developed by Daniel McFadden, which won him the Nobel Prize in economics. Both these approaches were enabled by rapid advances in computing and GIS technologies, together with the proliferation of the Internet, which brought high-performance computing to individual desktops.

The **suitability score** approach originated from landscape analysis, particularly through Ian McHarg's most celebrated book, *Design with Nature* (1969). McHarg popularized the "overlay method" in which transparent map layers, each containing information about one specific land aspect (slope, soil type, land use, vegetation,

Land classification A system for categorizing land according to the use and physical attributes of that piece of land. Several such schemes are now used; the most common being the USGS Anderson Land Classification Scheme.

Discrete choice models Models examining choices between two or more alternatives such as choosing between different modes of transport or different residential locations.

Suitability score A score assigned to different plots of land according to its suitability for development.

floodplain, drainage, etc.), are overlaid to identify suitable areas that can be developed with minimum disruption to the natural environment. This approach became one of the primary functions of GIS, especially in the formative years of ESRI's ArcGIS mapping and analytics platform.

The overlay approach was also the foundation for the **Land Evaluation Site Assessment** (LESA) technique adopted by the US Department of Agriculture (USDA) in evaluating the suitability of land parcels for agriculture, conservation, or development. This technique essentially assigned scores to a set of land attributes (land evaluation or LE) and site characteristics (site assessment or SA), which were individually weighted according to their importance for a designated use. These scores were then aggregated to determine which parcels were most suitable for that use.

McHarg's land suitability approach was among the first forays of urban models into the domain of sustainability. This approach has provided the methodological foundation of Environmental Impact Assessments (EIA) mandated by the National Environmental Policy Act of 1969 (NEPA) for evaluating large public projects (McHarg and Steiner, 1998). The EIA continues to be a principle component of sustainability practices given its comprehensiveness and its orientation towards the health of humans and ecosystems. Modeling techniques used in EIA include spatial overlay as discussed in *Design with Nature*. These models typically combine different layers of information about areas important for maintaining ecological services (such as watersheds, bio-habitats, and vegetation) so that the site can be developed in the most environmentally benign manner.

The basic tenets of the land suitability approach adopted for the urban context is implemented in an urban model called "What If?" developed by Richard Klosterman of *What If?, Inc.* (Asgary, Klosterman, and Razami, 2007; Klosterman, 2008). The approach incorporated development capability of different predetermined spatial units of land reflected in a suitability score. While the technique relies on a significant amount of user input, the process is primarily based on assumptions about growth, public policies, and decision rules supplied by the users, mostly as weights for physical and locational characteristics of parcels that are suitable for particular developments or for conservation. Klosterman emphasizes that "What if" is not a forecasting tool but a planning support tool that shows *What* would happen *If* a) specific development policies are enacted; b) growth assumptions prove to be true; and 3) the user-supplied suitability scores are appropriate and reasonable (See Figure 2).

A variation of the suitability approach was implemented in California Urban Futures (CUF-1) model developed by John Landis and his team at Berkeley (Landis,1994, 1995). Designed to determine suitability for development, this model was based on "profitability" rather than physical and environmental constraints assumed by the user. In CUF-1 as in What If?, the units of land were synthetically created from land attributes; hence they have no legal existence. These parcels are constructs generated from spatial intersection and/or union of various land attributes such as zoning, slopes, density, distance from transport infrastructure, and others. Landis called them developable land units or DLUs. These DLUs ranged in size from one to several hundred acres.

In CUF-1, each DLU was assigned a profitability score based on estimated costs of development in that land unit. Only residential developments were

Land Evaluation Site Assessment A point-based approach for rating the relative importance of agricultural land resources based upon specific measurable features. The land evaluation component measures soil quality, while the site assessment component evaluates the site's importance of agricultural activities in relation to conservation or development potential.

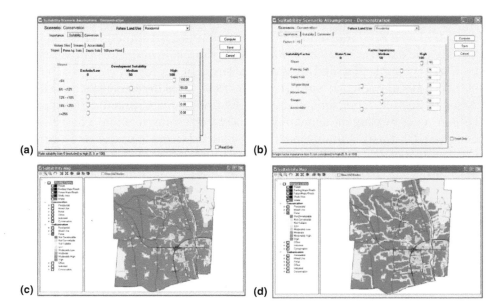

Figure 2 *a-d: Using "What If?" to generate suitability maps*. 2a) Top left: user interface showing residential suitability set to be high for slopes below 6%, medium for slopes 6-12%, and 0 for all other slopes. 2b) Top right: user interface showing the relative importance of various factors considered for residential suitability. 2c) Bottom left: suitability map of retail enterprises under a suburbanization scenario. 2d) Bottom right: another suitability map for retail under a conservation scenario that protects environmentally sensitive areas near streams

Source: Klosterman, R.E. 2008. A New Tool for Planning: The What If? Planning Support System. In Chapter 5 of R. K. Brail edited *Planning Support Systems for Cities and Regions*. Cambridge, MA: Lincoln Land Institute

a-d: Planning Support Systems for Cities and Regions, Lewis D. Hopkins, International Journal of Geographical Information Science, 3/1/2011, Taylor & Francis, reprinted by permission of the publisher (Taylor & Francis Ltd, http://www.tandfonline.com)

considered in this first version. The future residential growth in the region (calculated separately for counties and cities) was then allocated according to profitability rankings of the DLUs. A subsequent process examined the growing DLUs adjacent to the urban areas to determine if they should be incorporated within the city boundaries. The process stopped when all residential growth was allocated to the cities and unincorporated areas in the county.

Although CUF-1 adopted a market-based approach, it assumed that profitability is independent of demand characteristics and purely a function of costs of residential development. In other words, the initial estimate of growth was not affected by demand parameters, which would presumably impact congestion levels and cost of living and thereby moderate growth (Landis and Zhang, 1998). Also, given that no other types of land uses were considered, other types of development (commercial, retail, industrial, etc.) could not compete for land parcels (DLUs). Thus, CUF-1 was quite limited in its scope and ability to capture land market dynamics.

The second generation California Urban Futures model, CUF-2, made several improvements over its predecessor. First, it incorporated and predicted different land use types including commercial, industrial, and single- and multi-family residential. This version of the model predicted land use transitions from vacant to each of the above mentioned land-use types and also redevelopment from one use to another. Second, the spatial unit of analysis was now 100m x 100m (1 hectare) grid cells instead of the DLUs of CUF-1. Third, it generated job forecasts for each 3-digit North American Industry Classification System (NAICS) sector through separate econometric models to determine employment growth. The employment growth parameters are then used to drive the demand in commercial and industrial land uses.

Finally and importantly, CUF-2 was among the first land use models to implement a statistical framework based on a theory called the "**random utility theory**," which allowed estimation of probabilities of discrete events. Discrete events are unique and exclusive activities within a set of possible activities that can occur. For example, a commuter can choose among a set of unique mode choices such as self-driven auto, carpool, or transit. Only one option (a discrete event), say, self-driven auto, is finally chosen, at which point the others, carpool and transit, are rejected (exclusive). Land use change is also a similar discrete event since each type of land use is unique and exclusive. Once a group of parcels are designated as "residential," we would (at least in theory) eliminate all other uses. CUF-2 implemented a statistical technique known as "**multinomial logit**" to estimate the probabilities of land use change to any one of the discrete land use types or between them. This form of modeling based on random utility theory using multinomial logit techniques is now pervasive in the current generation of urban and transportation models.

Random utility theory A theory that allows modeling of preferences for multidimensional goods such as housing, transportation, etc.

Multinomial logit Predictive analysis used to describe data and explain relationships between variables with more than two levels.

The New Era of Urban Environmental Models

The evolution of urban and environmental models has accelerated since the new millennium. Advances in computation allows complex dynamic simulations at fine spatial and temporal scales. In other words, we can now look closely at how neighboring land uses and activities are influencing a specific parcel and how this influence is changing at regular temporal intervals (yearly, quarterly, etc.). As discussed previously, the random utility theory has been instrumental in expanding the range of competing activities we can model as well. A good example of such a highly disaggregated and spatially as well as temporally dynamic model that is becoming a "gold standard" in metropolitan planning organizations is *UrbanSim* (Waddell, 2002; 2000).

UrbanSim has been in active development since the late-1990s, first at the University of Washington and now at the University of California at Berkeley. In 2005, it was completely re-engineered to adopt a more extensible and modular platform, the Open Platform for Urban Simulation (OPUS), based on the Python code. In 2016, the newest implementation, a cloud-based platform, was launched. It is maintained by UrbanSim, Inc. (http://urbansim.com).

UrbanSim

UrbanSim introduced several innovations in the field of urban environmental models. First, it is a highly disaggregated model with each household, business,

Agent-based modeling A class of computational models that simulate the behavior of autonomous agents in relation to each other and observe the patterns that emerge from such interaction.

and developer behaving as a "decision-making agent." The interplay of these individual's decisions leads to the final outcome of their location in space. This form of bottom-up modeling is also known as **"agent-based" modeling**. UrbanSim explicitly models each household's decision to locate in or relocate to a neighborhood in the metropolitan area. Similarly, every business is identified and location and relocation decisions estimated.

Second, new developments in the real estate sector are predicted by modeling the change in land prices. Previously, few urban models included real estate prices in their estimation. Third, the changes in travel behavior and its impact on locational and development decisions are explicitly incorporated in the modeling platform by interfacing UrbanSim with a travel demand model. Therefore, UrbanSim is usually run in tandem with a separate travel demand model given that this feature is not included in the current platform.

Finally, the UrbanSim OPUS platform offered a graphical user interface (GUI) that showed all the models and datasets in one place, provided drop-down, user selectable commands to run the models and showed GIS-based maps of the output. It also included a module to "evolve" the households as they age over the period of the simulation. That is, households were no longer configured as static entities over 20- or 30-year horizons, but grew and/or split due to marriage, educational choices, dissolution of marriages, new births, and deaths.

The cloud-based version extends the field even further. The newest UrbanSim models use local data for each metropolitan area with parameters estimated using advanced statistical methods to reflect local conditions (See Figure 3). Tools to

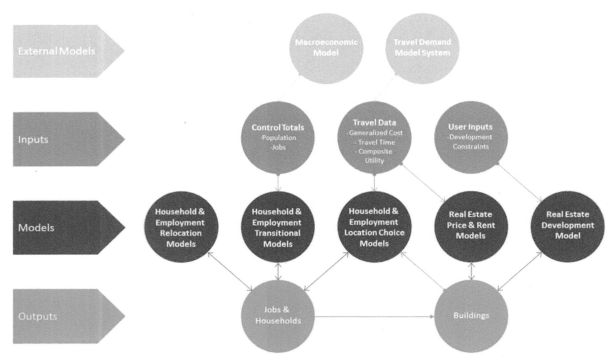

Figure 3 UrbanSim flowchart
Source: UrbanSim. Accessed 5/12/2017 at http://www.urbansim.com/urbansim/

automatically search for the best-fitting model specifications and calibrate the models to observed data at a more aggregate level are provided. Census block models for the US have been preconfigured and are ready for use. For other metropolitan areas, options can be implemented at a parcel level or zone level. Once the models are built and calibrated, transportation and land use inputs are uploaded and edited into UrbanSim scenarios using visual interfaces. The cloud platform enables simultaneously simulation of multiple scenarios. Smaller regions typically run in less than 30 seconds per year, while larger regions (e.g., 3–5 million population) may take 2–4 minutes per year to simulate. A significant number of indicators are available to better understand the simulation results at different levels of geography and to evaluate the scenario from a policy perspective. The online platform also allows users to visualize the indicators as 2D or 3D maps, charts, and tables.

In addition to these subscription-based models, UrbanSim, Inc. provides an open source Urban Data Science Toolkit (UDST). UDST is a growing portfolio of Open Source tools developed and maintained by UrbanSim Inc. and hosted on GitHub.com (https://github.com/UDST/urbansim). Contributors from academia, planning organizations, businesses as well as individuals support development and Urban Data Science. Most elements are implemented in the Scientific Python stack. UDST is based on the same constructs that UrbanSim is built from with various components designed to integrate seamlessly with one another. It focuses on specific problems related to urban spatial analysis. UDST projects include:

- *UrbanSim* - A platform for simulating urban real estate markets and their interaction with transportation.
- *ORCA* - A generalized framework for data processing and orchestration to support UrbanSim, ActivitySim, and other types of modeling.
- *ActivitySim* - A platform for simulating Activity-Based Travel (initial phase of development).
- *Pandana* - A fast network accessibility engine for computing accessibility metrics.
- *Spandex* - Spatial Analysis and Data Extraction.
- *Synthpop* - A Population Synthesizer.
- *ChoiceModels* - A library of flexible discrete choice models, including Multinomial Logit, Nested Logit, Mixed Logit, and Latent Class Models.
- *UrbanAccess* - A library to obtain, clean, merge and analyze GTFS Transit Networks and OSM networks for pedestrian and transit accessibility (UrbanSim Urban Data Science Toolkit, http://www.urbansim.com/udst/).

Several other models have been developed and tested in different parts of the world. Some, such as the Oregon Statewide Integrated Model (SWIM2) and IRPUD (Dortmund, Germany), are similar to UrbanSim in their approach in terms of their level of spatial disaggregation and the explicit modeling of individual agent behavior. Other modeling approaches include **cellular automata** (such as CLUE and SLEUTH) and system dynamics (similar to Forrester's model described earlier). Many previous studies have categorized and evaluated urban land use/transport/ environmental models based on attributes such as the level of disaggregation, the ability to incorporate dynamics, methodological approach, and whether they

Cellular automata A collection of "cells" on a grid where each cell is affected by the condition of its neighboring cells according to a set of predetermined rules.

are region specific or generic. Readers are directed to the reviews by Haase and Schwarz, (2009); Agarwal et al., (2002); and Wegener (2004) for more in-depth examination of different model characteristics.

Modeling with Cellular Automata

Cellular automata (CA) consists of an array of similar "cells" that are arranged in a 2-dimensional lattice where each cell is affected by changes in the state of its neighbors. Cells in the automata have different discrete states (e.g., dead or alive; residential, commercial, industrial, or vacant). Also, the state of a cell changes based on the aggregate changes in adjacent cell states according to a predetermined rule. Imagine, for example, a neighborhood with a group of identical houses and one of the houses in the middle of this neighborhood is abandoned and falls into disrepair. Very soon the households in the houses immediately adjacent to it begin to resent this eyesore and move out of the neighborhood. They underprice their houses since buyers will demand a discount to have the abandoned property next door. This, in turn, affects other houses adjacent to the first wave of home sales and they drop in value as well. This devaluation will continue till the values are low enough for an investor to buy up several of the properties and redevelop them to the latest standards, which then raises home values again. Cellular automata models can capture this process of decline and redevelopment of housing.

One of the most celebrated examples of a land use model based on cellular automata is SLEUTH. The acronym is based on the six attributes that the model uses to drive urban growth—slope, land use, exclusion, urban extent, transportation, and hillshade. These six data types are provided as input tables for a gridded map, where each grid represents a "cell." Based on several years of information on these parameters, five rules of transition for each cell is determined. These transition rules are based on a) diffusion (controlling overall scatter of growth); b) breed (likelihood of new settlements being generated); c) spread (growth outward and inward from an existing cluster); d) slope (a resistance parameter and threshold beyond which development does not happen); and e) roads (an attraction parameter). The process of figuring out the appropriate parameters for the five transition rules is called *calibration*. The calibration process actually proceeds in three sequential stages where 13 different goodness of fit measures are evaluated at each stage and the best values for the transition rules extracted. The final calibration parameters are used for simulating future changes in land use.

Although SLEUTH is among the most popular, there are many others that have contributed to the development of CA models (see Santé et al., 2010 for a review). Recent developments in CA modeling have relaxed the simple rules (that the future state of each cell depends on its current state and the state of its neighbors given a set of transition rules) and introduced more complexity to make them amenable for urban modeling. The more recent crop of CA models has introduced techniques for 1) accommodating irregular sized cells that can also be defined in 3 dimensions; 2) defining complex neighborhoods that are larger than just the adjacent cells; 3) nesting cells within larger neighborhoods with different rules based on neighborhood types; 4) making transition rules more complex and variable in space and time; and 5) allowing different time frames (and time steps) for different types of cells. These advances have brought CA models into the mainstream of urban modeling that tends to be more disaggregated with more exogenous influences and diverse competing uses.

Advancing Sustainability through Urban Models

While patterns of transportation and land use can capture a substantial component of a community's sustainability, it is by no means the complete picture. Particularly, the social and ethical components of sustainability could be missed if not included in assessments. The social aspect takes on a significant role in our aspirations for sustainability given that it is one of the three dimensions of sustainability discussed by Poveda, Campbell, Lozano, and others (Campbell, 1996; Lozano, 2008; Poveda, 2017). The three dimensions or "pillars" of sustainability —social, environmental, and economic—have been operationalized with the help of different indicators.

The social dimension reminds us that sustainable practices need to benefit all and not just a section of the population. Therefore, equity and justice are values that are embedded deeply in all conceptualizations of sustainability. Urban models can be designed to capture these equity aspects. This is usually accomplished by explicitly including costs and/or benefits of policies for different groups of people categorized by income, race, ethnicity, or other social divisions.

Consider the following example of a city that intends to become more sustainable by building a light rail transit system through many of its high-density residential corridors. Light rail transit offers comfortable, congestion-free, and reasonably rapid transportation to many of its popular destinations. If many automobile trip-makers switch to using light rail transit, pollution and GHG emissions will decline, improving the environment. However, the convenience of the light rail could potentially increase the desirability of the residential areas close to the stations, thereby increasing rents and housing values. Low–income renters would then be disadvantaged and perhaps displaced by higher income groups. Social sustainability, under this scenario, would decline.

As this example points out, operationalizing truly sustainable outcomes requires thinking through complex dynamics that incorporate both positive and negative aspects. While certain dimensions of sustainability are boosted in particular projects, other dimensions may be compromised. The challenge is to find solutions that maximize the net benefits across all three dimensions of sustainability. Urban models can help show how each of these dimensions will be impacted by different sustainability initiatives. The models also show how the outcomes can be improved and managed, and harms mitigated prior to implementing expensive projects that often have long-term consequences for the community.

The scenario of disparate sustainability outcomes of light rail transit was highlighted in a publication about the Valley Metro Light Rail project in Phoenix, Arizona before it began operation. The UrbanSim modeling environment (discussed earlier) was used to simulate light rail impacts on the number and type of households along the transit line between 2008 and 2015 (See Joshi, et al., 2007). One particularly startling result, which would not be apparent without modeling the dynamics of household location choice with changes in transportation options,

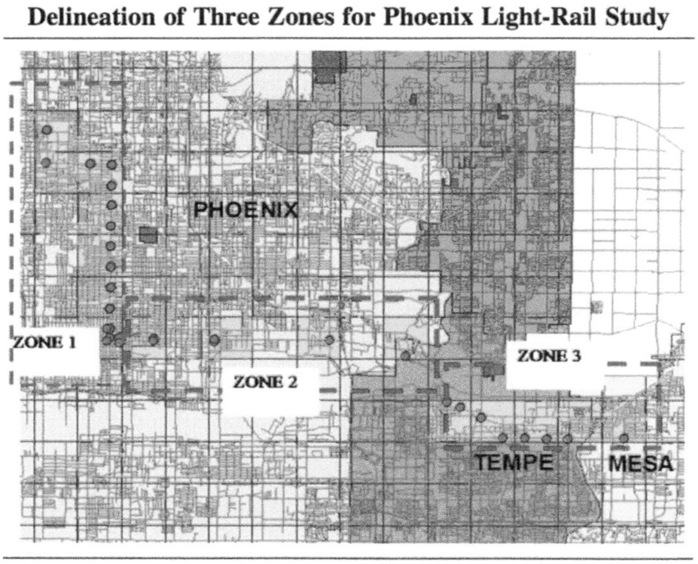

Delineation of Three Zones for Phoenix Light-Rail Study

Figure 4 Delineation of Light Rail Zones in Phoenix

Source: Joshi et al. 2006, page 102

Simulating the Effect of Light Rail on Urban Growth in Phoenix: An Application of the UrbanSim Modeling Environment, Himanshu Joshi, Subhrajit Guhathakurta, Goran Konjevod, et al, Journal of Urban Technology, 8/1/2006, Taylor & Francis, reprinted by permission of the publisher (Taylor & Francis Ltd, http://www.tandfonline.com)

was revealed by this simulation. Contrary to expectations, gentrification occurred in one area.

The researchers who developed the model to test the implications of Phoenix's light rail were specifically interested in understanding what would happen to three areas around the planned light-rail stations (See Figure 4). The three areas, or zones, represented three different types of developments with distinctive characters, mixes of use, and history of development. Zone 1 comprised some of the oldest areas in Phoenix including the downtown business and uptown arts districts. Zone 2 represented most of the industrial corridor around the airport and included neighborhoods with high concentrations of low-income and minority households. The neighborhoods around Arizona State University (ASU) with many small commercial establishments and high-density student housing comprised a large section of Zone 3. With the three zones identified, the UrbanSim modeling environment was used to generate scenarios of future numbers and types of households with and without the proposed light rail. By comparing the two scenarios, the researchers could assess the implications of the light rail on households' future location choices.

Researchers found that for Zones 1 and 2, more households located around station areas in the scenario with the light rail than without. This finding is expected as land close to transit is typically in high demand, resulting in more intensive use of land and higher housing densities. However, in Zone 3, the model showed that housing densities around station areas declined in the scenario with light rail transit compared to the one without. On further scrutiny, the researchers noticed that Zone 3 households were projected to be of a higher income level with a higher percentage of White households than the scenario without light rail. The authors of the study noted "Zone 3 will be the most affected area with the introduction of light rail, partly due to gentrification" (Joshi et al., 2007: page 107).

The anticipated gentrification of Zone 3 is especially unwelcome for the student community close to the ASU campus. The area is currently among the most job-rich areas in the metropolitan region and introducing light rail transit makes it even more desirable to up-market households. The scenario indicated that it can be expected to transition from more rental-based housing to more ownership-based housing. This would put additional strain on the ASU student community, which already has difficulty finding cost-efficient, short-term housing. Thus, as Zone 3 became more economically and environmentally sustainable, serious concerns about social sustainability would emerge. Many of the expectations of this model, especially about transition of housing types, have been realized.

Given that sustainability is a multidimensional concept, we are constantly faced with decisions about enhancing overall sustainability without compromising the other dimensions, especially the social dimension. Furthermore, ethical issues related to balancing the various dimensions of sustainability cannot be resolved by models that use strictly objective parameters. Contextual issues of culture, history, and social characteristics are critical in understanding how sustainable approaches should be designed. Urban models cannot provide an all-encompassing answer that unequivocally points to a decision. Instead, the models generate useful information about future scenarios as a frame of reference. In the case of Phoenix light rail project, ASU and the city of Tempe were dedicated to maintaining affordability. Hence, several plans were put in place to build more student housing as well as develop and maintain affordable housing in Zone 3.

The story of the Phoenix light rail reminds us that sustainable outcomes cannot be assumed even if a project is labeled sustainable based on its effect on the environment and the economy. The most difficult aspect of enhancing sustainability is to improve social sustainability together with the other dimensions. This effort requires substantial planning and a commitment to discover and address potentially harmful outcomes. Urban models can help inform this process and provide the critical information necessary to balance multiple dimensions of sustainability.

Where are we Headed?

We are now witnessing the confluence of several technological and social trends that are revolutionizing the way we interact with our physical surroundings and with each other. High-speed communication networks have integrated many urban functions such as transport, service delivery, and emergency management. Devices that sense locations, see objects, measure the ambient environment (temperature, noise, air quality, other activities) are everywhere. You are probably have one such device in your possession, a smartphone. For instance, it is now possible to track the flow of people and crowds through different parts of the city with the help of cellphone records. This can be complemented with information about the most significant events that are happening by capturing the data from social media like Twitter feeds.

This new era will change some of our fundamental notions of sustainable development. The assumptions about patterns of household and job location and their implications for travel will be tested when alternative fuel vehicles and new transportation systems are widely adopted. If indeed every single-family residence becomes both a producer and a consumer of energy (e.g., solar) and shares excess energy in real time through smart grids, low-density sprawling areas may have a surfeit of energy compared to more dense areas in the central city. In contrast, new types of materials and new designs as well as technologies could make dense central cores cooler and more energy efficient. Moreover, advanced communication and networking technologies can help in controlling ambient temperatures and water use with the help of sensors that bypass the pitfalls of human behavior. The resulting impact on city form and function from a combination of different transformative technologies that are just beginning to make their appearance is difficult to ascertain. Carefully designed urban models that consider the assumed behavior of multiple agents can provide a reasonable account of what lies ahead for urban residents.

Urban models are designed to portray the future state of a region by showing how its essential functions are going to change and how these functions may be accommodated in space and time. Where people live and work, how they travel, what buildings they use, and how they spend their leisure time are important questions for determining the sustainability of a region. By projecting social behavior and conditions of urban life into the future, urban models also allow us to estimate the amount of energy, water, and open space that will be needed. In addition, they can inform us about the quality of air, water, and land we will inherit. However, urban models are not omniscient, that is, they only project what we can imagine. It is only our imagination and our ability to form mental models that ultimately allow us to contemplate and envision urban sustainability. Urban models are the tools to make our mental models correspond to reality.

Chapter 7

Resilience: An Innovative Approach to Social, Environmental, and Economic Urban Uncertainty

Stephen Buckman and Nelya Rakohimova

Resilience as a New Approach

The rapid explosion in urban population in the second half of the 20th and first half of the 21st century has brought uncertainties, complex challenges, and levels of risk for urban regions not experienced any time in history. Risk, previously considered primarily a non-urban issue, is becoming more and more of a concern for urban decision makers. Urban areas are characterized by their "(a) scale, (b) densities, (c) inhabitants' livelihood strategies, (d) economic systems and resource availability, (e) governance systems, (f) public expectations, (g) settlement structures and form, (h) likelihood for compound and complex disasters, and (i) potential for secondary impacts on surrounding rural areas and regions" (Wamsler, 2014: 4). As Wamsler points out, urban risk is unique and important to study because of the potential impacts on large populations and the services embedded in these regions.

To address these risks and challenges, city officials together with urban and regional researchers are exploring resiliency and sustainability approaches. Resiliency, much like sustainability, focuses on interdisciplinary systems thinking and new approaches to deal with future uncertainty. It offers a useful starting point for understanding the mechanisms that hinder or enable cities to cope with structural change, disruptive events, situations of crisis, and the ability to recognize slow changes and find ways to manage these.

The simplest definition of resiliency is the ability to adapt and cope with uncertainty. It is built on hope and reduces fear. It instills confidence and strength when dealing with change (Newman et al., 2009). Within an urban context, resiliency aims to "restore the historical functions of cities as places where citizens can find safety and protection from disasters and environmental change" (Wamsler, 2014: 10).

This does not mean, however, that urban resiliency looks to revert to what was, but rather it seeks to change and adapt to potential impacts by stewardship of urban infrastructure and governance to build adaptive mechanisms. Adaptation is key to resiliency in that change is viewed as inevitable. Thus, for a community to be resilient and, for that matter, sustainable, it must have the ability to embrace change and move forward.

Resiliency is often considered a cornerstone of sustainability, as the sustainability of any system requires resilience. Enhancing a city's resilience not only makes it more sustainable, but also has the potential to provide a higher quality of life in economic, social, and environmental terms as resilient communities are strong and connected (Beatley, 2009). For that reason, the United Nations General Assembly included resilient cities in its 2016-2030 Sustainable Development Goals (SDG)— *Goal 11: Empower inclusive, productive and resilient cities* (Assembly, 2015).

In this chapter, we will review and discuss the historical components of resiliency, which grew out of ecological and engineering resilience and the importance of adaptation to resiliency thinking in terms of urban systems. We will also describe the various levels of resiliency, including economic, social and environmental capital as well as adaptive governance as ways to deal with shocks and slow burns to a system. Lastly, we will discuss how certain cities rank when it comes to resiliency.

Historical Background, Sustainability and Key Concepts

Resilience has been studied across disciplines ranging from environmental research to materials science and engineering, psychology, sociology, and economics. It has been defined as "the ability to recover quickly from illness, change, or misfortune. Buoyancy. The property of a material that enables it to assume its original shape or position after being bent, stretched, or compressed. Elasticity" (Smith et al., 1998).

The concept of resilience first emerged in ecological studies, research, and scientific journals during the 1960s and early 1970s. A primary work from this era was ecologist C.S. Holling's *Resilience and Stability of Ecological Systems* (1973). Holling defined resilience as a "measure of the persistence of systems and of their ability to absorb change and disturbance and still maintain the same relationships between populations or state variables". He found that high variability actually enables systems to absorb changes in ecological processes, random events, and disturbances "before its controls shift to another set of variables and relationships that dominate another stability region." (Folke, 2006: 254). Resiliency was born out of engineering as well (Holling, 1973; 1996).

From an engineering perspective, resilience centers on how quickly a system (i.e., a city's infrastructure or electrical system) can return to equilibrium after a disturbance. From the ecological perspective, resilience looks at how a system adapts and persists. These notions rely heavily on the idea of a hazard or disturbance offsetting the balance of a system creating the need to reconcile that balance. The primary difference is that the engineering definition assumes a fixed state while the ecological definition acknowledges the existence of multiple **equilibria** and the ability of a system to change into alternative stability domains. It is within this understanding and definition of adaptation that we can use the "idea" of resiliency as an intervention.

The concept of resilience has also been applied to individuals', communities', and larger societies' ability to recover after disturbances or long-term debilitating processes. In this context, it is essential to understand that social and ecological

Equilibrium The condition of a system in which competing influences are balanced, resulting in no net change. Ecological equilibrium is a point or period in time which the state of an ecological system is at climax, wherein it stops to grow or decline.

systems operate within the same dynamic system, that they are invariably interdependent and constantly co-evolving. These **social-ecological systems** are systems that include societal (human) and ecological (biophysical) subsystems in mutual interaction (Gallopín, 1991). Social-ecological systems can be specified for any scale from the local community and its surrounding environment to the global system constituted by the whole of humankind and the ecosphere. These systems provide the biophysical foundation and ecosystems services for social and economic development.

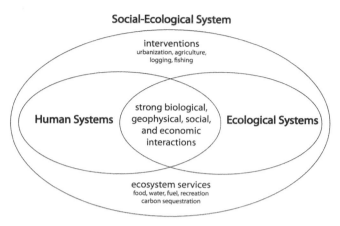

Figure 1 The Social-Ecological System

Evolutionary resilience, or socio-ecological resilience, questions the idea of equilibrium, advocating that systems change over time with or without a radical shock as in instances of prolonged drought. Thus, not only do disasters shock the system but slow, evolving change, or a "slow burn," such as climate change can shock the system as well. This "slow burn" is not confined to environmental shocks, however. It also encompasses areas of economic and social impacts.

For instance, the evolution of neoliberal market conditions have weakened the capacity of urban regions to socially adapt to change as market ideologies control the debate. As the market becomes more entrenched in our daily lives, people begin to think of themselves as being governed less by others and more by themselves (Hall & Lamont, 2013). Thus social resiliency, which has been shown to be a key measure in times of recovery becomes important. Social resiliency is about how members of community sustain their well being in the face of uncertain challenges. In this sense, Hall and Lamont (2013) see resiliency "not as the capacity to return to a prior state but as the achievement of well-being even when that entails significant modifications to behavior or to the social frameworks that structure and give meaning to behavior" (p. 13).

In essence, the key to all resiliency thinking is uncertainty. This idea of resiliency does not see the world as an orderly system but rather as a chaotic and complex changing system, which challenges the utility of relying on past forecasting to determine and plan for uncertainties (Davoudi et al, 2012). This empirical concept of complex systems is called the **adaptive cycle**. This model exposes the mechanisms that can support or prevent resilience in systems. It shows that change is a part of urban systems and that internal and external shocks can change the local structure and function of neighborhoods, districts, or entire urban agglomerations.

Generally, the pattern of change in adaptive cycles is addressed in a sequence of four phases. First is a rapid growth phase, during which resources are exploited. Second is a conservation phase in which resources are accumulated and stored. The third is a collapse phase in which resources are released, and the fourth is a phase of reorganization. After reorganization, the cycle starts anew (Resilience Alliance, 2014). The growth and conservation phases together constitute a relatively long developmental period with fairly predictable, constrained dynamics; the release and reorganization phases constitute a rapid, chaotic period. Resilience research

Social-ecological system A system that includes societal (human) and ecological (biophysical) subsystems in mutual interaction; ecosystems, from local areas to the biosphere as a whole provide the biophysical foundation and ecosystem services for social and economic development.

Adaptive cycle A model that can be used to describe a variety of phenomena such as ecological succession where an ecosystem goes through exploitation, conservation, release and reorganization.

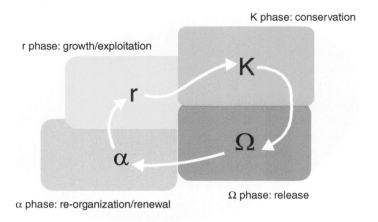

Figure 2 The Adaptive Cycle
Source: Resilience Alliance, 2014
From Panarchy by Lance Gunderson and C.S.Hlling. Copyright © 2002 Island Press. Reproduced by permission of Island Press, Washington, DC.

must focus on these phases to understand the nature of vulnerabilities and shock and their prevention or possible opportunities for reorganization if release phase already has occurred (Figure 2).

To illustrate the adaptive cycle in an urban context, we look at the history of Detroit, Michigan. Detroit experienced dynamic and diverse growth in the late 19th and early 20th centuries when the city was a rapidly expanding center for a wide variety of manufacturing, including the automobile. This was its growth phase. The city consolidated in the early and middle decades of the 20th century as manufacturing became more inflexible and less diverse, and resources became more concentrated and tightly interconnected. This was its conservation phase. The breakdown of the auto-dominated economy during the second half of the 20th century due to business cycles, class tensions, racial unrest, and mobile capital pressures represents the release phase of the adaptive cycle. The results of this collapse were unemployment, vacancy, poverty, pollution, and violence. In January 2013, Detroit released *The Detroit Future City Strategic Framework* formally beginning its reorganization and regeneration phase (Detroit Future City, 2016).

Urban and Community Resilience

Detroit's transformation from a rapidly expanding manufacturing center to its current state of redefining itself demonstrates the extremes of the adaptive cycle. Other metropolitan areas and medium- or small-sized cities face myriad concerns from a variety of shocks and disturbances that are different than in the past. For example, modern cities are highly vulnerable to natural and human-made disasters. According to the 2007 UN Report entitled *Mitigating the Impact of Disasters,* there was a fourfold increase in the number of recorded natural disasters from 1975 to 2006. Human-made disasters saw a tenfold increase from 1975 to 2006, with the greatest rates of increase being in Asia and Africa. In light of this, urban planning needs to understand the factors shaping the socio-spatial aspects of cities

Table 1 System Shocks

	Example
Origin	Natural: earthquakes, weather events Economic: market shocks, employment, recessions, poverty Biological: diseases Social: demographic, riots, revolutions Technological: industrial accidents Political: change government, wars
Lasting time	Minutes: earthquakes, landslides Days: hurricanes, riots Weeks: stock market crashes, weather events Month: housing prices, revolutions Decades: famine, droughts, demographic change
Frequency	Chronic: weather events Frequently: hurricanes Rarely: industrial accidents
Scale of Impact	Neighborhood: riots, diseases City: heat island effect Region: weather events Global scale: climate change, peak oil

and the institutional structures for managing them to prevent or recover from these shocks and disturbances that are occurring with alarming greater frequency.

In the resilience framework, shocks are defined by origin, lasting time, frequency, predictability, and scale of impact (See Table 1). Natural disasters have received the most attention from scientists over the last four decades, as they are especially destructive in urban areas where there is a concentration of people and resources. Worldwide statistics reveal an increasing number of disasters causing more than 3.3 million deaths and 2.3 trillion dollars in economic damages (Gencer, 2013).

The term "natural disaster" is problematic, however. Urban areas are not disaster and risk prone by nature; rather their socio-economic structural processes equate to risk. Furthermore, population movement and population concentration substantially increase *vulnerability*. Natural shocks and disturbances only become disasters when coupled with vulnerable human conditions such as inadequate governance systems, social inequity, outdated urban planning, and economic inequality. Compounding these problems are lasting periods of disturbances, or slow burns, which are the most challenging for urban and community planning and development. These lasting disturbances, including long-term demographic change, economic decline, and environmental degradation, can be addressed during the planning process as they are part of the urban adaptive cycle.

Aging and poverty are good examples of such disturbances. For instance, as the "baby boomer" generation (born between 1945-1964) retires, there will be an increasing strain on social welfare services and the already stressed Social Security system. With the majority of baby boomers living in cities and the anticipated Social Security shortfall not covering basic needs, urban communities potentially face major problems in the next decades with a significant percentage of older adults living in poverty.

In addition, cities are socially vulnerable when people, organizations, and societies are not able to withstand the adverse impacts from multiple stressors to which they are exposed. This vulnerability is often a result of social inequalities that lead various groups to be susceptible to harm and inhibit their ability to respond (Cutter et al., 2003). Thus, social vulnerability is not only measured by exposure to hazards, but also in the sensitivity and resilience of the system to prepare, cope, and recover from such hazards (Turner et al., 2003).

Flooding from Katrina in New Orleans
Source: @Zack Frank/Shutterstock Inc.

These disturbances and shocks occur at different spatial scales as well. Consequently, urban planning must address different levels of spatial development such as the local effects of global environmental change. Thus, cities and communities facing the same disturbances often experience completely different impacts. For example, coastal cities historically have an economic advantage as trading centers but are vulnerable to current climate change causing significant sea level rise. Cities also have critical infrastructure vulnerabilities, such as transportation and energy delivery systems, which can cause serious problems if disrupted.

Operating across multiple scales, policy makers deal with uncertainties at both the regional and local level. *Uncertainty* is an element of any disaster as impacts cannot be quantified or are completely unknown. Political and cultural globalization also add greater uncertainty. To cope with these uncertainties, cities need a robust approach to decision making. Policy makers and planners must take into account potential weak spots and system failures. They also need to prepare for a wide range of futures rather than focusing on optimal design solutions for the current state of the city. At the same time, there is increasing difficulty in relying on extrapolation of past events when the physical and human conditions that contextualized those events is continually being re-shaped by local and global forces. When decision makers feel disempowered by uncertainty, there is a danger that investment in disaster preparedness and mitigation will be left outside of urban development strategies (Pelling, 2003). To overcome this, scenario planning, which empowers decision makers with worst- and best-case options, and anticipatory governance can be used to foresee and plan for change as well as address uncertainty (Quay, 2010).

How a City or Community can be Resilient

A key component of resiliency research is learning how a city or community can enhance its resilience to uncertainty. Key approaches can be categorized into four major themes—*stability, recovery, transformation,* and *adaptation* with the common thread being the ability to withstand and respond positively to stress or change (Adger, 2000; Resilience Alliance, 2007; Maguire and Cartwright, 2008). Stability relates to a system's efforts to endure a shock and its consequence. Recovery refers to a system's ability to 'bounce back' from a change or stressor and return to its original state. Transformation is concerned with renewal, regeneration, and reorganization to create opportunities for innovation and development (Folke, 2006). It is especially applicable in cities and communities where a system is already degrading and efforts to get it back into a desirable stable phase are no longer possible without innovative policies, financial support, and system change. Of these, adaptation has received the widest attention in the academic and practitioner communities.

Adaptation can be both reactive and proactive. Reactive, or autonomous, adaptation represents a response to a stress that has already occurred. However, reactive adaptation does not always end well as it can be too late to take some actions. Proactive adaptation anticipates future stresses and builds the capacity for taking measures to lessen the negative impacts from these future events. This enables cities and communities to *"bounce forward"* and prevent or mitigate some shocks and disturbance. Such approaches depend on one's ability to envision what the future might look like. They are also influenced by one's ability to have learned from past experiences, particularly what worked (and what did not) in similar circumstances. Moreover, adaptation must be multi-faceted and take a holistic approach. There is not just one aspect of a community that must be resilient, but rather the whole community must be resilient. Thus, adaptation needs to embrace economics, the environment, socio-cultural dynamics, and civic institutions.

Adaptation is closely connected with the concept of *adaptive capacity*. Adaptation includes actions taken to reduce vulnerabilities and to increase resilience; adaptive capacity is the ability to take those actions. In this sense, both adaptation and adaptive capacity relate to the ability of social and political institutions to think and act towards anticipated events and reduce vulnerability. Urban and community resiliency is concerned with adapting to change, not trying to control it.

Magis (2010) stated that community resiliency "is the existence, development, and engagement of community resources by community members to thrive in an environment characterized by uncertainty, unpredictability, and surprise" (401). It is important to note that this definition involves all community members, not just decision makers and planners. Thus, resiliency and adaptive capacity are best conceptualized by how a community as a whole develops its social, economic, and environmental capital. More importantly, how this capital is shared throughout the community will determine its success.

Cities and communities with high adaptive capacity are resilient. They are able to re-configure themselves without significant declines in crucial functions relative to primary productivity, social relations, and economic prosperity. This is an ever-evolving process. It is not set in stone and requires frequent reassessment to avoid losing adaptive capacity. A loss of adaptive capacity, and thus resiliency, results in missed opportunities and constrained options during periods of reorganization and renewal.

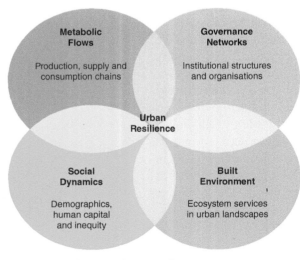

Figure 3 Urban Resilience Pillars
Source: Adapted from Resilience Alliance, 2007. A Resilience Alliance Initiative for Transitioning Urban Systems towards Sustainable Futures. Camberra: CSIRO.

Adaptive Capacities of Cities and Communities

Establishing a strong, resilient community requires a strong foundation. There are four pillars that form this base: **economic capital**, **social capital**, **environmental capital**, and *adaptive governance*. While each of these areas are often thought of as separate, they work holistically and reinforce one another. If one area is insufficient or neglected, the weight must be taken up by one of the other areas or the community and its systems will collapse. This is especially true when communities and systems are under stress.

Ideally, a community should be equally strong in all four areas allowing it to be 100 percent resilient. In reality, this is never the case. There is always a give and take between these pillars. Resilient communities are flexible enough to bolster those pillars undergoing flux and adapt to both near- and long-term changes.

Economic Capital

It is easily understood that the economy has an important role to play within community resiliency. The stronger a community's economy, the more it can rebound from disaster as it is likely to have the economic resources to rebuild, retrofit, and restore the important functions that keep it moving. The role of economic capital for community resiliency entails making sure a community's wealth increases its citizen's standard of living for the present and the future. This is done though a few key ways.

First, the level of economic resources is important. The levels of corporate taxation, property tax, business turnover as well as other standard economic variables must be stable enough to support the community in times of both prosperity and change. A second important aspect is the degree of equality in the distribution of resources. A resilient economic structure does not have huge disparities between the poorest members and the wealthiest members. Hence, the more polar the level of wealth in a community, the less economically resilient that community is. Lastly, the scale of diversity of economic resources is important. The more diverse an area's economic resources, the easier a community can weather an economic storm or natural disaster. One only needs to look at the City of Detroit and the collapse of the auto industry to see how the lack of economic diversification can create significant vulnerabilities in an urban area (Sherrieb et al., 2010).

Abandoned Michigan Train Station: Symbol of the power and decay of Detroit. (Photo by Stephen Buckman).

Social Capital

Another important area for community resiliency is its social capital. This is especially important in times of distress such as post-disaster recovery. While supported by federal subsidies, local communities take on the brunt of on-the-ground recovery efforts not the government. Social capital involves the level of citizen participation, place attachment, and sense of community. It is premised on the fact that communities have many non-economically driven resources based on its social, spiritual, cultural, and political foundations (Magis, 2010). As Putnam (2001) and others have shown, the stronger a community's social capital is, the more balanced a community is in terms of mutual aid and non-governmental support. Likewise, the weaker a community's social capital is, the more susceptible to collapse that community is.

There are three types of social capital that are important in terms of community resiliency. The first is bonding, which entails the close ties and cohesion of groups within a community. The second is bridging, which entails the loose ties that groups in a community have with each other. The third is vertical linkage between groups, which binds lower and higher socioeconomic groups. This is important for poorer community members to be able to tap into resources (Magis, 2010). Thus, social capital is about social support, social participation, and community bonds (Sherrieb et al., 2010). It provides the underpinnings for building community resiliency by promoting social resilience.

Social resilience, in essence, is the ability of individuals and groups to cope with and adapt to social, political, and environmental change and disturbances (Wilson, 2012). A socially resilient community enables adaptive capacity by engaging all of its members in the ever-evolving process of dealing with change. By ensuring social resilience, decision makers and planners take a proactive position rather than a reactive one. A prime example of both economic and social resiliency in action is the "Local First" movement.

Local First is an economic and social campaign that communities undertake to keep resources local. This entails employing local labor, eating from local farms, and buying from local companies. These activities circulate money locally, create a greater pride in one's local area, build connections with the local community, and insulate communities from economic shocks outside of the community. An excellent example of this campaign is Local First of Western Michigan, which encourages local consumption in the Western Michigan region. Their mission is "to foster the development of an economy, grounded in local ownership, which functions in harmony with our ecosystem, meets the basic needs of our people, encourages joyful community life, and builds wealth" (Local First). The economic impact can be seen in Figure 4, which shows the impact to the local community in economic terms of buying locally.

Environmental Capital

Original ecosystems play an important role in preventing various shocks and disturbances, especially natural disasters. Yet, natural landscapes and some of their features are often ignored during the planning process. This can increase community vulnerability. For example, the disappearance of wetlands and the

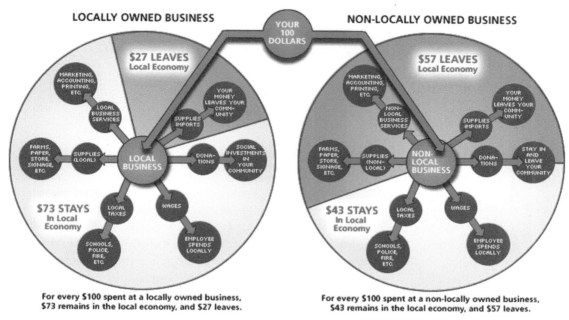

Figure 4   Impact of Locally Owned Businesses in Western Michigan

Source: Local works: Examining the impact of local business on the West Michigan economy a civic economics study for local first grand rapids Michigan.

barrier islands south of the City of New Orleans was identified as a major factor in the devastation caused by Hurricane Katrina because it failed to offer a natural barrier to coastal flooding.

In addition, urban landscapes are subject to a rapid rate of change, continuing disturbances, and complex interactions between various processes. This together with fragmentation caused by development patterns affects the capacity of these landscapes to generate ecosystem services that sustain urban quality of life. For instance in many cases of flooding, the original shape of a river valley and the river's historical path was ignored or altered during urban development. Curitiba in Brazil offers an example of development that takes the natural flow of its rivers into account. Curitiba's Seasonal Parks were designed using the ecosystem's natural adaptive capacity. The parks protect the city from flooding while offering recreational areas for citizens and tourists with facilities such as restaurants, amusement parks, and other services that offset the economic costs of park construction.

The built environment and infrastructure are also important to a city's adaptive capacity. For instance, the

Opera de Arame: Wire framed opera house in a former quarry as part of the Curitiba park system. (Photo by Stephen Buckman).

physical location of roads, railways, airport, etc. has a significant influence on the flow of commerce and people in and out of cites, which is important during periods of crisis. New urban development should focus on enabling flexibility in the built environment to account for the changing needs and requirements of urban populations. It should provide opportunities for preventing and avoiding harm during various kinds of disturbances. New, innovative means of addressing urban complexity are needed to negate major vulnerabilities.

Adaptive Governance and Collaborative Planning

Flexible, integrated, and holistic forms of governance are innovative ways of dealing with the urban complexity. The emerging challenges of urban development require networks and institutions that are able to capture and share knowledge in a transparent way; adapt to social, ecological, economical, and political changes; and build capacity for long term observation, monitoring, and perspective. Local, regional, and international governance and institutional structures need to increasingly take into account collaborative participatory approaches, including adaptive co-management, community self-management, and the development of academic initiatives (Resilience Alliance, 2007b).

Adaptive co-management is an emerging approach for urban governance where the aforementioned characteristics are combined. Adaptive management assumes natural resource management policies and actions are not static but adjusted based on the combination of new scientific and socio-economic information. Adaptive co-management combines this learning dimension with the linkage dimensions of collaborative management in which rights and responsibilities are jointly shared. Multiple stakeholders participate in learning, developing a shared understanding, and establishing goals, objectives, and management decisions.

Adaptive co-management promotes place-specific governance approaches in which strategies are sensitive to social and ecological feedback and oriented towards urban resilience and sustainability (Olsson et al., 2004). Thus, adaptive co-management allows for open flows of information for a well-functioning recovery

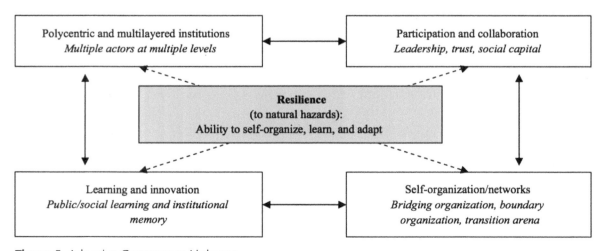

Figure 5 Adaptive Governance Linkages

process during and after disasters. When individuals see that that there is a well working government that communicates freely and openly with the community, individuals and NGOs have a better sense that the governance and community structure will properly deal with adversity (Norris et al., 2007).

Communication and empowerment are important before an actual disaster event, however. One of the most effective ways to achieve this is through open participation within the governance process. An example of this is the collaborative planning process often undertaken when constructing a master or disaster plan. In contrast to a traditional top-down approach to planning, collaborative planning allows citizens to have open dialogue with the planning commission and city council. In essence, collaborative planning allows for multiple viewpoints to come to the table.

As Margerum (2011: 6) states, "collaboration is an approach to solving complex problems in which a diverse group of autonomous stakeholders deliberate to build consensus and development networks for translating consensus into results." What is especially valued in planning of this nature—at least by scholars promoting collaborative planning—is candid and explicit discussions that break down power hierarchies (Brand et al., 2007). Toward that end, collaborative planning involves four key variables to achieve success: stakeholders who spend time understanding the problems; a deliberative process between stakeholders that allows everyone to debate the issues; consensus among stakeholders and the public; and networks that can translate consensus into results (Margerum, 2011). This form of planning helps to empower disenfranchised members of the community and allows for transparency, which opens veins of communication and is a key aspect of community resilience.

The collaborative urban planning process.
(Photo by Stephen Buckman).

Resilience Indicators, Rankings, and Typologies

Measuring urban resilience is not an easy task. Yet, it is important find and understand the gaps where governance, infrastructure, and community solutions are needed. It is also important to find ways to quantify resiliency in terms of which areas are more resilient than others. Quantifying oftentimes qualitative data allows policymakers and communities to understand where they are resiliently strong and weak. An effective way to quantify this data is through the use of indicators.

An indicator is a measurement of some phenomena over time and space that aids decision makers, and has policy and societal implications. It can be defined as "any form of information that allows a decision maker to quickly and effectively determine the state of a system or what is being measured" (McAslan, 2015: 235). Indicators take on a specific segment of analysis, and therefore, come in many forms. There are economic, community, and sustainability indicators, among others. Objective indicators (i.e., economic) are concerned with concrete data and subjective indicators (i.e., community) that are based on perceptions and opinions (McAslin, 2015). The goal of the analysis determines the indicator(s) to be used.

It is important to distinguish between indicators and indices here as the two are often used together, yet they are different in scope. The difference is one of scale. An indicator is concerned with one specific measure while an index is concerned with a group of measurements or more precisely a group of indicators. For example, a sustainability index would be concerned with numerous issues of sustainability such as recycling, greenhouse gas emissions, water usage, etc. while an indicator would only be concerned with one of those variables.

An example on how indicators and indices can be used to show the resiliency of a region is the Building Resilient Regions Network's Resilience Capacity Index (RCI; http://brr.berkeley.edu/rci/). The index is a single statistic summarizing a region's score on 12 equally weighted indicators—four indicators in each of three dimensions. These dimensions include Regional Economic (income, quality economic diversification, regional affordability, and business environment), Socio-Demographic (educational attainment, lack of disability and poverty, high level of health insurance), and Community Connectivity attributes (civic infrastructure, metropolitan stability, homeownerships, and voter participation). The RCI reveals strengths and weaknesses of 361 US metropolitan areas as of March 2011, allowing regional leaders to compare their region's capacity profile to that of other metropolitan areas.

According to the RCI, metropolitan areas in the northeastern and midwestern US tend to have higher index scores than those in southern and western states. For instance, Rochester, MN; Bismarck, ND; and Minneapolis-St. Paul, MN-WI were ranked highest in terms of overall resilience. Merced, CA; McAllen, TX; and College Station, TX were ranked lowest. High RCI tends to be found in slow-growing regions where metropolitan stability, regional affordability, homeownership, and income equality is typically higher. Low RCI is found in metropolitan areas with rapid population growth.

Having the capacity to be resilient, however, is no guarantee that in the face of a stress a region will be able to effectively respond and recover. Higher capacity indicates that a metropolitan region has factors and conditions that should position it for effective post-stress resilience performance. Likewise, lower capacity does not necessarily mean that a region will falter in the face of a stress. Rather, it indicates that the region lacks factors and conditions thought to position it for effective post-stress resilience performance. The Grovesnor Group, an international development and management company, quantified "the resilience of 50 of the worlds most important cities" (Barkman et al., 2014; 3).

In *Resilient Cities: a Grosvenor Research Report*, resilience comes from the interplay of vulnerability and adaptive capacity. These were defined through five main themes for vulnerability—climate, environment, resources, infrastructure, and community characteristics—and five for adaptive capacity—governance, institution, technical and learning features, planning systems, and funding structures. Calgary, Toronto, and Vancouver were found to be the least vulnerable while New York, Toronto, and Los Angeles were found to be the most adaptive. Specifically, the report looked at the vulnerability to physical impacts of climate change; pollution and overconsumption; access to basic needs such as energy, food, and water; housing, transportation and basic utilities; and social equity and cultural issues such as access to education, health facilities, and crime standings, among others. Adaptive capacity included government transparency and leadership; long-term visioning, including disaster management planning; technological innovation; and funding access and borrowing capacity.

What both the RCI and Grovesnor Group rankings show is that there is a certain set of qualities that makes a resilient system. These are exemplified by the Rockefeller Foundation's 100 Resilient Cities' initiative. The initiative identifies seven key qualities of a resilient system. These include **reflecting** on the past to understand uncertainty and change; designing **robust** urban areas that are able to withstand hazard event impacts without significant strain or loss of function; creating **redundant**, diverse systems that decrease impacts while meeting community needs; adopting **flexible** strategies that enable systems to change in response to various circumstances; leveraging the city's **resourceful** people and institutions to find different ways to achieve goals during times of stress; promoting **inclusive** policies that broaden the engagement of communities to include vulnerable and underrepresented groups; and coordinating **integrated** governance and planning to ensure that city systems meet the needs of all stakeholders and deal with multiple issues (100 Resilient Cities, 2016).

Building on these qualities, the Rockefeller Foundation framework specifies four "dimensions" and 12 "drivers" or indicators for analyzing a city's resiliency. The dimensions are health and wellbeing; economy and society; infrastructure and environment; and leadership and strategy. Health and well-being indicators include "meets basic needs" to ensure minimal human vulnerably; "supports livelihoods and employment" opportunities for all residents; and "ensures public health services" to provide adequate safeguards to human life and health. Economy and society indicators include "promotes cohesive and engaged communities" that build a collective identity and mutual support networks; " ensures social stability, security and justice"; and "fosters economic prosperity" to enhance the availability of financial resources and contingency funds at the community level. Infrastructure and environment indicators include "enhances and provides protective natural & man-made assets," "ensures continuity of critical services," and "provides reliable

Figure 6 Resilient Categories, Indicators, and Qualities Chart
Source: Da Silva, J., & Morera, B. (2014).

communications and mobility" to reduce physical exposure and vulnerability. Leadership and strategy indicators include "promotes leadership and effective management," "empowers a broad range of stakeholders," and "fosters long-term and integrated planning" to facilitate a holistic vision shared by all sectors and stakeholders (The Rockefeller Foundation, 2015). Figure 6 charts these qualities, indicators, and categories to demonstrate how they work together to inform a holistic vision.

While indicators and indices allow for regions to be judged on their level of resiliency, the question of what a resilient region looks like and how those regions differ remains. In other words, what types of urban form or structure excels in what areas? Typologies can be used to help address this question.

The use of typologies to compartmentalize regions is a mainstay of urban planning especially as it pertains to sustainability. In essence, typologies combine indices to create a "type" that can then be used to place urban regions in context. For example within sustainability planning, urban regions can be placed into four typologies: neotraditional development, urban containment, compact city, and the eco-city, as defined by Jabareem (2006). Each of these typologies is based on how a city incorporates and utilizes seven key variables: compactness, sustainable transport, density, mixed land uses, diversity, passive solar design, and greening (Jabereem, 2006). Thus if one was to talk about city "A" as a compact city and city "B" as urban containment, it would be understood what the strengths and weaknesses of each city are based on its typology.

While the use of typologies in sustainable urbanism is quickly becoming the norm for urban regions, creating resiliency typologies comes with a whole new set of concerns. These include "resilient to what, resilient for whom, and resilient how" Most resiliency literature discusses resiliency in terms of environmental resiliency. However, resiliency has concerns beyond the environment such as the economy, society, and governance. Therefore, creating a resilient urban typology must take into account the various aspects of the resiliency discussed to put regions into context.

Davidson et al. (2016), in accordance with a detailed overview of the literature, created three distinct typologies for resilient regions. Type 1 is basic resilience or "static," type 2 is "adaptive resilience," and Type 3 is "transformative resilience." The authors also defined the relationship of each typology's elements. For instance, Type 1 elements "are related in reducing disturbances and maintaining system status quo, while the elements of Type 2 adaptive resilience are related to adapting to change but maintaining structural function. Conversely, the defining Type 3 transformative resilience element, transformability, involves a transition, either purposeful or unintended, from the status quo and replacement of adaptation as the lead change response" (Davidson et al., 2016: 34).

While typologies allow for categorization, regions will not necessarily fit neatly into one type or another. They may overlap or may shift from one to the other over a temporal scale. What can be understood from these typologies and the basic premise for creating resilient typologies is a general makeup of a metropolitan region. Furthermore, typologies allow for a larger understanding of what regions are moving in what directions. If a policy maker understands that their region falls into a Type II "adaptive" typology, they can determine what might be needed to stay in that typology or move to another typology.

It is important to understand that indicators, indices, and typologies are not definitive labels but rather tools that help policy makers and planners make decisions. These measurements allow for various variables, both quantitative and qualitative in nature, to be presented in a way that is easily digestible. Thus, these tools can help make regions more resilient by understanding its strengths and weaknesses.

Lessons Learned

This chapter introduces the concept of resiliency as it pertains to urban systems. Key lessons from in this chapter include:

1. That community resilience deals with both abrupt shocks such as natural disasters (i.e. Hurricane Katrina) and "slow-burn" events such as the impacts of climate change (i.e., increased temperature, prolonged drought) and economic impacts such as the gradual loss of a manufacturing base.

The long-term impacts of slow-burn events are just as important and oftentimes harder to plan for than natural disasters.

2. Urban environments are socio-spatial systems that are unpredictable and thus must be looked at in terms of bouncing forward not back. Adaptation and the adaptive cycle constitute a series of changes that rebalances the system enabling it to start anew and are key to understanding resiliency. Resilient cities exist in a fluid state that relies on change as its life source.

3. Resiliency is composed of various pillars that together strengthen the larger whole. The key pillars can be categorized into four major platforms— stability, recovery, transformation, and adaptation. The common thread is the ability to withstand and respond positively to stress or change.

As stated throughout this chapter, resiliency thinking is beginning to, and will continue to, play a greater role in the way we plan for and adapt to changing economic, social, and environmental impacts in our cities. The static model of city planning no longer holds true, as cities become more complex and unpredictable. Resiliency thinking allows for urban regions to address and adapt to change and stressors.

Chapter 8

Understanding Climate Risk: Mitigation, Adaptation, and Resiliency

Bjoern Hagen

Introduction

Global **climate change** is one of the most important science and societal issues of the 21st century. Climate change is often perceived as a global issue; however, its impacts occur at the national and local scale as well as globally (Pittock, 2009; NRC, 2010). This does not exclude the US. Extreme weather events are increasing in frequency and scale, impacting various regions and sectors across the country and many areas are facing climate conditions never experienced before. Other impacts such as ongoing drought and increasing temperatures in the southwestern US have led to an earlier start of, and longer lasting, wildfire season. Prolonged droughts have also increased the competition for limited water resources among people and ecosystems. In other regions, such as the Northeast, Midwest, and Great Plains, data show that heavy rainfalls (which frequently exceed the capacity of infrastructure systems such as storm drains and sewer systems) have increased over the past century. This has led to a demonstrable uptake in flooding events, land erosion, and landslides (IPCC, 2013).

The majority of the scientific community is in agreement that human behavior and current urban patterns, especially automobile usage, are key factors in the rapid increase of the average global temperature in recent decades (Calthorpe, 2011) Cities cover less than one percent of the earth's surface but carry disproportionate responsibility for **greenhouse gas (GHG) emissions**, which constitute the leading cause of global climate change. Most of the world's energy consumption either occurs in cities or is a direct result of the way cities function. Cities are not only a major contributor to climate change; they also play a major role in solving current and future challenges stemming from climate change. Urban sustainability, therefore, must take a critical role in discussions about climate change causes, impacts, and solutions.

In recent years, there has been growing acknowledgement among scientists and policymakers that many climate change challenges can be met through the design, development, and redesign of urban space and structure. Making cities more sustainable will not only reduce the causes of climate change (**mitigation**) but also

Greenhouse gas emissions Greenhouse gases are gases in the atmosphere that absorb and re-emit solar radiation back to the earth's surface. The four most common greenhouse gases (GHGs) are carbon dioxide (CO_2), methane (CH_4), halocarbons, and nitrous oxide (N_2O).

Mitigation Measures that reduce the causes of climate change. The majority of mitigation strategies aim to reduce greenhouse gas emissions.

Resiliency In the context of this chapter, resiliency refers to the ability to cope with potential negative impacts of climate change.

Adaptation Adjustments and measurements undertaken by natural or human systems in response to current and possible future impacts of climate change.

Institutional capacity The ability of an institution, such as the federal government, to perform functions, solve problems and set and achieve objectives that will reduce the threats and impacts of climate change.

High levels of uncertainty Climate change is characterized by high uncertainties regarding the types and severity of future impacts due to the complexity of climate science itself, future human behaviors and decisions, and internal climate system processes.

Global Warming The rapid increase of global temperatures in recent decades caused by natural factors, natural process, or human activities.

improve our **resiliency** toward the effects of climate change that can no longer be avoided (**adaptation**). This chapter presents not only the issues and science behind climate change but also examines the importance of urban sustainability through mitigation and adaptation policies and applications as well as building **institutional capacity** to make more effective decisions under **high levels of uncertainty.**

Climate Change

Climate is the most significant component of the world as we know it. Landscape, plants, and animals are greatly influenced by long-term climate conditions and urban areas are often affected by short-term climate fluctuations. Prior to the introduction of irrigation and the start of industrialization, climate determined food supplies, trade, trade routes, and where people could live. In general, the term 'climate' is the typical range of weather and its variability experienced at a particular place (Archer & Rahmstorf, 2010). People speak of 'climate variability' describing the irregularities of weather at a particular location from one year (or decade) to another. Changes over longer time scales are referred to as 'climate change'. Although, modern technology allows people to live in places where it was impossible to live earlier, local climate should still be a pivotal factor when determining an appropriate design for buildings and urban areas.

It is well known that climate can change over time. Yet, the precipitous acceleration of the rate of change and the observability of the **global warming** trend recorded in the last few decades is alarming. Aside from some natural variations, scientists concur that these rapid climate changes are mostly caused by human activity or, in other words, are anthropogenic, as they result from human influence on nature (IPCC, 2013a). As shown in Figure 1, worldwide surface

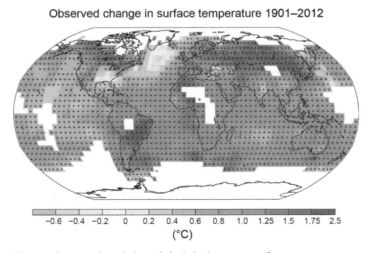

Observed change in surface temperature 1901–2012

Figure 1 Observed annual and decadal global mean surface temperature anomalies from 1850 to 2012 and map of the observed surface temperature change from 1901 to 2012

Source: Intergovernmental Panel on Climate Change, 2013; figure SPM.1, Panel B

temperatures have been increasing since 1901. Each of the last three decades has been warmer than any other decade since 1850 and records show that the current top 15 warmest years all occurred since 1998. According to the National Aeronautics and Space Administration (NASA), the average combined land and ocean surface temperature increased by 0.86 degrees Celsius (33.55 °F) between 1880 and 2015 (NASA, 2016). Simultaneously, climate data indicates that since 1950, the number of unusual and extremely cold days and nights is decreasing, whereas the number of abnormally warm days and nights is increasing.

Figure 2 shows the temperature trends by continent, global averages between 1910 and 2010, and compares these observations with simulated climate change emphasizing the impact of human behavior on the significant temperature increases

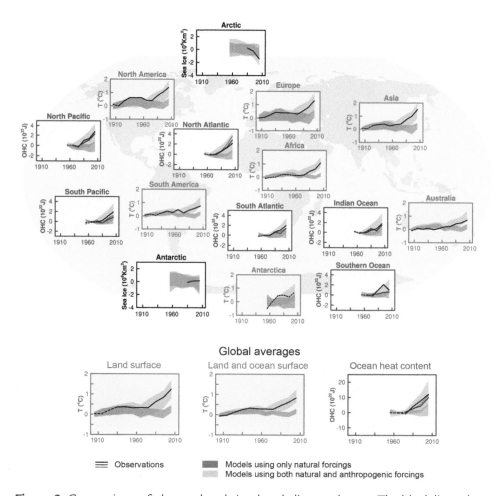

Figure 2 Comparison of observed and simulated climate change. The black lines show the actual measured temperatures. The blue bands show the expected temperatures if climate would only be influenced by natural causes. The pink bands show possible trends of temperatures if human actions are acknowledged as causes for climate change
Source: Intergovernmental Panel on Climate Change, 2013; figure SPM.6

Climate models Sophisticated computer programs that quantitatively represent the interactions of the atmosphere, oceans, land surface, ice, and human behavior. There are myriad climate models available, ranging from relatively basic models that focus on Earth's heat balance to very complex and detailed simulations that aim to show possible future impacts of global climate change under a variety of different assumptions.

over the last 50 years. The black lines show the actual measured temperatures. The blue bands show simulated temperatures from various **climate models** assuming that only natural causes or forcings are impacting climate conditions. The pink bands show the spread of model outputs due to human actions, such as GHG emissions, as causes for climate change. Consistently, the observed temperatures overlap with the simulated temperatures that presume anthropogenic climate change, suggesting that global temperature trends, especially since 1950, cannot be explained through natural causes alone.

Research shows a larger than 95 percent level of confidence (IPCC 2013) that the majority of the observed increases in global surface temperatures since 1951 have resulted from human-induced GHG emissions. Overall, from 1950 to 2000, the warming trend was around 0.13 degrees Celsius (32.23 °F) per decade, almost twice as much as in the previous century (IPCC, 2007). This trend continued into the beginning of the 21st century and is expected to accelerate even further in the future. Worldwide, the years between 2000 and 2009 were the warmest ever measured (NASA, 2010). More recently, the National Oceanic and Atmospheric Administration (NOAA) concluded that 2015 was the warmest year on record (NOAA, 2015). Figure 3 summarizes significant climate anomalies and events in the year 2015.

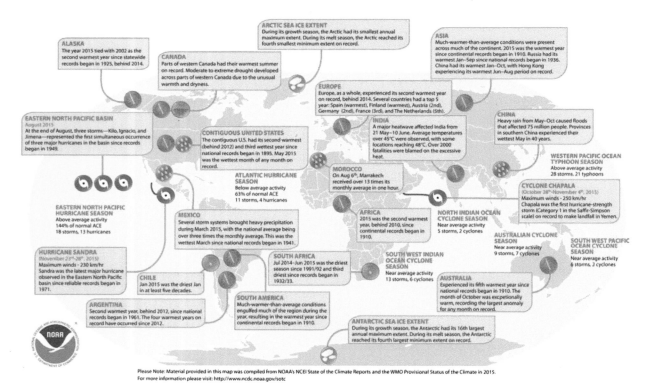

Figure 3 Selected Significant Climate Anomalies and Events in 2015
Source: NOAA, 2015

Science Behind Global Climate Change

Since 1979, when the **National Academy of Sciences (NAS)** first raised concern about global warming, the body of knowledge and the amount of scientific data documenting this phenomenon has grown. The year 1988 marked the start of the Intergovernmental Panel on Climate Change (IPCC), founded by the World Meteorological Organization (WMO) and the United Nations Environment Programme (UNEP). Today, the IPCC is considered the leading institution for the assessment of climate change. Its mission is to monitor the worldwide scientific research regarding climate change.

With the help of thousands of scientists, the IPCC regularly assesses the available scientific information relevant for improving the understanding of climate change and its possible environmental and socioeconomic impacts. Participating scientists are divided into three "Working Groups" that focus on 1) the scientific assessment of today's research regarding global climate change (IPCC, 2013a); 2) the potential impacts of climate change to socioeconomic and natural systems and how they can be reduced (IPCC, 2014a); and 3) evaluating options for avoiding the causes of global climate change (IPCC, 2014b). The results are summarized and published in specific chapters of the *Assessment Reports of the Intergovernmental Panel on Climate Change.*

> **National Academy of Sciences** A private nonprofit institution whose members act are advisers to the nation on science, engineering, and medicine.

IPCC Assessment Reports

The IPCC has released five Climate Change Assessment Reports since 1988. The first report (1990) did not find sufficient scientific proof to demonstrate a relationship between human behavior and global climate change. Nevertheless, the report projected that by the year 2000, the connection between human actions resulting in GHG emissions and global climate change would be made. The second report, released in 1995, concluded that evidence of human-induced climate change had already been found, and that these findings surfaced five years earlier than projected by the first report. The third report (2001) supported this claim and provided more evidence that temperature increases over the past 50 years were linked to GHG emissions. In 2007, the fourth report finally stated that (with a likelihood of 90–99%) global climate change is driven by human-caused emissions of heat-trapping gases and projected serious environmental damages can be expected in the future. The working group reports for the fifth assessment were released between September 2013 and April 2014. These documents use strong language emphasizing the urgency to take action against global climate change. Furthermore, the latest assessment report concludes with a 95% certainty that human behavior (human-induced GHG emissions) stands as the main reason for global warming since 1950.

Causes of Global Climate Change

The basic principle of the earth's climate is that the sun's energy entering the atmosphere is reflective and a considerable proportion of this incoming radiation is reflected back to space by snow, ice, and clouds. This energy is mostly transmitted as visible or ultraviolet light. The sunlight reflected away from earth is referred to as the earth's **albedo** and does not "deposit" energy. If the exchange between

> **Albedo** A measurement used to determine how much solar energy is reflected from earth back to space.

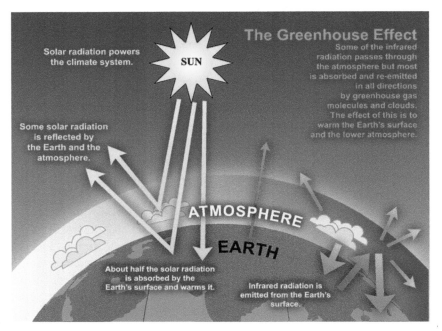

Figure 4 The Greenhouse Effect

Source: Climate Change 2007: Working Group I: The Physical Science Basis from the IPCC Fourth Assessment Report FAQ 1.3, Figure 1. Page 98

incoming and outgoing energy becomes unbalanced, meaning that less energy is reflected back to space than in the past, then the earth's surface and atmosphere temperatures change. The lower the albedo, the greater the heat absorption on earth. Various factors that can change the earth's temperature are called climate-forcing agents; the strength of these factors is called **radiative forcing**.

Human-caused GHG emissions have played a significant role in the changes of the earth's temperature in recent decades. This phenomenon, called the **greenhouse effect**, is illustrated in Figure 4. Greenhouse gases such as CO_2 and methane (CH_4) function as **heat trapping gases**, preventing the earth's albedo from reflecting the sun's energy back into space. Rather, those gases trap the energy inside the earth's atmosphere and send parts of it back to the surface. As a result, the atmosphere and the earth's surfaces warm up more and cool down less over time.

Greenhouse Gas Emissions

The four most common GHGs released by humans are carbon dioxide (CO_2), methane (CH_4), halocarbons, and nitrous oxide (N_2O). Overall, GHG emissions have increased 70 percent between 1970 and 2004, with carbon dioxide being the largest contributor (IPCC, 2007). Carbon emissions alone have increased by about 90 percent (EPA, 2016) as illustrated in Figure 5.

Systematic measurements of the concentration of CO_2 in the earth's atmosphere began in the 1950s. Since then, CO_2 concentrations have been rising at an accelerating rate. For example, CO_2 concentrations rose 20% faster between the years 2000 to 2004 compared to the 1990s. The majority of the emitted CO_2 has been captured and stored by the ocean, resulting in increased levels of seawater

Radiative forcing The strength of climate forcing agents in changing the earth's temperature.

Greenhouse effect The phenomenon that prevents the earth's albedo from reflecting the sun's energy back into space. GHGs trap the energy inside the earth's atmosphere and sends parts of it back from the surface. As a result, over time, the atmosphere and the earth surfaces warm up more and cool down less.

Heat trapping gases Gases that prevent the sun's energy from being reflected back into space, trapping the energy inside the earth's atmosphere and sending parts of it back to the surface.

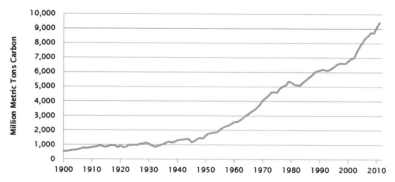

Figure 5 Global Carbon Emission from Fossil Fuels, 1900-2011
Source: United states environmental protection agency, 2016

acidification. In addition to its high percentage in the earth's atmosphere, CO_2 has a notably longer lifespan and persistency factor than other GHGs.

Greater CO_2 concentration in the atmosphere is, to a large extent, due to an increase in human-caused emissions from fossil fuel combustion, deforestation, and cement manufacturing. Burning of fossil fuels, mostly by private automobiles, is the largest single source of CO_2 emissions (IPCC, 2013). The transport sector alone accounts for about 23 percent of the overall CO_2 emissions from fossil fuel combustion (International Transport Forum, 2010). Another significant source of CO_2 emissions is the building sector. In the US, buildings account for approximately 40% of total CO_2 emissions (USGBC, 2012). The significant emissions result from their high electricity consumption, which is generated by burning fossil fuels, such as coal or natural gas. Deforestation causes high CO_2 emissions from burning or decomposing trees and soil carbon.

Methane (CH_4) is the second most frequent GHG. Compared to a CO_2 molecule the radiative forcing from a CH_4 molecule of methane is about 30 times stronger. However, CH_4 molecules lifespan of eight years is much shorter than CO_2. In addition, CH_4 concentrations in the atmosphere have not increased since 1993 (Archer & Rahmstorf, 2010). Natural and artificial wetlands, as well as oil wells, comprise the largest sources of methane emissions. Although methane concentrations are currently stable, methane sources are expected to increase due to thawing permafrost, another example of climate change impacts.

Halocarbons and nitrous oxide have significantly smaller impacts on climate change than CO_2 and CH_4. In fact, the concentrations of halocarbons in the atmosphere are declining as a result of international efforts to protect the ozone layer (NOAA, 2005). Looking at the global emissions from all different GHG sources by sector, electricity and heat production as well as agricultural, forestry, and land use are the economic sectors with the highest amount of emissions. As illustrated in Figure 6, almost 50 percent of all global GHG emissions are linked to these sectors.

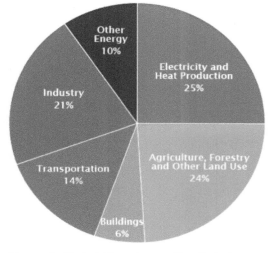

Figure 6 Global Greenhouse Gas Emissions by Economic Sector. *Source*: EPA, 2016.

Current Climate Change impacts

An average temperature increase of 0.86 degrees Celsius (33.55 °F) since the beginning of the 20th century might sound insignificant. However, the impacts of climate change are already visible in the US and globally. Increases in air and water temperature reduced frost days. A higher frequency and magnitude of heavy rainfall, a rise in sea level, reduced snow cover, glaciers, permafrost, and sea ice are also observed. Such changes can affect human health, water supply, agriculture, coastal areas, and the natural environment. One recent conclusion is that in many areas of the world, global climate change impacts are occurring faster than once expected (Pittock, 2009).

Sea Level Rise and Ice Sheets

Sea Level Rise The phenomenon of rising sea level refers to the fact that the mean high tide levels have increased constantly over the past decades.

Thermal expansion The process of ocean water expanding as it gets warmer is called "thermal expansion". In terms of climate change sea water expands in volume due to increasing temperatures leading to measurable sea level rise.

The increase in ocean temperatures and the melting of ice sheets are direct results of global climate change and the main contributors to ocean expansion and **sea level rise**, observed since the beginning of the twentieth century. The process of ocean water enlarging as it gets warmer is called "**thermal expansion.**" Considering that the oceans are on average 3,800 meters deep, even an average expansion of one hundredths of one percent would result in the ocean rising by 38 centimeters — posing a significant risk to coastal cities and their inhabitants.

The impact of ice sheet melting is a far bigger issue. If all ice sheets melted entirely, the sea level could rise by as much as 70 meters. This would change the world's coastal landscapes forever. Currently, data shows that sea level has risen 19 centimeters since the beginning of the twentieth century (IPCC, 2013). More important, recent studies summarized in the latest IPCC reports strongly suggest that the rate of sea level rise has been accelerating. Between 1901 and 2010, the global average sea level rise was 1.7 millimeters (mm) per year, the annual average between 1971 and 2010 was 2.0 mm, and the annual average increase since 1993 ranges from 2.8 to 3.6 mm.

The two ice sheets with the greatest potential impact on sea level rise are Greenland and Antarctica. Both ice sheets are already decreasing. If this trend continues, it could only be a matter of decades before the Greenland ice sheets are completely melted and the Antarctic ice sheets become unstable. The melting of the Greenland ice sheet alone could raise sea level by seven meters. Measurements in Greenland show that the ice closest to the sea is already melting on the surface and creating meltwater ponds and water streaming towards the open sea. Sea level rise is not the only outcome of melting ice, however (Archer & Rahmstorf, 2010). The loss of surfaces that reflect sunlight back into space decreases the earth's surface albedo, adding more heat to the earth's surface (See earlier discussion of this chapter.)

Compared to ice sheets, mountain glaciers and ice caps contain significantly less water but melt much more quickly under increasing temperatures. Glaciers have been retreating since the 18th century, but the rate of melting has been increasing since the 1970s. This trend is shown in Figure 7.

The four photographs of the Muir Glacier in Alaska, or what remains, were taken at four different times. The two pictures on the left were taken in August 1941 (upper) and September 2004 (lower). The pictures on the right were taken in

Figure 7 Retreat of the Muir Glacier in Alaska from 1941 to 2004. Left: August 1941 (upper) September 2004 (lower); Right: September 1976 (upper), September 2003 (lower)

Source: Retreat of the Muir Glacier in Alaska http://www.bartholomewmaps.com/fragileearthnew/FE%20172- 173%20Muir%20Glac.jpg

September 1976 (upper) and September 2003 (lower). Over 65 years, the glacier retreated more than seven miles and the resulting runoff created a mountain lake. The glacier's substantial shrinking illustrates the dramatic impact an increase in temperature can have. Furthermore, mountain glaciers and snow packs store winter precipitation and release it slowly over the summer. This provides a fresh water source when it is needed for agricultural irrigation. With glaciers retreating and snow packs declining, fresh water from mountain streams could decrease as could downstream storage of water supplies (See Chapter 10).

The melting of permafrost soils is another problem that increases the concentration of GHGs in the atmosphere. Arctic permafrost underlies almost one fifth of the planet's land surface and contains methane hydrate. As long as it is frozen, methane hydrate does not present any danger for the environment. But once the ice thaws, the methane hydrate converts to CH_4 and becomes a very potent heat trapping gas.

Precipitation and Drought

In addition to increasing global temperature, climate change impacts precipitation patterns. Unlike temperature, which has increased almost everywhere on the planet, precipitation is increasing in some parts of the world and decreasing in others. The warmer the air becomes, the more water it can store. This water is then released during colder days. This can lead to storm floods and heavy damage in areas where the infrastructure is not able to handle the release of large amounts of water in short amounts of time. The map in Figure 8 shows the areas where heavy rainfalls have increased or decreased.

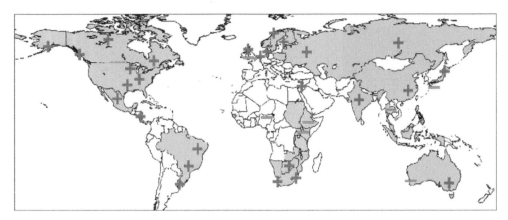

Figure 8 Changes in Heavy Rainfall around the World both Increases and Declines
Source: Climate Change 2007: The Physical Science Basis. Working Group 1 Contribution to the Fourth Assessment Report of the intergovernmental Panel on Climate Change, FAQ 3.2. Figure 1. Cambridge University Press

For example, heavy rainfall increased on the east coast of the United States but decreased in Africa, where food shortages and hunger are already a major concern. In 2012, Hurricane Sandy caused extraordinary rainfalls on the east coast resulting in floods, heavy property and land damages, and even human casualties. In total, Hurricane Sandy caused 117 deaths in the US and 69 in Canada (CNN, 2013). The total economic loss caused by the hurricane in New Jersey for travel and tourism spending alone was estimated at $950 million by the US Department of Commerce (2013). In addition, the New Jersey State Government concluded that it would cost approximately $29.5 billion to repair the damages caused by Hurricane Sandy.

Besides the increases in frequency and magnitude of heavy rainfall, seasonal changes in precipitation are occurring. Precipitation in some regions has decreased in the summer but increased in the winter, resulting in increasing risks of flooding during the winter and drought in the summer. These changes are especially important to land ecosystems and the agricultural sector. Farmers are concerned about seasonal rainfall changes impacting their growing and harvesting seasons. Heavy rainfall is already delaying spring planting in some areas of the US, jeopardizing the livelihoods of farmers. Flooding of the fields during the growing season causes low oxygen levels in the soil, which destroys crops and increases the likelihood of root diseases. In addition, research suggests that increasing temperatures will most likely reduce livestock production during the summer season.

Seasonal changes may also impact areas that rely heavily on tourism and winter sports. In some areas, precipitation that used to fall as snow during the winter is now falling as rain. Consequently, the reduction in the snowpack not only shortens the winter sports season but also reduces the water runoff during the summer when water is most needed for agriculture. We find this occurrence in the Southwestern US, where there is concern about long-lasting drought conditions.

There are different ways to define and measure the severity of droughts. The most common measurement is the **Palmer Drought Severity Index (PDSI)**,

Palmer Drought Severity Index The most common measurement used to define and measure the severity of droughts. The index considers not only the monthly amount of precipitation but also the regional average temperatures. Drought severity is shown in terms of minus numbers and excess rain by plus numbers.

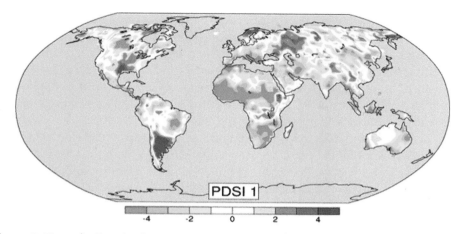

Figure 9 Drought Severity from 1990 to 2002 according to the Palmer Drought Severity Index. Negative numbers indicate drought; positive numbers indicate excess rainfall

Source: Intergovernmental Panel on Climate Change, 2007; FAQ 3.2, figure 1

which considers not only the monthly amount of precipitation but also the regional average temperatures. As shown in Figure 10, droughts became more severe between 1990 and 2002 according to the PDSI. In this index, the severity of the drought is shown in terms of minus numbers indicating drought and plus numbers indicating excess rain. Although heavy rainfall has increased in some areas, dry areas have more than doubled in size since the 1970s (IPCC, 2007).

Human Health, Food Insecurity, and Ecological Refugees

In addition to environmental impacts, global climate change can also cause or intensify health and social issues. The US Global Change Research Program's (USGCRP) Interagency Crosscutting Group on Climate Change and Human Health (CCHHG) recently published its 2016 Climate and Health assessment. The report indicates that global climate change is increasing "exposure to elevated temperatures; more frequent, severe, or longer-lasting extreme events; degraded air quality; diseases transmitted through food, water, and disease vectors (such as ticks and mosquitoes); and stresses to our mental health and well-being" (page 2). These issues impact both developing and developed countries. In the US, for example, data indicate that ticks spreading Lyme disease, the Anopheles mosquito carrying Malaria, and other viruses are spreading northward due to warmer temperatures. Warmer temperatures are also contributing to an increase in pollen allergies, as they cause the pollen season to start earlier in the year. For those interested in this topic, please see the following Internet sites:

- World Health Organization: http://www.who.int/globalchange/en/
- Environmental Protection Agency: http://epa.gov/climatechange/effects /health.html
- US Global Change Research Program: http://www.globalchange .gov/explore/human-health

Food security The ability to obtain reliable access to a sufficient quantity of affordable, safe nutrition

Another problem area exacerbated by global climate change, which is becoming more and more visible, is **food security**. The United Nations Food and Agriculture Organization (FAO) warned in 2008 that climate change will negatively impact all aspects or dimensions of food security. Those dimensions consist of food availability, food accessibility, food utilization, and food system stability (FAO, 2008). Although food systems in all countries will be impacted, the world's poorest and most food insecure are the most at risk.

The Global Food Security Index is another measurement developed to assess food insecurity (Global Food Security Index, 2016). The main categories from which the index is compiled are a) affordability, b) availability, and c) quality & safety. According to the 2016 rankings, 89 of the 113 countries included in this measurement have improved their food security over the last five years. However, the rankings also show that several developing nations still struggle with providing sufficient infrastructure, political unsettlement, and food inflation – all of which pose significant barriers to reaching a satisfying level of food security. In the latest rankings, the top ten nations are all developed countries with the US being on top followed by Ireland, Singapore, Australia, Netherlands, France, Germany, Canada, United Kingdom, and Sweden; the ten countries with the lowest scores are all developing countries with Madagascar coming in at 104th followed by Malawi, Burkina Faso, the Democratic Republic of Congo, Haiti, Mozambique, Niger, Chad, Sierra Leone, and in Burundi coming in 113th. For those interested in this topic, please see the following Internet sites:

- Food and Agriculture Organization of the United Nations: http://www .fao.org/climate-change/en/
- Global Food Security Index: http://foodsecurityindex.eiu.com/
- US Department of Agriculture http://www.usda.gov/oce/climate_change /FoodSecurity.htm

According to the IPCC and the World Bank, the environmental impacts of climate change will also lead to social and economic challenges in the areas of poverty reduction, growth, and development. There is relatively little research on this topic with most being focused on country-level challenges, but two of the most significant are anticipated to be climate refugees and national security. According to the Global Governance Project in its *Global Environmental Governance Reconsidered* (2012), climate refugees are a "subset of environmental migrants" who are forced to move "due to sudden or gradual alterations in the natural environment related to at least one of three impacts of climate change: sea-level rise, extreme weather events, and drought and water scarcity." In terms of security, climate change threatens human welfare as well as countries' sovereignty and national interests. For those interested in this topic, please see the following Internet sites:

- The World Bank: http://www.worldbank.org/en/topic/climatechange
- United Nations High Commissioner for Refugees (UNHCR): http://www .unhcr.org/en-us/environment-disasters-and-climate-change.html

Future Impacts of Global Climate Change

Climate Models and Scenarios

The **Kyoto Protocol** aimed to reduce GHG emissions by 5.2% below 1990 emission levels by 2012. However, worldwide GHG emissions are still increasing. This is, in part, due to the fact that future climate change is already built into the system as a result of past GHG emissions that take decades to disappear from our atmosphere. Therefore, ongoing impacts will occur despite efforts to reduce GHG emissions. Research has found a strong likelihood that extreme weather events and sea level rise will continue to increase and droughts will become longer and more severe regardless of current actions. Additional impacts are expected to include major alterations in oceans, ice, and storms as well as massive dislocations of species, pest outbreaks, and major shifts in wealth, technology, and societal priorities (Stern, 2006). There is a wide array of climate models available, ranging from relatively basic models that focus on the earth's heat balance to very complex and detailed simulations that show these potential future impacts of global climate change under a variety of different assumptions.

The models compute outcomes or **scenarios** based on different assumptions regarding possible future amounts of GHG emissions, policy selections, behavior and actions, and other aspects that might impact future climate trends. It is important to understand that these models do not predict the future, rather they simply offer possible future scenarios. The future of climate, to a large degree, depends on human behavior, which is impossible to predict. The scenarios, however, do provide important data to decision-makers, allowing them to make better-informed, long-term decisions that impact the future. Figure 10 shows

Kyoto Protocol Initially adopted in December 1997, the Kyoto Protocol is an international treaty with the goal to reduce GHG emissions and prevent further increases in the global temperature. Today, more than 190 countries have signed the treaty.

Scenarios Possible future circumstances computed by climate models. They are based on different assumptions regarding possible future amounts of GHG emissions, policy selections, behavioral actions, and other aspects that might impact future climate trends.

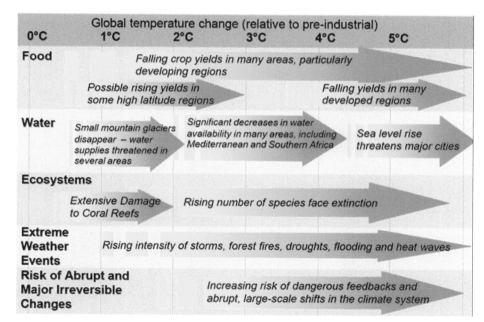

Figure 10 Project Impacts of Climate Change according to Specific Rises in Temperature

Source: Stern, 2006; figure 2

climate change impacts by sector as temperatures increase based on different scenarios. Although there are short-term benefits in some areas, the scenarios indicate significant negative consequences in all sectors over the long run.

Future Global Temperatures

According to the 5th IPCC Assessment Report (2013), the global mean surface temperature is likely to increase by 0.3°C to 0.7°C between the years 2016 and 2035 compared to temperatures measured between 1986 and 2005. As shown in Figure 11, projections reaching further into the future suggest average temperature increases ranging from 0.3°C to 1.7°C all the way to 2.6°C to 4.8°C by the year 2100 depending on the emission scenario. Moreover, the IPCC concludes with a very high degree of certainty that there will be more hot weather extremes in the future and less cold temperature anomalies.

Future Precipitation Patterns

In addition to rising temperatures, changes in precipitation rates will also impact society and the natural environment. A secure water supply is fundamental for our food supply and the livelihood of plants and animals. Yet, possible future precipitation changes are much harder to predict than other climate change impacts. By and large, current climate models operate on a large spatial scale, making it very difficult to capture important regional differences in rainfall. Therefore, the uncertainties regarding possible future trends of rainfall extremes or droughts can be quite large.

Despite this uncertainty, all models anticipate future droughts, heavy rainfall, and floods (IPCC, 2013). Specifically, they indicate that wet areas will become wetter and dry areas will become even drier. Furthermore, as mean surface temperatures increase, extreme precipitation events will most likely intensify as well. Climate models show that heavy rainfall will become stronger and more frequent, especially over most of the mid-latitude landmasses and wet tropical regions. Monsoon seasons are likely to start earlier and last longer in many regions. Droughts are

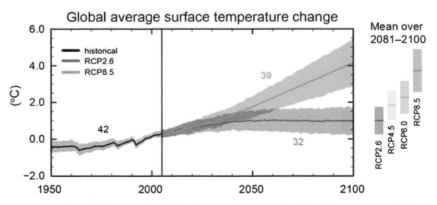

Figure 11 Projected Change in Global Annual Mean Surface Temperature Relative to 1986-2005

Source: Intergovernmental Panel on Climate Change, 2013 ; figure SPM.7 Panel A

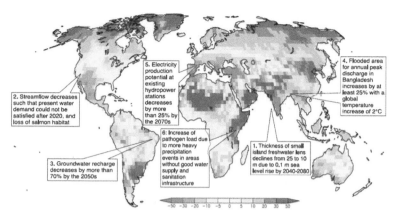

Figure 12 Possible threat to water security worldwide in the future. Blue shows increased runoff, red decreased runoff in percent.
Source: Intergovernmental Panel on Climate Change, 2007 ; figure 3.8

forecast to amplify in Australia, the eastern parts of New Zealand, as well as in the Mediterranean, central Europe, and Central America. In terms of snowfall, decreases in the length of the snow season can be expected in most of Europe and North America. The wide range of possible impacts is summarized in Figure 12.

Future Impact on Water Security

One of the most significant climate change impacts on human society is water security (IPCC, 2007). Regions in the Mediterranean, Southern Africa, Western Australia, and Southwestern US will likely face serious future droughts. Moreover, water stress in regards to quality and availability will increase, affecting up to two-thirds of the global land area. In turn, this will affect food security and water quality adversely impacting human health. According to the 2007 IPCC report, by the year 2050 one to two billion people could suffer from droughts and decreasing water quality.

Uncertainty in Climate Change Projections

All future climate change impact scenarios are characterized by uncertainties. Uncertainty arises from internal processes in the physical climate system, such as changes in vegetation, variations in the earth's orbit around the sun, or volcanic eruptions. It also arises from human decisions and behaviors, which is very unpredictable and influenced by attitudes towards quality of life and wealth. Future GHG emission trends will depend heavily on human decisions and behaviors, which, in turn, will influence future climate change. Among these important influences will be the development and availability of new technologies, the implementation of different environmental policies, and their level of public acceptance and support.

Given these uncertainties, decision makers find it challenging to anticipate and employ appropriate adaptation strategies to counter future climate change impacts. Traditional approaches such as making decisions based on "worst-case" scenarios do not translate readily to the highly complex and uncertain issue of climate change.

Advanced scenario planning Advanced scenario planning presents a flexible framework that allows decision makers, to develop long-term strategies based on many different possible scenarios. Advanced scenario planning and includes methods such as aggregated averages, risk assessments, sensitivity analysis of factors or decisions driving the scenarios, identification of unacceptable or worst case outcomes, and assessment of common and different impacts among the scenarios.

Anticipatory governance framework Anticipatory governance relies on the development and analysis of a range of possible scenarios, rather than a forecast or selection of a single scenario. It presents a new model for decision making while dealing with high uncertainties and consists of the anticipatory future steps and feedback creation of flexible adaptation strategies, monitoring and action.

Instead, a more flexible framework is required that allows decision makers to develop strategies based on many different possible scenarios with feedback loops. This approach is referred to as **advanced scenario planning** and is a key component of the **anticipatory governance framework** (Quay, 2010) (See chapter 11).

Anticipatory governance is "a system of institutions, rules and norms that provide a way to use foresight for the purpose of reducing risk, and to increase capacity to respond to events at early rather than later stages of their development" (Fuerth, 2009, p.29). It presents a new decision-making model that accounts for high uncertainties and consists of anticipatory future analysis and feedback creation for flexible adaptation strategies, monitoring, and action. Anticipatory future analysis is based on advanced scenario planning and includes methods such as aggregated averages, risk assessments, sensitivity analysis of factors or decisions driving the scenarios, identification of unacceptable or worst-case outcomes, and assessment of common and different impacts among the scenarios. Due to the uncertainties surrounding climate change and the changing impacts over time, the final steps 'monitoring and action' demand that policy makers and decision makers revise adaptation strategies on a regular basis.

Climate Change Governance and Strategies

Mitigation and adaptation strategies are considered the two main policy responses to global climate change. However, they are not independent. In fact, mitigation and adaptation are driven by the same set of problems; therefore, the more mitigation that takes place the less adaptation will be needed, and vice versa. Mitigation requires comprehensive changes in numerous aspects of society and the built environment to cope with the effects of global climate change impacts that are unavoidable. Adaptation includes significant political intervention, behavior change, and support for climate strategies and policies in the next decade to confront the causes of climate change and reduce the potential negative consequences. Successful implementation of these strategies, particularly for improving the resiliency of urbanized areas, will be one of the major societal challenges in the twenty-first century.

International Treaties and Frameworks

On the international scale, more than 150 countries signed one of the most important and well-known policy frameworks following the 1992 Earth Summit in Rio de Janeiro. The initial goal of the "United Nations Framework Conventions on Climate Change" (UNFCCC) involved the reduction of GHG emissions to a level that would prevent any additional negative impacts on the climate system (UNFCCC, 1998). Focusing on climate change, mitigation, and adaptation while also acknowledging issues of social equity and sustainable development, the framework entered into force in 1994. This, in turn, led to the Kyoto Protocol in 1997 (UNFCCC, 1998a). The developed countries that signed the Kyoto Protocol committed to reduce GHG emissions by an average of 5.2 percent below 1990 levels between 2008 and 2012 (United Nations, 1998). The reduction target for the US was set at seven percent. However, the US signed but did not ratify the treaty and withdrew its support in 2001, never meeting its goal. The US did not ratify the treaty primarily due to economic concerns, arguing that the protocol would harm the economy and result in the loss of jobs (NBC, 2005).

In addition to the Kyoto Protocol, the UNFCCC also established yearly climate summits among its members called the "Conferences of the Parties" (COP), which have led to numerous international agreements for contending with climate change. Major milestones where reached in 2007 during COP 13 in Bali, Indonesia that led to the "Bali Road Map" and the "Bali Action Plan" (http://unfccc.int/key_steps/ bali_road_map/items/6072.php), and outlined a negating process with the goal to achieve a legally binding climate treaty (UNFCCC, 2008; 2008a). Two years later in 2009 at COP 15 in Copenhagen, 114 countries signed the "Copenhagen Accord" re-emphasizing the need to cut GHG emissions and establish financial support mechanisms for mitigation efforts in developing countries (UNFCCC, 2009).

In 2010, COP 16 in Cancun, Mexico led to the "Cancun Agreements" (http:// unfccc.int/meetings/cancun_nov_2010/items/6005.php), which included recognizing the efforts by developing countries to reduce GHG emissions (UNFCCC, 2011a). The "Cancun Adaptation Framework" improved the planning and implementation processes of adaptation projects and actions to further reduce emissions caused by deforestation and forest degradation (UNFCCC, 2011). In addition, countries continued negotiations around making emission reduction pledges official and discussed extending the Kyoto Protocol past the year 2012, which marked the end of the first commitment period (2008-2012). The 2011 climate change conference (COP 17) in Durban, South Africa focused mostly on the continuation of the work started with the "Bali Roadmap" and the "Cancun Agreement" (http://unfccc. int/key_steps/durban_outcomes/items/6825.php). The major outcome was the decision to adopt a new, universal, legally binding agreement on climate change no later than 2015 during the COP 21 in Paris, France (UNFCCC, 2012).

With the end of the initial phase of the Kyoto Protocol in 2012, next steps were a major talking point at COP 18 in Doha, Qatar. The Doha Amendment extended the Kyoto Protocol for another eight years, beginning January 1, 2013 (UNFCCC, 2013). Compared to the first commitment period, however, the second phase was appreciably more limited in scope due to the lack of participation of US, Canada, Japan, and Russia, and the fact that developing countries like China (the world's largest emitter), India, and Brazil were not subject to emissions reductions. Despite these limitations, COP 18 laid the groundwork for more extensive action. This capacity-building decision platform is known as the Doha Climate Gateway (http://unfccc.int/key_steps/doha_climate_gateway/items/7389.php; UNFCCC, 2013a).

The 2013 meeting, COP 19, took place in Warsaw, Poland. During this conference, consensus was reached between the members to financially support developing countries in their efforts to reduce GHG emissions and adapt to impacts already occurring through the Green Climate Fund and Long-Term Finance, among others (UNFCCC, 2014; http://unfccc.int/key_steps/warsaw_outcomes/items/8006. php). Countries confirmed their commitment to reach a climate agreement by the COP 21 conference to be held in Paris, France in 2015, but the summit was highly criticized for the lack of urgency amongst negotiators. COP 20 was held in Lima, Peru in 2014. This conference and the resulting "Lima Call for Action" set the stage for COP 21 (UNFCCC, 2015). Conference outcomes included developing elements of the new agreement as well as setting the ground rules for submitting contributions. The conference also made important strides in recognizing the importance of adaptation as well as mitigation of GHG emissions (http://unfccc.int/meetings/lima_dec_2014/meeting/8141.php; UNFCCC, 2015a).

These efforts led to over 180 countries signing the Paris Agreement at COP 21 (http://unfccc.int/files/essential_background/convention/application/pdf/english_paris_agreement.pdf; UNFCCC, 2016). The agreement marks the first serious attempt supported by most UN member nations to resolve global climate change since the 1997 Kyoto Protocol. As of October 5, 2016, over 60 countries ratified the agreement, surpassing the requirement that the agreement only becomes legally binding if ratified by at least 55 countries covering a minimum of 55 percent of global GHG emissions. Unlike the Doha Amendment, the US, Canada, Japan, and Russia signed the Paris Agreement with the US and Canada ratified it on September 3 and October 5, 2016, respectively. As a result, the Agreement entered into force on November 4, 2016 and full implementation will begin in 2020.

The aim of the Paris Agreement is to limit GHG emissions to the degree necessary to hold the increase in global average temperature "well below 2 degrees Celsius above pre-industrial levels and to pursue efforts to limit the temperature increase even further to 1.5 degrees Celsius" (UNFCCC, 2016a). Compared to the Kyoto Protocol, emission reduction targets are required for both developed and developing countries but each country decides how to reach its reduction goals. The agreement also includes regular reviews, progress reports, and financial aid for poorer nations to help them cut emissions as well as cope with already occurring negative climate change impacts.

Since its adaptation, the newly elected US President Donald Trump announced in June 2017 that the country will withdraw from the Paris Climate Agreement. Despite this announcement, many US cities, states and private sector organizations state that they will continue to strive for the targets set by the agreement. Thus, it is not clear what the US' role will be in resolving climate change and fostering a clean energy economy.

Urban Efforts and Networks

On the national scale, many cities have come to terms with the fact that something needs to be done about climate change and are developing and leading initiatives rather than relying on the US Government, international community, or comprehensive global treaties. Cities are often considered a significant part of the climate change problem as they consume major amounts of energy and are the main producers of GHGs creating climate change risks in the first place. The spatial concentration of industry and transportation, as well as domestic and commercial buildings assign cities a key role in how energy is produced and GHGs are emitted. Furthermore, cities are often located along coastal regions or rivers, making them more vulnerable to climate change impacts. Rapid urbanization also increases the vulnerability of cities as urban centers are already overstressed and new city dwellers, particularly the poor, are often relegated to live in high-risk areas. In developing countries, many poor individuals are forced to construct their homes in informal settlements on floodplains, in swamp areas, or on unstable hillsides, all of which lack necessary infrastructure and basic services (Rosenzweig et al., 2011).

Cities are both the victim and perpetrator of climate change and many are actively engaged in finding solutions (See Chapter 13). Overall, cities and urban areas consume about 75 percent of the world's energy and are responsible for up to 75 percent of GHG emissions. They are also centers for innovation with the potential to develop the new technologies, and urban policies and strategies needed to address local climate change impacts. Municipalities have a significant say in aspects of urban planning, building codes, public transportation, and the supply of energy, water, and waste services that determine vulnerabilities and the production of GHG emissions. As a result, city governments are in an advantageous position to implement climate change mitigation and adaption strategies.

To improve local strategies and learn from experiences in different places, cities have begun to organize themselves in transnational municipal networks. One of these networks is the "C40 Cities Climate Leadership Group", a network of the world's largest cities willing to address climate change (http://www.c40.org). Among the network's tasks are raising finances, sharing knowledge, and providing additional partners and expertise to its member cities to engage in specific climate change-related projects. For example, C40 networks share knowledge and support one another with respect to rapid transit systems, climate risk assessments, sustainable urban development, and sustainable solid waste systems. Another well-known transitional city network with similar goals and approaches is the "Local Governments for Sustainability" (ICLEI), which was established in 1990 (http://www.iclei.org/). ICLEI has member cities in over 85 countries ranging from mega-cities to small and medium-sized towns. Within the US, the lack of political action on the federal level has led to the formation of the US Conference of Mayors Climate Protection Agreement (http://www.usmayors.org/). By signing the agreement, city mayors commit to take three decisive actions against climate change. First, meet the GHG reduction targets outlined by the Kyoto Protocol for their community. Second, lobby at the state and federal levels to implement more mitigation policies and reduce GHG emissions bellow the benchmark set by the Kyoto Protocol for the entire US. Third, work toward the implementation of a national emission trading system. So far, over 1,000 mayors have signed the Climate Protection Agreement (The US Conference of Mayors, 2014).

Mitigation

Mitigation addresses the core cause of human-induced climate change, namely the large amount of energy consumption and resulting GHG emissions. The concept of mitigation is clearly understood by scientists and decision makers. As a result, various international treaties signed by many countries, such as the Kyoto Protocol (1997) and the Paris Agreement (2016), have set GHG reduction goals and strengthened international cooperation in the fight against climate change. The fact that GHG emissions are easy to assess and can be monitored quantitatively has led to the development and implementation of numerous mitigation strategies. The following sections address three major sectors of energy consumption, especially in urban environments, and discuss how mitigation strategies can reduce their GHG emissions.

Transportation Sector

The transportation sector is among the fastest growing areas of energy use. In the US, it is one of the largest contributors to the nation's GHG emissions, constituting 26 percent of total emissions in 2014 (EPA, 2016). The heavy reliance on combustion engines fueled by fossil fuels has led to a significant increase in GHG emissions over the past decades. Despite increasing gas prices, many consumers still prefer large vehicles with powerful combustion engines. Thus, the free market has little impact on the people's travel behavior and the mode of transportation they choose. In contrast, mandatory regulation regarding fuel economy has been much more effective in reducing GHG emissions while also encouraging research towards higher fuel efficiency (IPCC, 2007). As a result, new technologies have emerged that make automobiles up to 40 percent more efficient compared to cars relying entirely on fossil fuels. Technologies such as hybrid and electric vehicles, turbo diesels, and biofuels will decrease GHG emissions even further as these technologies become widely available.

Strategies that provide technological alternatives to fossil fuels are not the only measurements available for mitigation of climate change, however. Research suggests that well-designed communities and sustainable transportation planning can reduce transportation sector GHG emissions by decreasing the need to drive and the overall amount of vehicle miles traveled. Underlying strategies include mixed land use, implementation of compact development patterns, creation of walkable environments, and allocation of transportation alternatives such as public buses or light rail systems. These strategies are being embraced by a number of initiatives including Transit-Oriented Development and Complete Streets.

Transit-Oriented Development (TOD) focuses on reducing automobile traffic and GHG emissions by building and revitalizing urban communities around centralized transit systems, specifically "fixed-guideway" or light, heavy, and rapid rail systems. According to the Transit Oriented Development Institute (2016) components include walkability, train stations and public space as central components, and mixed-use (office, residential, retail, civic) nodes within walking distance. TOD also promotes alternative means of transit, including bicycling and other modes of public transportation. In the US, a 2014 Federal Transit Administration Research report by the Center for Transit Oriented Development noted that transit ridership increased by 36 percent from 1995 to 2008 and investment in new fixed-guideway transit systems increased from 27 to 40 or 48 percent (pg. 1). Thirteen new regions added transit systems, consisting of 292 new train stations between 2000 and 2010. In addition, 589 stations were added to existing systems (pg. 5).

The "complete streets" approach is also gaining momentum across the US and the world. For instance, the National Complete Streets Coalition, launched in 2004,

Figure 13 Copenhagen's bicycle culture reduces GHG emissions and promotes a healthy lifestyle. (*Source*: 337279727)

reports that over 730 US local, regional and state agencies have implemented over 950 policies nationwide (Smart Growth America, 2016). The essence of planning for complete streets is that all users are guaranteed safe, easy-to-use rights of way. This includes providing access to businesses, workplaces, and community amenities for all people regardless of age, ability, income, or race. An excellent example of this approach is taking place in Copenhagen, Denmark.

The city incorporates public transportation, pedestrian, and bicycle access to promote healthy living and reduce transportation GHG emissions (Figure 13). In 2015, it was named the #1 most bike-friendly city in the world by the Copenhagenize Index. It edged out Amsterdam, which held the top spot for the last two years, thanks to an increase in ridership to 45% of its population (Copenhagenize Design Co., 2015). Copenhagen promotes its bicycle culture with approximately 400 km (249 miles) of cycle paths, including bike lanes, a network of cycle superhighways (Denmark.dk, 2016), and bike bridges (The Blog, 2016). For a more in-depth discussion on transportation, see Chapter 12 *Sustainable Transportation*.

Building Sector

Changing building energy can also reduce GHG emissions significantly. Therefore, the building sector provides low-cost opportunities to reduce CO_2 emissions. Many energy-saving efficiency measures for heating, cooling, and lighting buildings are being implemented. For example, energy efficiency can be improved and heating costs reduced by improving the insulation of buildings, sealing leaks, and using energy-efficient windows. Depending on the climate, these measures alone have the potential to decrease heating costs by up to 90 percent (Archer & Rahmstorf, 2010).

Reducing emissions from air conditioning and decreasing cooling costs is also relatively simple. This can be achieved through architectural design and siting choices, such as orienting the long axis of a house east-west, so that wall areas receiving hot morning and afternoon sun are minimized. Another energy-saving measure is to use reflective materials and light colored surfaces, so that heat is reflected away from the building. Providing shade through landscaping is also an effective way too cool buildings and reduce cooling costs by up to 40 percent. In recent years, concepts, such as the Smart Growth and New Urbanism, and green building rating systems, such as Building Research Establishment Environmental Assessment Method (BREEAM) and Leadership in Energy and Environmental Design (LEED), have been developed and introduced into planning policies to establish a sustainable development pattern for neighborhoods and communities.

Smart Growth is based on ten basic principles that build strong communities, encourage economic growth, promote healthy behaviors, and protect the natural environment (Smart Growth Network, 2014). The principles are:

- Mix Land Uses
- Take Advantage of Compact Building Design
- Create a Range of Housing Opportunities and Choices
- Create Walkable Neighborhoods
- Foster Distinctive, Attractive Communities with a Strong Sense of Place
- Preserve Open Space, Farmland, Natural Beauty, and Critical Environmental Areas
- Strengthen and Direct Development Towards Existing Communities
- Provide a Variety of Transportation Choices
- Make Development Decisions Predictable, Fair, and Cost Effective
- Encourage Community and Stakeholder Collaboration in Development Decisions

New Urbanism is similar to Smart Growth in that it promotes many of the same principles, such as mixed land use, walkability, and compact urban design. However, it focuses more specifically on the human element of public policy and development practices, emphasizing equity and diversity. New Urbanism looks at three distinct aspects of urban life—the Region, consisting of the Metropolis and surrounding cities and towns; the neighborhoods, districts, and corridors that make up the Metropolis; and the blocks, streets, and buildings that residents live and work in (CNU, 2014).

Green building rating systems have been used since 1990 when BREEAM was introduced in the UK. These systems assess myriad aspects of planning, designing, constructing, and operating buildings. Materials used, energy efficiency, land use, pollution and waste remediation, accessibility to public and alternative transportation modes, and water use are some of the common aspects considered during assessment. Assessment can occur at any life-cycle stage of a building—during construction, while in use, or when being renovated. In the US, the US Green Building Council (USGBC) has been developing rating systems to define and measure existing sustainable buildings and new developments under the LEED program since 1993.

, Currently, there are five LEED rating systems a) Building Design and Construction, b) Interior Design and Construction, c) Building Operations and Maintenance, d) Neighborhood Development, and e) Homes. In order to receive LEED certification, projects that fall within the five different rating systems need to earn a certain number of credits for meeting different sustainable design categories and performance benchmarks. For example, development projects such as master planned communities are awarded points based on the materials used, water efficiency, energy use, site location, level of access to public transit, and others. In total, the LEED program offers four types of certification (certified, silver, gold, and platinum) depending on the points scored on the rating system (USGBC, 2014).

BREEAM USA, launched in June 2016, focuses specifically on the country's more than 5 million in-use commercial buildings. The primary difference between LEED's Existing Building (LEED-EB) and BREEAM In-Use programs is that there are no prerequisites with the In-Use platform, enabling all commercial buildings to participate. It also provides a benchmarking mechanism so building managers can track their progress and is science based rather than consensus driven. It is administered through a partnership of BuildingWise, a US-based LEED certification consultant, and BRE, the U.K. Building Research Establishment group. It provides a sustainability assessment framework aimed at maximizing resource efficiency and facility use, occupant satisfaction, and economic benefits (BREEAM-USA, 2016).

Industry Sector

The industrial sector is the third major source of energy use and a significant contributor to global climate change. In the US, the EPA's *Inventory of U.S. Greenhouse Gas Emissions and Sinks* determined that industry was responsible for roughly 2000 million Metric tons or 21 percent of direct CO_2 emissions in 2014. Compared to the transportation and housing sector, the annual increase in emissions from the industrial sector is relatively low at only 0.6 percent. Total emissions account for both direct (fuel and energy consumption, leaks) and indirect (energy generation by power plants for use by others) emissions (direct and indirect). Industry's total CO_2 emissions (direct and indirect) in 2014 were 29 percent, making it the largest contributor of greenhouse gases of any sector (EPA, 2016a).

Within the industrial sector, the metal and chemical production industries are the most energy intensive and account for 85 percent of the sector's total GHG emissions. The highest potential for reducing emissions is to offer the steel, cement, and petroleum industries incentives to use cleaner industrial processes or implement stricter environmental laws. The latest IPCC report argues that emission cuts of up to 40 percent are possible if companies were charged $20 for each ton of CO_2 or any other GHGs emitted into the atmosphere (IPCC, 2014a). There are other actions that individual plants can undertake as well. These include energy efficiency measures, alternative fuels, recycling, and training and awareness (EPA, 2016a).

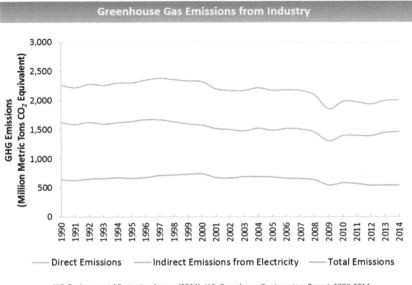

Figure 14 **Industry GHG Emissions for the United States between 1990 and 2014**

Source: https://www.epa.gov/ghgemissions/sources-greenhouse-gas-emissions#industry

Adaptation

Adaptation strategies focus on avoiding or reducing harmful impacts caused by global climate change. They aim to increase resilience or decrease vulnerability to current and anticipated impacts. Even if current mitigation measurements prove to be successful in reducing GHG emissions, global climate change is already occurring and further climate change is built into the system from previous emissions. This makes the design and implementation of adaptation strategies a necessity to reduce the impacts that can no longer be prevented and to prepare for possible future threats.

Unlike mitigation strategies that focus primarily on global and national scales and require international cooperation, adaptation strategies are implemented at regional and local scales. Another difference is that adaptation strategies are subject to high uncertainties because their effectiveness depends on the accuracy of regional climate and impact projections. Moreover, there are no global or national targets and schedules for adaptation projects. They are strictly determined by regional and local agencies based on perceived and actual impacts.

Adaptation Strategies and their Impact

Reducing greenhouse emissions takes time, and benefits will not be fully realized for decades. Adaptation measures, on the other hand, have much shorter timelines. Due to the strong links with development initiatives and the fact that implementation occurs mostly on local or regional scales, adaptation efforts and their results are much faster and more visible when compared to mitigation measures. Furthermore, the efficiency of adaptation strategies is less dependent on the actions of others and does not need international agreements. Since adaptation is primarily a local and regional problem, appropriate strategies must be evaluated on site.

It is impossible to minimize the economic costs of climate change impacts, protect human health and welfare, and limit harm to infrastructure, ecosystems, and biodiversity without the successful implementation of adaptation strategies. In the case of sea level rise, floods, and droughts, adaptation measures protect settlements or enable populations to relocate. In the case of water shortages, securing new supplies or the implementation of conservation measures will become necessary. Adaptation measures also play an important role in the agricultural sector. Farmers and ranchers might adapt to climate change impacts by changing crops, raising different livestock, or relocating. Public health measures include protection for vulnerable populations from increasing heat-related events and shifts in disease and insect patterns.

Real world examples of adaptation measures include coastal defense planning and storm surge barriers, as shown in Figure 15, and early warning systems to prepare for sea level rise and storm floods. Other regions are starting to prepare for water shortages by creating water storage systems, implementing conservation measures, and building seawater desalination plants. Additional measures include land management planning and zoning law updates to prevent soil erosion; shifting ski slopes to higher altitudes in regions dependent upon winter tourism; and preparing emergency plans to deal with future heat waves or other extreme weather events.

© Gertje/Shutterstock.com

Figure 15 Storm Surge Barrier in Zeeland, Netherlands Built after the Storm Disaster in 1953.

Local Climate Change Action Plans

A majority of the world's energy consumption either occurs in cities or is a direct result of the way cities function (e.g. through transport of goods to points of consumption in cities). Nevertheless, the concentration of resources in cities can be a useful weapon in fighting climate change. Cities are in an excellent position to lead the way for other municipalities to follow, as they are often centers of new thinking and policy innovation. In recent years, many cities and states have developed **Climate Change Action Plans** (or Climate Action Plans) to mitigate and adapt to climate change.

Climate Action Plans and their recommendations present a great opportunity. They provide a framework to change current development patterns and establish a sustainable way of living reduce the vulnerability to climate change, and increase the adaptive capacity of communities. These plans help states and cities identify and evaluate feasible and effective policies to reduce their GHG emissions via a combination of public and private sector policies and programs.

The first generation of Climate Action Plans focused mainly on improving municipal operations in terms of energy use and GHG emissions with the most prominent strategies as follows:

- Creating building codes and standards that include practical affordable changes that make buildings cleaner and more energy efficient;
- Conducting energy audits and implementing retrofit programs to improve energy efficiency in municipal and private buildings;
- Installing more energy–efficient traffic and street lighting;
- Implementing localized, cleaner electricity generation systems;
- Developing rapid transit and non-motorized transport systems;
- Using clean fuels and hybrid technologies for buses, rubbish trucks, and other city vehicles;
- Implementing schemes to reduce traffic, such as congestion charges;
- Creating waste-to-energy systems at landfills; and
- Improving water distribution systems and leak management.

Today, these plans are also addressing adaptation strategies and jurisdiction-wide mitigation policies such as land-use planning (Millar-Ball, 2010). By the end of the last decade, at least 141 local US jurisdictions had developed Climate Action Plans. A very good example of a Climate Action Plan that considers mitigation and adaptation strategies was Chicago's 2008 plan.

After consulting with dozens of experts, an internationally recognized research advisory committee, and numerous Chicago business, labor, civic, and environmental leaders, the city set goals to achieve a 25 percent reduction of the 1990-level GHG emissions by 2020, an 80 percent reduction of the 1990-level GHG emissions by 2050, and prepare the city for the effects of global climate change (City of Chicago, 2008). The plan outlines 5 main strategies, 26 actions for mitigating GHG emissions, and 9 actions to prepare for climate change. It identifies key opportunities to meet the emission goals by improving residential, commercial, and industrial building energy efficiency as well as upgrading power plants and improving their efficiency. The plan also emphasizes the need to reduce GHG emissions by decreasing the amount people who drive and improving vehicle fuel efficiency. Various strategies include supporting principles of transit-oriented

Climate change action plans Many cities and states have developed "Climate Change Action Plans" to mitigate and adapt to climate change. These plans help to identify and evaluate feasible and effective policies to reduce GHG emissions through a combination of public and private sector policies and programs. In addition, these plans provide a framework to change current development patterns and establish a sustainable way of living in the future, which reduces the vulnerability to climate change and increases the adaptive capacity of communities.

development and policies encouraging car sharing and carpooling, improving fleet efficiency, and achieving higher fuel efficiency standards. In terms of adaptation, the five strategies focus on the management of possible heat waves, the protection of air quality, the preservation of plants and trees, the engagement of the public and local businesses, green urban design, and innovative cooling. In addition to the environmental benefits of these actions, Chicago's Climate Action Plan also points to other benefits such as the potential for thousands of new jobs once the policies are implemented.

Portland, Oregon is also well known for its efforts to become more sustainable and prepare for climate change. In 1993, Portland became the first local government in the US to adopt a climate action plan. Because the plan addressed issues such as public transportation and energy supply, for which other entities hold ultimate decision-making authority, the process and final plan were highly collaborative and involved a broad network of partnerships. Thus, the *Local Action Plan on Global Warming* was developed under the guidance of a steering committee and extensive input from residents, businesses, non-governmental organizations, and public sector representatives. The plan identified strategies to reduce GHG emissions in six areas: land-use planning, transportation, energy efficiency, renewable energy, solid waste and recycling, and urban forestry. Activities targeted both city government operations and community-wide initiatives, building on a tradition of natural resource stewardship and the high value Portland residents place on local quality of life. The overall goal was to reduce GHG emissions in Multnomah County, Oregon to 10 percent below 1990 levels by 2010 while minimizing costs and maximizing co-benefits (City of Portland, 2001).

In 2007, Portland City Council and the Multnomah County Board of Commissioners adopted resolutions directing staff to design a strategy to reduce local carbon emissions 80 percent by the year 2050. The resulting 2009 Climate Action Plan guides future city and county efforts and provides a framework for the region's transition to a more sustainable and climate-stable future. The visions of the plan include:

- Each resident lives in a walkable and bikeable neighborhood that includes retail businesses, school, parks, and jobs.
- Green-collar jobs are a key component of the thriving regional economy, with products and services related to clean energy, green building, sustainable food, and waste reuse and recovery providing living-wage jobs throughout the community.
- Homes, offices, and other buildings are durable and highly efficient, healthy, comfortable, and powered primarily by solar, wind, and other renewable resources.
- Urban forests, green roofs, and swales cover the community, reducing the urban heat island effect, sequestering carbon, providing wildlife habitat, and cleaning the air and water.
- Food and agriculture are central to the economic and cultural vitality of the community, with productive backyard and community gardens and thriving farmers markets. A large share of food comes from farms in the region and residents eat healthily, consuming more locally grown grains, vegetables, and fruits.

Chapter 9

Sustainable Agricultural Systems for Cities

Rimjhim M. Aggarwal, Carissa Taylor, and Andrew Berardy

Introduction: The Challenge of Feeding Cities

As the world becomes increasingly urbanized, with cities encroaching on former agricultural land and farmers moving to cities, the question of who will feed urban residents and how has become critical. The world today is very different from nineteenth-century Europe when productivity in agriculture was rising. To feed the influx of people moving to cities, the rate of production for nearby farmlands was simply increased. In the twentieth century, farming in cities was discouraged or ignored as urban planners drew a clear distinction between rural areas as primarily agricultural and urban areas as primarily industrial and residential. The same model was carried to other regions of the world by colonial rulers, and it was believed that farming had no place in the building of modern cities. Despite this, several cities in Europe continued to have significant land devoted to farming.

Now, we live in a world where the majority of people live in cities, far from the primary sources of food production. The whole process, from growing food to bringing it to consumers, has become highly energy intensive. Furthermore, there are large ecological impacts associated with each step of the lifecycle process, from the local to the global scale. Besides these production issues, there are emerging issues on the consumption side such as what we eat and why. Rising rates of food deprivation and obesity within even relatively wealthy urban areas have raised concerns about the **accessibility of healthy food** within cities. As our **food systems** become integrated globally, questions about food safety are regularly raised, and we seem to be less and less in control of what gets onto our plate. In the US and Western Europe in particular, this feeling of disconnect and loss of control over something as intimate as the food we eat has motivated people to look for alternative ways of securing their food. In some cities in Africa and Asia where the rate of migration into urban areas is very high, widespread poverty and unemployment has driven residents to find ways of growing their own food to survive.

Accessibility of healthy food lrefers to physical proximity to retail stores selling healthy food and affordability of that food.

Food systems The set of interdependent processes, infrastructures, and enterprises involved in the production, processing, distribution, consumption, and disposal of food for a given population.

Given the centrality of food in our lives and its role in shaping resource use on this planet, it is clear that we cannot achieve sustainability without fundamentally restructuring our food system. By 2050, the world population is expected to increase to 9.1 billion, with around 70% of people living in cities. Feeding all those people will require increasing food production by 60%, according to the UN's Food and Agriculture Organization (Allen, Guthman, & Morris, 2006).

Together with increasing the quantity of food, it is also essential to pay attention to its quality. With growing concerns over obesity and other health issues, urban residents must have access to a healthy, wholesome diet. The twentieth-century model of keeping cities and agriculture distinct and separate has proven to be both ineffective and unrealistic. In fact, if we look closely, we realize that agriculture has always been practiced in and around cities in different forms (see Howard's Garden City insert below). The problem has been that urban agriculture is generally treated as a marginal or fringe activity and not systematically integrated within city planning and decision making.

Box 1: Ebenezer Howard's Garden City

A need for closer integration between agriculture and cities is not a new concept. In the late 19th century, concerned with the downsides of capitalism and the increasing urban-rural divide, English stenographer Ebenezer Howard sought to develop an approach to urban planning that would help re-connect people to their agricultural land and surrounding ecosystems. In his 1898 publication, *Garden Cities of Tomorrow*, Howard explained that the current rural-urban divide was untenable. Towns provided employment but squalid living conditions, while the country was healthy and beautiful but lacked modernization and job opportunities. Thus, he argued "town and country must be married" and this must be done in a carefully planned way. He called his utopian idea the "garden city."

Each household would receive a plot of land enough for a small garden, and would be located in close proximity to a community park or open space which also would contain some sort of public amenity such as a theater, museum, hospital, or library. A central city park would also contain the city's major shopping and business center. All this would be bounded by an agricultural greenbelt, which could provide the city with much of its food, and rail would connect garden cities to one another. Under his urban planning model, each garden city would be small – bounded to 6000 acres and no more than 32,000 people. Consider for a moment how small this is. New York City is 193,000 acres and has over 8 million residents. Even in Phoenix, Arizona's "little" suburb of Tempe is far larger – with over 25,000 acres and more than 158,000 residents.

The first city built under the garden city model was Letchworth, England – established in 1904. Other examples of towns and suburbs designed using "garden city" principles include Hampstead Garden, England; Fairfield, AL;

Kincaid, IL; Goodyear Heights, OH; Greenbelt, MD; Yorkship Village, NJ; Forest Hills Gardens and Sunnyside Gardens in Queens, NY; and Beloit and Kohler, WI (Gillette, 2010). Ultimately, the garden city movement was not widely adopted, though many of the communities that were created were considered successful. Some blamed the lack of political support for new modes of urban planning and the rampant culture of individualism. Others suggested that Howard's designs were too rigid to account for the evolving and diverse nature of modern cities (Gillette, 2010; Batty, 2008).

References

Batty, M. (2008). The Size, Scale, and Shape of Cities Science 319, 769–771 (2008).
Gillette, H. (2010). Civitas by Design: Building better communities, from the garden city to the new urbanism. Philadelphia, PA: University of Pennsylvania Press.

In the context of urban sustainability, we need to ask a different set of questions regarding the relationship between cities and agriculture. Following are questions we pose that planners, city officials, and others interested in urban sustainability should be asking. In your vision of a sustainable city, how would the food system be designed to meet the diverse needs of city residents? Does it make sense to bring food production sources closer to consumers? If so, to what extent is agriculture an appropriate use of city space and other potentially scarce resources such as water? Thinking beyond food provision, what other services (or disservices) does agriculture provide within cities? How can we better incorporate the beneficial services that agriculture provides within urban development? Addressing these questions requires us to fundamentally rethink what sustainable agriculture means in an urban context and how we can transition toward it. This is an exciting new area of research and action, where we have more questions than answers.

The rest of this chapter is organized as follows. We begin by discussing how sustainable agriculture has generally been defined. Then we describe three different agricultural systems that currently provide for the food needs of cities. Each system has its own unique characteristics and we critically examine the sustainability implications of each of these systems. Next, we broaden the discussion to think about agriculture as not only providing food but also other kinds of services and disservices. Lastly, given this broader understanding, we discuss how we might integrate agriculture into urban planning and development to better design future cities.

What Is Sustainable Agriculture?

Many definitions of the term sustainable agriculture exist today. One particularly useful and overarching definition represents the legal definition of Sustainable Agriculture addressed by Congress in the 1990 "Farm Bill"

"An integrated system of plant and animal production practices having a site-specific application that will, over the long term:
- Satisfy human food and fiber needs;
- Enhance environmental quality and the natural resource base upon which the agricultural economy depends;
- Make the most efficient use of nonrenewable resources and on-farm resources and integrate, where appropriate, natural biological cycles and controls;
- Sustain the economic viability of farm operations; and
- Enhance the quality of life for farmers and society as a whole."

Embedded within this definition we can clearly see the core elements of sustainability: meeting social, environmental, and economic needs over the long term. Achieving agricultural sustainability, however, is much easier said than done and may involve making difficult tradeoffs. For example, providing low-income populations with affordable fruits and vegetables may come in direct conflict with maintaining economic viability of farms. In the next section, we discuss the sustainability of the different agricultural systems that provide food for cities.

Food supply chain
A network of activities related to food moving from production to consumption.

Lifecycle approach
A holistic view which takes into account the entire process associated with a product or service from beginning to end, including the inputs and outputs involved as well as how it is disposed if applicable.

Agricultural Systems for Feeding Cities

Much of our food takes a very complex path from farm to plate. Examining the **food supply chain** from a **lifecycle approach**, we understand that the process begins with production on a farm. From there, food is processed, packaged, and distributed either directly to consumers or to retailers. Consumers purchase the food and then prepare and consume it, disposing of the waste. This process sounds simple enough, but at each step along the way, there are different stakeholders, decision makers, and regulations involved.

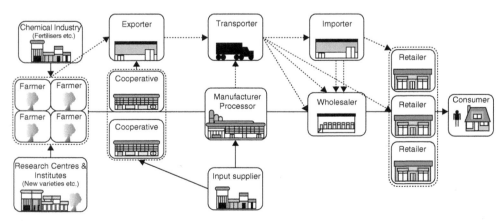

The Food Supply Chain.

From Supply Chain Management: An International Journal by Georgios I. Doukidis, A. Matopoulos, M. Vlachopoulou, et al. Volume 12, Issue 3, Pages 177–186 (May 08, 2007). Copyright © 2007 Emerald Group Publishing Limited.

A food system is the chain of activities connecting food production, processing, distribution, consumption, and waste management as well as the associated actors and regulatory and institutional environments. The sustainability implications of the three dominant systems that provide food for cities, **industrial food systems, organic** food systems, and local food systems, are examined below and compared in terms of the sustainability criteria outlined above.

© Studio 1a Photography/
Shutterstock, Inc

The Industrial Food System

Today, most of us participate in what we call the industrial food system. This is the system we all know and love (or hate). So, let's take a moment and un-package it. How much do we really know about all the things we put in our grocery cart?

Over the past century, the food system has undergone a dramatic shift. Development of high-yielding hybrid crops, widespread irrigation, application of synthetic fertilizers and pesticides, and the increasing use of machinery to replace human and animal labor all contributed to an increase in agricultural yields. Farms focused on efficiency, with each specializing in producing a limited set of crops to streamline their efforts and maximize yield. These practices spread throughout the globe in what is now known as the "**Green Revolution**."

Four-fifths of the world's food production now comes from **industrial agriculture** (American Farmland Trust, 1997). This industrial system was initially heralded as widely successful. New techniques doubled the yields of many staple grains and increased the amount of food available for the developing world's growing population. This brought the rate up from below 2,000 calories per person in the early 1960s to more than 2,500 calories per person by the mid-1980s (Arnould & Thompson, 2005). Would this have been possible without industrial agriculture? It would be hard to imagine feeding the 54.5 percent of the world's population, or over 4 billion people, without some sort of mass food production. However, from a perspective of sustainability, we need to ask what the broader environmental, social, and economic impacts of the industrial food system are. What are the advantages of this system? What are the disadvantages?

Economic Impacts—Rise of Large Farms and Agribusiness

Green Revolution technologies were expensive and required substantial capital investment upfront. Despite widespread government subsidies for agricultural inputs such as fertilizers, pesticides, and water, the additional costs were too much for many small farmers. In the 1950s and 1960s, the number of farms in the US was reduced by half. Meanwhile, large farms thrived and consolidated with the average US farm size nearly doubling (Barney & Worth, 2008; Barrett, 2010). These trends have continued. Since 1982, there has been a 40% decrease in the number of small and mid-sized farms (Barrett, 2010). As large farms increased in size and market share, they outgrew the regional and local markets and began integrating horizontally (i.e., with other farms) or vertically (i.e., with processors,

Industrial food system The way we eat and how the food is created in mass quantities to supply that demand.

Organic Foods produced not using synthetic pesticides or chemical fertilizers, additives or solvents.

Green Revolution A series of changes to global agricultural practices from the 1930's to 1960's that focused on the use of fertilizers, pesticides, and selected crop varieties to increase crop production, especially in developing countries.

Industrial agriculture The system of chemically intensive food production developed in the decades after World War II, characterized by large-scale single-crop farms and animal production facilities.

Agribusiness All of the business and the range of activities involved in food production including farming, seed supply, farm machinery, chemicals, fertilizers, pesticides, transportation, processing, distribution, and retail sales.

Ecological footprint Measures the amount of biologically productive land and water area required to produce all the resources an individual, population, or activity consumes as well as the land area needed to absorb the waste they generate, with consideration for prevailing technology and resource management practices.

Food miles Distance food travels from production to consumer.

Eutrophication Excessive nutrients in a lake or other body of water, frequently due to runoff from the land, which causes a dense growth of plant life and death of animal life from lack of oxygen.

Biodiversity Biodiversity is the variability among living organisms from all sources, including terrestrial, marine, and other aquatic ecosystems and the ecological complexes of which they are part; this includes diversity within species, between species, and of ecosystems.

Food insecure State of lacking access to enough food to support an active, healthy lifestyle at all times.

distributors, retailers) so they could operate at national or international scales. Over time, this integration has led to the rise in **agribusiness**—large multinational corporations that now control much of the food system worldwide.

Environmental Impacts and the Ecological Footprint

With this expansion, the global food system's **ecological footprint** has increased significantly. Industrial, or conventional, food systems require extensive use of nonrenewable fossil fuel resources for use in agricultural chemical production, on-farm machinery, processing, packaging, and transportation. In the US, agriculture is estimated to account for approximately 17% of the country's fossil fuel consumption. Recent studies suggest that US food travels approximately "4200 miles total in its life cycle"(Born, & Purcell, 2006) and the average American meal contains ingredients from at least five different countries (Bryld, 2003). "**Food miles** contribute to a highly inefficient food system, which produces 1 unit of food for every 7.3 units of fossil fuel energy input." (Bryld, 2003) Global estimates of total agricultural greenhouse gas (GHG) emissions range from 20–22% of all anthropogenic GHG emissions (Canfield, Glazer, & Falkowski, 2010). Agriculture also accounts for over two-thirds of water extracted from lakes, rivers, and aquifers worldwide. (Cofie, van Veenhuizen, & Drechsel, 2003).

From 1960 to 2000, US use of synthetic nitrogen fertilizers increased 800%. Yet, it is estimated that approximately 30–80% of the nitrogen applied to crops is never taken up by the plants themselves. Much of it ends up in streams and lakes, where it can lead to **eutrophication**. Eutrophication encourages algae to bloom due to the sudden influx of nutrients. When the algae die, the bacteria that decompose them consume large amounts of oxygen, triggering a chain reaction of dieoffs that can ripple across the entire aquatic ecosystem. For instance, fertilizer runoff from farms along the Mississippi River flows into the Gulf of Mexico, generating a "Dead Zone" of approximately 5,000 square miles (Colding, 2007). Excessive and improper use of agricultural chemicals has also led to loss in **biodiversity** and severe impacts on human and animal health. Biodiversity has also suffered due to a shift to industrial agriculture, which focuses on genetically altered, limited diversity seeds grown in large monocultures.

Food Security, Nutrition and Health

Despite a 25% increase in the amount of food available per person worldwide and a doubling in US food production, many go without adequate food. In 2015, nearly

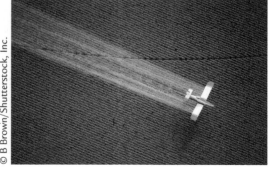

© B Brown/Shutterstock, Inc.

13% of US households were **food insecure** – meaning that they did not have access to enough food for an active, healthy lifestyle at all times throughout the year (Conner, Colesanti, & Smalley, 2010). The urban poor and minorities feel the worst of the effects. While the Green Revolution was able to increase the total amount of food available, it could not address people's inability to afford or otherwise access that food (Connor, 2008; Conway, 1997).

Growing concerns about obesity and other diet-related diseases, have led to research on how food intake may be influenced by the access that consumers have to different food outlets (i.e., supermarkets, convenience stores, and fast food outlets) in their neighborhood. Recent research has indicated that in some low-income and minority neighborhoods, residents have limited access to affordable nutritious food because they live far from a supermarket or large grocery store and do not have easy access to transportation. These areas are referred to as food deserts (Powell, Slater, Mirtcheva, Boa & Chaloupka, 2007; Glanz, Sallis, Salens, & Frank, 2007; Larson, Story, & Nelson, 2009;). Box 2 provides further details on the varied definitions of food deserts, their prevalence within the US, their impacts, and the various policy and non-policy options being explored to address the problems they pose.

Box 2: Food Deserts—Definitions, Prevalence, Impacts, and Action

The term 'food desert' was reported to be first used in the early 1990s in Scotland by a resident of a public housing sector scheme (Cumins and Macintyre, 2002). Since then a number of different definitions have been proposed. The US Department of Agriculture (USDA), has defined a food desert as "a census tract with a substantial share of residents who live in low-income areas that have low levels of access to a grocery store or healthy, affordable food retail outlet" (USDA, 2014). The USDA defines low access as "based on the determination that at least 500 persons and/or at least 33% of the census tract's population live more than one mile from a supermarket or large grocery store" (USDA, 2014).

> **Food deserts** Low-income neighborhoods in rural and urban areas that are more than one mile from a grocery store. This equates to 23 million Americans, including 6.5 million children.

A study conducted by USDA in 2008-2009 estimated that 23.5 million people (or 8.2 percent of the total US population) lived in food deserts (USDA, 2014). For residents in these areas, convenience stores and other small grocery or corner stores may be more common than supermarkets. These smaller stores generally stock energy-dense unhealthy foods and little or no produce. They often charge more for the healthier foods that are available as well (Walker et al., 2010). Poverty, lack of access to transportation, poor development of public transportation, high crime rates, and low awareness about nutrition further compound this problem.

There are several different theories on how food deserts have come into existence. The most popular explanation pertains to the changes in demographics in large US cities between 1970 and 1988, when economic segregation became very prominent. This period saw the migration of affluent households from the inner city to suburban neighborhoods. This led to a decline in median income in the inner city (Walker et al., 2010). This was also the time at which supermarket chains were growing. These chains found the suburban areas more attractive because of larger availability of space and greater demand and moved with the affluent households. This left a void in the inner city as smaller stores had been driven out of business when supermarkets had first moved in. Other theories view food deserts as yet

(Continued)

Box 2 Food Deserts—Definitions (*Continued*)

another manifestation of the process of deprivation of inner city neighborhoods which can be traced back to structural problems of poverty and racial segregation, declining demand for low skilled workers, high crime rates, and zoning laws.

What can be done to mitigate or eliminate food deserts? At the policy level, The Food, Conservation, and Energy Act of 2008, also known as the 2008 Food Bill, implemented by the USDA prioritized the need to address domestic food accessibility and nutrition issues. The subsequent 2014 Farm Bill authorized funding for the Healthy Food Financing Initiative (HFFI) to provide start-up grants and affordable loan financing for food retailers, farmers' markets and cooperatives that sell and deliver healthy goods to "food deserts" (CDC, 2014). However, while the bill sought to improve the retail environment for nutritious food, it also cut, by around $8 billion, the food stamp program that 15 percent of Americans relied on to purchase food, thus limiting its effectiveness (FRAC, 2014).

Several states have enacted legislation to attract full-service grocery stores and supermarkets to underserved communities and improve the quality of the food sold at small corner stores (FRAC, 2014). In addition to the direct health benefits of such initiatives, communities may also realize indirect economic benefits, including job creation and community-wide revitalization. An often-cited example of these direct and indirect benefits is the The Pennsylvania Fresh Food Financing Initiative, which helped develop supermarkets and other fresh food outlets in 78 underserved urban and rural areas, through a successful statewide public-private initiative (Treuhaft and Karpyn, 2010). While improving access to healthy food is critical, it is not sufficient by itself. Change in behavior and habits towards consumption of healthier food is also required.

Case Study: Phoenix Metropolitan area

USDA has developed a web-based Food Locator Tool (http://www.ers.usda.gov/data/fooddesert), which enables users to view a map of census tracts that qualify as food deserts. Users can scan the map, zoom into an area, and download statistics on population characteristics of a selected tract. Using this tool, you can map the food deserts in your area. For example, the map below was constructed using this tool. It shows food deserts in Maricopa County, which encompasses Phoenix and other neighboring towns that comprise more than half of Arizona's population. Based on the Locator's statistics, approximately 57 percent of the population in Maricopa County has low access to a supermarket or large grocery store.

The availability and access to food, or food security is an issue that Arizona is addressing through assistance programs, such as Supplemental

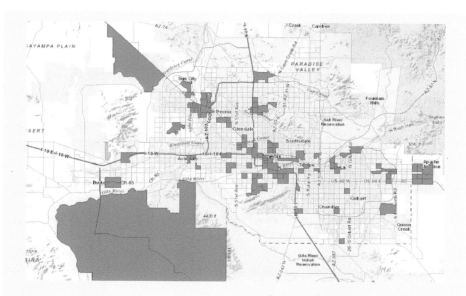

Nutrition Assistance Program (SNAP), Nutrition Assistance (formerly the Food Stamp Program), Coordinated Hunger Program, and the Emergency Food Assistance Program. Other initiatives being promoted by community activists as part of a holistic community food system include:

- encouraging farmstands and community-supported agriculture programs;
- increasing the stock of fruits, vegetables, and other healthy foods at neighborhood corner stores or small groceries;
- growing food locally through backyard and community gardens;
- improving transportation to grocery stores and farmers' markets; and
- promoting education about nutritional value of different food options.

References

Centers for Disease Control (CDC) (2014). State Initiatives Supporting Healthier Food Retail: An Overview of the National Landscape. Retrieved July 7, 2014 from http://www.cdc.gov/obesity/downloads/healthier_food_retail.pdf.

Cummins S, Macintyre S. (2002). "Food deserts - evidence and assumption in health policy making." British Medical Journal 325(7361): 436–8.

Food Research and Action Council (FRAC, 2014). Farm Bill 2014. Retrieved July 7, 2014 from http://frac.org/leg-act-center/farm-bill-2012/.

Treuhaft S. and A. Karpyn (2010). The Grocery Gap: Who Has Access to Healthy Food and Why It Matters. PolicyLink and The Food Trust.

United State Department of Agriculture (USDA) (2014). Food Deserts. Retrieved July 7, 2014 from https://apps.ams.usda.gov/fooddeserts/foodDeserts.aspx.

Walker, Renee E., Christopher R. Keane, Jessica G.Burke (2010). "Disparities and access to healthy food in the United States: A review of food deserts literature." Health and Place 16: 876–884.

Food waste

While some people struggle with food insecurity and food deserts, a completely different problem, food waste, occurs where there is excess that is spoiled, thrown away, or otherwise not consumed. Food waste is one of the biggest challenges for the sustainability of the US food system, as about 30% of food produced is wasted. Estimates are that nearly one-third or more of food is actually wasted because it is not eaten, spoils en-route, spoils with the end-user, spoils during storage or is simply discarded by processors/pickers because it is not the perfect specimen or is approaching an arbitrary sell-by date (Conway, 1997; Conway, 1998).

Long-Term Resilience

Long-term resilience
The ability of a system to maintain its functionality over an extended period of time despite impacts from external stressors or events.

Reliance on distant food sources can undermine local resilience by increasing regional vulnerability to shocks from fluctuating markets, mass-scale food contamination, rising fuel costs, and breakdowns in the global transportation system. It can also decrease communities' capacity to respond to such stresses. Global food prices rose by 80% between 2006 and 2008, significantly affecting the household budgets of urban dwellers (Dimitri & Oberholtzer, 2009). Due in part to the large separation between consumers and producers, people are often unaware of the impacts of their purchasing decisions. This makes change difficult. There have been some attempts at agricultural reform, however. One of these reforms is the organic food movement, which seeks to address some of the negative effects associated with the industrial food system.

Organic Food

The organic food movement began as a much more holistic and transformative endeavor than is sanctioned by current organic certification processes. Sir Albert Howard (1873–1947) is generally regarded as the founder of this movement. But his philosophy of agriculture was much broader than the organic-inorganic debate he has become known for today. He advocated proactive measures to prevent soil erosion, raising mixed crops (polyculture), integrating livestock farming with crop farming, recycling plant and animal waste back to the soil in the form of compost, and rainwater harvesting. Despite his call for a more holistic view of agriculture and his followers' later advocacy toward understanding the farm as an "organic whole," the debate over chemical applications was what galvanized the organic movement (Dubbeling, 2009). Barry Commoner's, *The Closing Circle: Nature, Man, and Technology*, and Rachel Carson's, *Silent Spring* that suggested the long-term hazards of pesticide applications for humans and other non-human living organisms, were instrumental in bringing these concerns to the forefront.

So, what exactly does "being organic" mean? According to the US Department of Agriculture's National Organic Program (NOP) regulations, organic crops are grown and processed

How Sustainable Is Organic Food?
Organic agriculture is grown without the pesticides or petroleum/sewage-based fertilizer being used.

© Malivan_Iuliia/Shutterstock, Inc.

without most conventional pesticides, petroleum-based fertilizers, sewage-sludge-based fertilizers, genetically modified organisms, or ionizing radiation (Eaton, Hammond, & Laurie, 2007). Animals raised organically must be fed only organic feed, receive no antibiotics or growth hormones, and have access to the outdoors.

For several decades, organic agriculture has served as the poster child of the broader sustainable food movement. In fact, many consumers equate organic with sustainability as far as food is concerned. In recent years, organic food has increased dramatically in popularity. In 1992, the US was home to fewer than 1 million organic acres; by 2008, it contained over 4.8 million acres of certified organic farmland (Economic Research Service [ERS], 2008). Consumers have been found to purchase organic products primarily for the following reasons: health, taste, environmental benefits, food safety concerns, animal welfare concerns, support for small and local farms, wholesomeness, agricultural heritage, and trendiness. So how well do our desires match up with reality? How sustainable is *organic*? Let's take a look at some of its key impacts to the environment, society, and the economy.

Reduction in Synthetic Chemicals—Pros, Cons, and Gray Areas

The core purpose of the organic certification process is to minimize the amount of harmful synthetic chemicals that are used in the production and processing of our food. Here is where organics truly meet sustainability goals. The rejection of synthetic fertilizers forces farmers to re-think the way they interact with soil. Soil health may be improved using organic fertilizers and **nitrogen-fixing cover crops** as well as innovative techniques such as low-tillage, crop rotation, and **inter-cropped polyculture**. Many organic farmers employ a combination of these procedures to maintain soil health, but proper management of nutrients still remains a challenge. For instance, recent research suggests that organic inputs tend to have adequate nitrogen but too little phosphorus and potassium, leading to a slow depletion of these key nutrients over time on organic farms (Economic Research Service, 2017).

Lower chemical pesticide use translates to fewer non-target plant and animal species being killed by potent chemicals. Studies show that biodiversity and species abundance and richness, in particular, tends to be significantly higher on organic farms (Edwards-Jones et al., 2008). Fewer pesticide applications also means less pesticide exposure for farmers and consumers. In the US, an estimated 300,000 **acute pesticide poisonings** occur each year. Worldwide, this figure may be as high as 26 million (FAO, n.d). Past studies have found that chemicals present in pesticides are known or suspected carcinogens, endocrine disruptors, and neurotoxins. Intuitively, these seem like good things to keep away from something so intimate to us as our food. Complicating the matter, however, is the fact that many organic products are grown in close proximity to nonorganic ones. This can result in pesticide-drift and contamination of otherwise organic products. It is also important to note that although under NOP organic regulations most synthetic substances are banned, there is a long list of allowed synthetic substances.

Organics—Energy Reduction?

Organic farms don't use fossil-fuel intensive pesticides and fertilizers, but depending on the crop and the level of mechanization on the farm, fewer pesticides can mean more mechanical weeding. Therefore, some of the savings may be lost.

Nitrogen-fixing cover crops Crops planted, also known as catch crops, to retain soil nitrogen that is then released back into the soil and absorbed by other nitrogen needing crops.

Inter-cropped polyculture Growing several different crops in one area creating individual growing patterns and higher return.

Acute pesticide poisonings Extensive use and exposure of pesticides and agro-chemicals in local agriculture and food production primarily in low- to middle-class communities due to the commercialization and globalization of agriculture.

Lifecycle assessment A method to account for the environmental impacts associated with a product or service. Originally used in a manufacturing context, it has been applied to agriculture to calculate associated environmental impacts.

A **lifecycle assessment** of the US food system as a whole indicates that only about 8% of the energy use of the entire system is embodied in the production of chemical fertilizers and pesticides. Some suggest that because organics avoid these chemicals, energy inputs per acre can be 30%–50% lower for organic crops (Feagan & Morris, 2009; Food Agricultural Conservation and Trade Act, 1990; Food Agriculture Organization, 2006).

Organics—More Nutritious?

Many perceive organic foods to be more nutritious. Is this really the case? Currently, the research is inconclusive. Studies tend to indicate that vitamin C concentrations are significantly higher in organic produce. However, a recent review of 39 studies comparing organic and conventional produce found that results can vary widely depending on the types of nutrients and the specific crop. For example, vitamin C concentrations were higher in organic tomatoes than conventional ones, but for carrots and potatoes, the reverse was true. Studies reveal no consistent, significant differences between conventional and organic products in terms of many other nutrients, such as vitamin B, vitamin A or beta-carotene (Glanz, Sallis, Salens, & Frank, 2007; Goland, 2002).

Organics—Lower Yield?

One critique levied against organic agriculture is that organic agriculture results in lower yields. A comprehensive meta-analysis that examined the relative yield performance of organic and conventional farming systems globally confirmed that, overall, organic yields are typically lower than conventional yields (Grace, Grace, Becker, & Lyden, 2007). They found, however, that these yield differences are highly contextual, depending on system and site characteristics, and ranging from 5% lower organic yields (rain-fed legumes and perennials) to 34% lower yields (when conventional and organic systems are most comparable). Further studies have found that agricultural diversification practices, multi-cropping, and crop rotations substantially reduce the yield gap when applied in only organic systems (Grey, 2000; Griffin & Frongillo, 2003).

Beyond the yield gap, lower yields may be compensated in part by savings on chemical fertilizers and pesticides. However, these savings may be potentially offset by the higher costs for seeds, labor, and machinery. Thus, price premiums are a key factor in ensuring that farms profit from their organic venture. Some research suggests that a premium of 10% above conventional products may be necessary to ensure comparable farm profits. This has not been a problem in the past because some consumers appear to be willing to pay up to 180% more for their organic products(Guthman, 2008; Hand & Martinez, 2010; Heckman, 2006).

More problematic is the lingering question of whether or not organic agriculture can feed the world. Critics argue that with 20% lower yields, up to 25% more land would need to be cultivated to supply the same amount of food. Environmental arguments for curbing the expansion of agricultural land aside (i.e., deforestation, biodiversity loss), include that there may simply not be enough land nor organic fertilizer available to feed the world organically (Heimlich & Anderson, 2001; Heller & Keolian, 2003). Additional research needs to be done to determine how much yields could be improved by changing on-farm management practices. Could a shift to organic agriculture increase agricultural land use and heighten food insecurity worldwide? It's possible, but it is also important to realize that food insecurity isn't just about availability of food—it's about the lack of stable access to healthy, usable food. This brings us to the next critique of organic agriculture, its inaccessibility due to cost.

Organic Food's High Price Tag: Are Industrial Organics a Solution?

People who try to fill their shopping cart with organic products soon realize that these products are much more expensive than their conventional counterparts. The higher cost and relative unavailability of organics significantly diminishes the movement's potential to meet the dietary needs of mid-to low-income populations. The good news is that many organic products have begun to infiltrate our mainstream supermarkets, at increasingly affordable prices. Much of this is due to the entry of large-scale farms into the organic market. These farms are able to achieve **economies of scale** and levels of efficiency that many small farms are incapable of, and as a result, can pass the savings on to their customers.

Economies of scale Efficiency based on large volume production that results in decreased cost per unit.

However, these large organic farms don't come without a cost. For many of us, the term "organic agriculture" conjures up images of small, pristine, family farms where farmhands work the good soil with nothing between them and the earth but their bare hands, or perhaps a shovel. For much of the food labeled USDA Organic, this is far from the truth. Organic farms that are 500 or more acres control over 60% of the organic US farmland. Furthermore, the size of organic farms is trending upward. In 1997, organic farms averaged 268 acres; in 2005, this figure was 477 (Heller & Keolian, 2003).

Many suggest that the creation of the "organic" label has done little to transform the industrial food system. This could be seen as both a blessing and a curse. Proponents of large-scale organic farming argue that by working within the industrial system, it is easier to convince farmers to adopt the practice. Critics argue that large-scale agribusiness has simply created a new forum in which business-as-usual agricultural practices may occur, albeit with less pollution and a lot more profit—a world of **industrial organics**. In 2008, approximately half of all organic purchases were made in mainstream supermarkets and big-box stores (Heller & Keolian, 2003). These stores require large volumes of consistent product and small-scale organic farms quickly get outcompeted.

Industrial organics Animals or crops that are raised or grown according to minimum standards for organic certification on a large scale, focusing on efficiency.

To summarize, organic regulations restrict chemical use, which means reductions in fossil fuel use and **embodied energy**. Overall, lower chemical use seems to translate to greater energy efficiency of organic crop production, better soil health, improved biodiversity, and less risk of human exposure to pesticides. Yet, there are tradeoffs. Yields are often lower in organic fields than conventional ones, thus requiring more land to produce the same amount of food. This becomes a sustainability problem since bioproductive land is decreasing at the global scale. For a variety of reasons, organic products are typically far more expensive than conventional ones, limiting their capacity to serve low-income populations. Finally, organic agriculture, for better or worse, does little to challenge the mainstream, globalized system of industrial agriculture in which small-scale farms lose out.

Embodied energy Energy used to produce/create a product during its entire lifecycle.

Thus, we need to ask ourselves if this is all we want from a sustainable agricultural system? What about other environmental issues such as water use, GHG emissions, soil erosion, nutrient runoff, deforestation, and biodiversity loss? What about farm livelihoods, the loss of small-scale farms across the US, and the domination of the industry by large corporate agribusiness? What about the inequities in our food system, rising obesity, rising costs of nutritious food, and the many hungry and food insecure in our midst? On all these issues, the organic movement as we see it today is silent. These are the principle questions of sustainable agriculture and cities for which we need answers.

Local Food

There is a growing argument that both the industrial and organic models of agriculture fail to address the bulk of the world's food system problems. Both models operate primarily within a structure of an intensely globalized agricultural system in which producers and consumers are distanced, and local regions remain highly vulnerable to global shocks. **Local food systems**, on the other hand, seek to minimize the distance from farm-to-plate and often use direct marketing approaches, where farmers sell directly to consumers. When talking about cities, local agriculture is often called **urban and peri-urban agriculture (UPA)**, which takes place inside the city itself or in the periphery. UPA includes horticulture, aquaculture, arboriculture, and poultry and animal husbandry. It can be found in the form of greenbelts around cities, community gardens, farming at the city's edge and on vacant inner city lots, fish farms, farm animals at public housing sites, municipal compost facilities, schoolyard gardens and greenhouses, restaurant-supported salad gardens, backyard orchards, rooftop gardens and beehives, window box gardens, and much more (see Box 1).

In the US, agriculture in metropolitan regions accounts for about 33% of total crop sales and 61% of the production of fruits and vegetables by acreage (Henneberry & Agustini, 2004). The **local food** trend is growing. In 2013, the US was home to more than 8,144 farmers' markets (ERS, 2017) and at least 2,500 Community Supported Agriculture (CSA) programs. Direct sales of farm products, including **direct-to-consumer** (DTC) and **intermediated marketing channels**, have increased from $812 million in 2002 to $6.1 billion in 2012 (Low et al., 2015).

What Is Local Food?

Defining this term is more complex than one might guess at first glance. Unlike organic or fair trade, there is no formal, overarching certification system that defines what makes a product *local*. The definition of local, therefore, varies from region to region and person to person. Geographical or political boundaries such as "state" or "county" are often used by academics, as datasets are readily available at this level. These types of definitions may also tend to be employed by retailers for marketing purposes (Hill, 2008). However, many consumers tend to conceive the term "local" in narrower terms expressed in "food miles" from their home—for example, a 50- or 100-mile radius (Hinrichs, Gillespie, & Feenstra, 2004; Hoefkens, 2009). Consumers also often inherently associate particular values with the term *local*. Often implicated in people's definition of local are attributes associated with *who* produced the food, *how* they produced it, or perhaps *where* the food was *purchased*. According to these definitions, large corporations and agribusinesses, or

Local food systems
Food production within a given geographic area that minimizes food miles and has increased use of direct marketing approaches.

Urban and peri-urban agriculture (UPA)
Growing, harvesting, producing, and consuming food in a city, town, or metropolis or its periphery.

Intermediated marketing channels Marketing opportunities in the local supply chain other than DTC transactions such as farmers selling to grocers, restaurants, or regional aggregators, i.e., food hubs. Channels also include direct sales to food services at schools, universities, hospitals, etc.

Direct-to-consumer
Marketing channels that engage consumers in face-to-face market transactions at roadside stands, farmers' maketes, pick-your-own and on-farm stores, and CSAs.

© Malivan_Iuliia/Shutterstock, Inc.

Rooftop Garden

farms without environmentally sound practices, may be excluded from consideration as truly local. The definition of local may be even further constrained to include only products that are bought directly from producers or those purchased from particular outlets branded as local (Hole et al., 2005).

Why Go Local?

Consumers buy local food for many reasons. Some of the most commonly cited include: improved quality, freshness and taste, environmental concerns, nutritional benefits, a desire to support local farms and economies, direct purchasing from farmers, obtaining organic products, and gaining a sense of knowing where their food came from (Hoppe, MacDonald, & Korb, n.d; Hunt, 2007). So, again, we must ask how well these expectations match up with reality. How sustainable is local food? Without clearly defined standards, the answer is often "*it depends.*" It depends on the specific practices of the particular farm or local food outlet in question and whom you are asking. So, let's take a look.

Is Local Food Sustainable?

Urban demand for local produce can provide profitable avenues for metropolitan area farms to maintain a sustainable livelihood. But, direct marketing isn't for everyone. Only 6.2% of farms in the US participate in direct marketing (Jarosz, 2008). So, why don't more farmers participate in local food systems? Research suggests a variety of reasons, most of which have to do with our mass conversion to industrialized, specialized agriculture. For many large farms, selling products exclusively at a local level is no longer feasible. They produce in quantities beyond what the local markets can consume. Their use of machinery, fertilizers, and pesticides allows them to produce large quantities with great efficiency. Working on a small-scale farm can be extremely hard work, with a great deal of physical labor involved. This may leave little energy for gathering their products, taking them to market and engaging with consumers and/or buyers. In addition, marketing directly to consumers is time-consuming and many farmers describe feelings of "burnout" associated with producing and marketing their products locally.

Competition with the 24-7, one-stop-shopping, highly subsidized supermarket experience is difficult as well. Americans are used to getting what they want, when they want it, and many don't want to be restricted to the seasonal produce that local farms have to offer. Because of this, local farmers have to spend extra time and effort educating the public and seeking out niche markets for their products. So, why don't small-scale farms sell their products at regional supermarkets? Supermarket chains—as well as large-scale foodservice, restaurants, and catering companies—typically only buy from large farms that can supply consistent and substantial quantities of products to all their stores (Kirchmann et al., n.d; Kirchmann, Bergstrom, Katterer, Andren, & Andersson, 2008). They tend to purchase products from regional food distribution farms because of market security and diversity of products.

Many people report that they buy local to support their local economy. Does buying local improve the local economy? Unlike supermarket purchases, where much of the profits "leak" to distant corporate headquarters and middle men, money spent on local food is more likely to go to a local farmer or retailer, who in

Local foods effect the local economy to a greater extent and more quickly than supermarket purchases. Local food outlets such as farmers' markets may create new job and volunteer opportunities, both directly and indirectly. One study calculated that for every person employed at an Oklahoma farmers' market, an additional 2.44 jobs were created throughout the state (Larson, Story, & Nelson, 2009).

turn, is more likely to spend it in the local region. It is estimated that for each dollar spent on local food, more than double this amount is re-circulated in the local economy (Kneafsey, 2010).

In addition to economic benefits, noncommercial forms of urban agriculture, including household and community gardens, play a particularly significant role in contributing to health and nutrition, especially in food-insecure communities. In developing countries, where poor, urban residents are estimated to spend 60–80% of their meager incomes on food, a few homegrown crops can go a long way to meeting nutritional needs while freeing up financial resources for other investments (Lin, Smith, & Huang, 2008; Low et al., 2015). Household gardens can also be vehicles of empowerment for women living in regions of the world that still discourage them from seeking a job or participating in economic activities. So, it would seem that for some farmers, and in terms of the local economy, a local food system can contribute a lot. But, what about the rest of the world? What happens when we choose to buy locally rather than import our products from somewhere else?

Defensive Localism—A Cautionary Tale

Defensive localism
Aversion to agriculture products not produced within the community.

Buying something produced locally sounds like a simple idea. But, as you start unpacking what it really means in terms of sustainability, things can get complicated. What if a better, more sustainable, product could have been sourced non-locally? Buying local for local's sake may be at best lazy sustainability, and at worst, irresponsible. Thus, scholars argue, we need to be careful not to fall into the "local trap"—assuming that local is better without looking more closely at the issues. Assuming that local is better can lead to a sort of **defensive localism**, in which nonlocal products are shunned without a second glance. Along with this can come an array of cascading effects, some potentially positive, others negative. As we have seen, there are local economic benefits to buying local. But, this may be at the expense of a job in another community elsewhere. Furthermore, the local scale may be just as susceptible to power imbalances, injustices, and irresponsible practices. In this sense, the local label can run the risk of appearing to embody a set of sustainability principles and values that inherently have, in fact, very little to do with the local scale (Lyso & Guptill, 2004; Mariola, 2008; Martinez et al., 2010). Let's take a deeper look at some of the other factors that need to be considered when thinking about sustainability.

Local Food—Better for the Environment?

The use of transportation that delivers food from farm to plate creates Food Miles, which average between 1,500 and 2,000 miles for what ends up on our plates.

Is eating local better for the environment? One widely used argument for local food is that it travels less distance from farm to plate. Many argue that fewer *food miles* entails less fossil fuel use and lower GHG emissions. While this seems to make sense intuitively, reality is much less straightforward. What *is* well established is that in the conventional system, food travels a long way to get to our dinner plate. Most studies peg this number at an average of 1,500–2,000 miles for US consumers (Martinez et al., 2010).

Things get complicated, however, when you start to factor in the *type* of transportation used to deliver food from farm to plate. Personal vehicles and small trucks used by farmers and consumers to transport food on a local scale have lower capacity, and therefore, transport fewer pounds of food per gallon of gas consumed. This is true even if these vehicles have relatively good fuel efficiency. It comes down to economies of scale. When carrying substantial amounts of food, large refrigerated trucks can actually be more fuel efficient on a per-pound basis.

When considering the energy impact of local food, transportation isn't the whole story either. In fact, the energy used in food miles to transport food from farm to plate only accounts for 14% of the overall energy consumption of the food system (McMichael, Powles, Butler, & Uauy, 2007; Milestad, Westberg, Geber, & Björklund, 2010). If we truly care about reducing the fossil fuel use and GHG emissions of our food purchases, we have to think about the big picture and perform a life cycle assessment.

Different regions and different farms may produce food with far less fuel consumption and far fewer GHGs emitted per unit of product than others, giving them a comparative advantage over local regions. This may be attributed to a variety of factors, such as soil type, climate, farm-management practices, and the nature of processing and packaging. The question becomes then, how much energy is embodied in the product as a whole by the time it hits your refrigerator? One study revealed that tomatoes purchased locally in the UK in the wintertime came from heated greenhouses and used more energy than would have been used to ship them from Spain (Miller, 2007). This demonstrates the importance of considering seasonality, crop yield, regional geography, and on-farm management practices before universally labeling local food as having lower energy use and a smaller carbon footprint. It must be stressed here that it is important to calculate energy use per unit of farm product produced. Some farms may be more productive than others, and therefore, though their energy consumption overall may be greater, they yield more crop per unit energy input.

It's also important to remember that energy use is just one of many indicators that we could use to assess the impact of a local food product on the environment as compared to a non-local option. There are many other aspects to consider. We must compare local and non-local options in terms of their use of other nonrenewable and scarce resources such as water and phosphorus and potentially detrimental products such as synthetic nitrogen-based fertilizers. We must compare them in terms of their impact on air, water, and soil quality, as well as overall biodiversity and ecosystem health. We must do all of this in terms of assessing not only the farm, but also the entire processing, packaging, and distribution **phases of the food system** as well.

Phases of the food system Production, processing, distribution, consumption and post consumption.

Local Food—Better for People?

Today, one of the key reasons that people in the US turn to local food systems is that, through them, they have a better sense of where their food comes from. Local food systems reconnect producers and consumers and heighten "agricultural literacy" in urban dwellers. They begin to understand what goes into their food, how it's grown, the seasonality of production, and the local environmental characteristics of their region. Because of this increased knowledge, consumers feel that they can make more informed choices when they buy or grow. Furthermore, direct contact with customers and the increased visibility of farms and their practices have been shown to be related to increases in on-farm biodiversity, and environmental and ethical farm management (Miller, 2007; Ministry of Agriculture Botswana, 2006; National Organic Program, 2008).

Local Inequities—Who Gets a Seat at the Local Table?

Who buys local food? Participation in local food outlets is very limited. A growing body of research suggests that local food systems primarily serve the urban elite and that low-income or minority customers are less well served (Nord, Andrews, & Carlson, 2003; Nord, Andrews, & Carlson, 2009). If local food is ever to be the answer to our food sustainability problems, its inequities must be addressed. The reasons behind low-income and minority consumers' lack of participation in local food markets are complex and not well understood. Much research is yet to be done in this field. One lens that can increase understanding of the issue is that of **food security**. The UN Food and Agriculture Organization (FAO) states that for food security to be achieved, food must be available, accessible, usable, and stable (Oberholzter, Dimitri, & Greene, 2005; Peters, Bills, Wilkins, & Fick, 2008). The US has more than enough quality food available, but what about the other factors? We'll address each in turn.

How accessible is local food to consumers? **Food accessibility** encompasses a number of factors—typically aspects of affordability, physical accessibility, and cultural acceptability. Many consumers cite the high cost of local foods as a major barrier to purchasing them. Providing food subsidies to low-income and other vulnerable groups through food stamps and food vouchers, and making these redeemable at local food outlets such as farmers' markets is one way of addressing the affordability issue. But, affordability is not enough to ensure food security.

Local food outlets such as farms, farm stands, farmers' markets, or CSA drop-off locations are not always physically accessible for consumers or access may vary from season to season. Markets must also provide culturally acceptable food. In terms of food, the culture with which we identify can unwittingly play a significant role in what we come to regard as acceptable food and how we utilize it (Pimentel & Pimentel, 2003; Pimentel et al., 2005).

To conclude, just like organic options, local food is not a miracle cure. On a regional scale, local food can bring many benefits. It stimulates the local economy, provides a profitable avenue for many small farmers, and can contribute to increasing resilience and food security. But, we have also seen that local food tends to be unaffordable and inaccessible for many urban residents. We have also discussed that although local food can reduce fossil fuel use, this is not always the case. Similarly, depending on the region, sourcing non-locally could mean savings in water inputs, fertilizer applications, pesticide use, or even improvements regarding issues of social justice, fair wages, or humane treatment of animals. In this light, it's not *local* in and of

Food security The condition in which all people, at all times, have physical, social and economic access to sufficient safe and nutritious food that meets their dietary needs and food preferences for an active and healthy life.

Food accessibility Access by individuals to adequate resources for acquiring appropriate foods for a nutritious diet.

© marlee/Shutterstock, Inc.

Farmers markets that are accessible for consumers are very important for continued sustainability.

itself that is important but rather what is actually happening in the process from seedling to dinner plate. Still, one can argue that it is much easier to know what's happening on the farm and in the distribution process when the farm is local—and, of course, this is one of the major reasons people buy from local farms.

Beyond Food: Ecosystem Services (and Disservices) of Urban Agriculture

Farms can offer urban regions much more than just food. Scholars increasingly recognize the many **ecosystem services** that agricultural lands provide. Essentially, ecosystem services are the benefits people obtain from a given ecosystem, and agriculture can be thought of as a managed ecosystem. Beyond food provisioning, these services include carbon sequestration; waste and nutrient cycling; water runoff management and groundwater recharge; maintenance of soil fertility; educational, recreational and economic opportunities; increased wildlife habitat and biodiversity; reduction in the urban heat island effect; and other microclimate and aesthetic improvements (Pimentel et al., 2005; Pimentel, Hepperly, Hanson, Douds, & Seidel, 2005; Pirog & Benjamin, 2003). We'll explore some of these concepts below as well as some of the potential *disservices* of urban and peri-urban agriculture.

Ecosystem services
The benefits provided by ecosystems. These include provisioning services such as food and water; regulating services such as flood and disease control; cultural services such as spiritual, recreational, and cultural benefits; and supporting services, such as nutrient cycling, that maintain the conditions for life on Earth

City Metabolism—Rethinking "Waste"

Currently, our cities operate in a linear manner. We bring in vast quantities of food and other resources, and emit vast quantities of waste, which is then processed in treatment plants before being released into waterways or being trapped in landfills. There is little active recycling of food products and by-products resulting in a slow, but persistent removal of nutrients—especially phosphorous—from agricultural lands (Ploeg et al., 2009). Phosphorus is a nutrient that is absolutely essential to plant growth. It is also a nonrenewable resource, harvested primarily in the form of rock phosphate. Farmers apply it to their fields as fertilizer to maintain soil fertility. Some of this is lost to runoff before the crops can take it in. The part that plants do uptake is shipped off to cities, embedded in our food. Nearly 100% of the phosphorus humans eat is excreted—mostly in urine. Eventually it ends up in our waterways or sewage sludge in landfills. Only 10% of this is estimated to recirculate back to agricultural lands—a serious problem when you realize that global phosphate rock resources will likely be depleted within the next few hundred years (Postel, 2000; Powell, Slater, Mirtcheva, Bao, & Chaloupka, 2007).

Because of this, many increasingly argue that it is crucial that we begin to think of the **metabolism** (inputs and outputs) of our cities as **circular** rather than linear (Pretty, 2008). Here, the development of closely integrated urban-agriculture systems could help. Sewage water, if properly treated, could be used to fertilize agricultural lands. In fact, many regions in China have had long-standing traditions of successfully recycling waste back onto cropland near urban areas using **aquaculture ponds**. Of course, recycling human waste into a form that can be applied to edible crops does not come without its own challenges. Improperly treated wastewater can increase exposure to diseases such as cholera, typhoid, giardia, dysentery, and more. Furthermore, urban soils may be contaminated

Metabolism The flow of inputs and outputs as well as resulting processes.

Aquaculture ponds Ponds installed and modified to produce grow and harvest fish, other animals, and irrigation of plants.

Growth of cities can destroy ecosystems.

with heavy metals from automobile exhaust and industry (Pretty, n.d). However, there are many examples of such ecological sanitation systems worldwide, which is promising for a move toward more closed-loop, regenerative urban-agriculture systems (Ricketts & Imhoff, 2003).

Wildlife Habitat & Biodiversity

Urban areas have become notorious for their destruction of ecosystems. As cities have grown, they have literally plowed down or paved over much of the world's prime agricultural land and biodiverse habitat(Swinton, Lupi, Robertson, & Hamilton, 2007; Taylor & Aggarwal, 2010).

In the US, much of this land now lies vacant. Urban agriculture provides an opportunity to provide cities not only with food, but also with much needed green space. Many studies show that due to their high plant diversity, backyard and rooftop gardens can provide key urban habitats for native pollinators, birds, and other wildlife (Topp, Stockdale, Watson, & Rees, 2007). However, urban agriculture doesn't always provide vibrant wildlife habitat. Chemical monocultures, heavy tillage, or farms producing vast amounts of nutrient runoff, for example, don't improve soil and ecosystem health or wildlife habitat; they degrade it. So again, we see that local, urban agriculture isn't always a good thing, but if practiced properly, it could be.

Microclimate Improvements

There are a number of ways in which urban agriculture can lead to microclimate improvements. Plants help minimize the movement of dust and air pollution. As plants respire, the water they've taken in evaporates, cooling the surrounding region. In desert climates, this urban heat island mitigation can play a critical role in making the city livable and reducing the amount of air conditioning required. Of course, the tradeoff here is water. For evaporative cooling to work, water is needed and this can be problematic for areas in which water is scarce (Trevwas, 2001).

Integrating Agriculture in Cities: Innovative Urban Design, Policy, and Community Efforts

In the above section, we discussed the various ecosystem services and disservices that agriculture potentially provides in cities. Given this discussion, a central urban design and policy question is whether and to what extent agriculture is an appropriate use of city space and scarce resources such as water. In most places, it has been observed that as a city grows and land prices rise, land held in agricultural production is out-competed by other uses. This is because the market values only the agricultural products (food and fiber). Other ecosystem services—such as microclimate regulation, green spaces, biodiversity, and cultural services—are excluded from most market value calculations.

As a society, we may value these other services for a healthy and resilient city. But, if these services are not calculated in market evaluations, then we either need some sort of policy support or alternative models of urban design and architecture, together with community efforts, that make urban agriculture viable. A wide range of policy instruments have been designed to manage urban growth and protect or promote agricultural production.

Agricultural zoning has been used widely by cities across the world to protect agricultural land use. Land use plans often indicate the areas within the city in which urban agriculture is allowed. Sometimes these plans may also include guidelines from planners on the types of urban agriculture that are permissible. In Botswana, for example, the City of Gaborone has set up poultry zones on land considered of low potential for other land uses (United Nations Environment Progamme [UNEP], 2011).

This vacant lot in an inner-city neighborhood has been transformed into a community garden.

In addition, in the US, legally recognized geographic areas called *agricultural districts* or *agricultural preserves* are designed to keep land in agricultural use. Agricultural preserves differ from exclusive agricultural zoning areas because enrollment in them is voluntary (United States Department of Agriculture (USDA), 2009). Farmers who join an agricultural preserve or district may receive a variety of benefits, such as differential tax assessment, which implies that they are taxed at a lower agricultural value rather than the higher values associated with developed uses (Williams, 2002).

Another popular approach, widely used not only by government agencies but also a large number of private land trusts, is the *purchase of development rights* (PDR). Under PDR, the landowner retains title to the land but voluntarily sells the development rights that prohibit future subdivision and development. Conservation easements are also being increasingly used to delineate environmentally vulnerable lands that then can be used for agriculture. Several municipalities are also exploring the option of promoting *multifunctional land use,* wherein food can be grown in combination with other urban functions such as recreation and city greening.

Another innovative policy option gaining ground in recent years is the use of vacant public and private land for urban agriculture. With the growing "sprawl" into the suburbs, the last couple of decades have seen a common pattern of inner-city neglect in most cities across the US. The Economic Recession of 2008, or the Great Recession as it is often called, has led to the depopulation and rise of abandoned properties in cities. For example, it is estimated that Chicago now has 70,000 vacant parcels of land. Urban agriculture can play a regenerative role by transforming these vacant lands, which are often neglected, to gardens that meet the food needs of urban residents.

The challenge here is negotiating a deal between the owners and potential users of these properties. Without a title or three- to five-year leases, the users risk losing their investment when the land is taken away for other purposes. The temporary nature of use may also make it difficult to secure other resources, such as water.

Urban settings are also better suited to small-scale and organic agriculture that serves the needs of the local community rather than industrial scale conventional agriculture. Space constraints and consideration of neighbors as well as the frequently higher value of urban land mean that a more practical approach is required. This includes easily manageable plots and local customers who understand the seasonality inherent in local farming.

All these policy options and challenges are leading to a great deal of creative thinking around the question of how elements of organic, industrial, and local food systems can be incorporated into cities. Architects are finding ways to fit food production into buildings, for instance, through rooftop gardens (which also provide cooling). Unused spaces such as riverbanks, median strips, schoolyards, and hospital lawns are being incorporated into a broader vision of creating green infrastructure within cities. New citywide coalitions are emerging to advance the goal of food security, both at the household and community level. Health and nutrition advocates are joining with community gardeners, university extension services, emergency food distributors, and city planners to design these new food systems. It is through these collaborations that the potential of urban agriculture in providing different types of ecosystem services can be realized.

Conclusions

In this chapter, we have reviewed several of the key food systems used in urban areas and discussed their sustainability implications. We have seen that much of the food that passes through US cities today comes not from the local region, but from farmlands and pastures hundreds or even thousands of miles away. In terms of sustainability, there may be some sound economic, social, and environmental reasons for sourcing food from far away—perhaps rain-fed agriculture is possible there, the land is more fertile, or the farm is large enough to achieve greater on-farm efficiency. But, there are costs as well. Buying into the globalized, industrial system may mean that more fossil fuels and chemicals were used to produce the food that arrives on our plates. Furthermore, relying solely on distant sources of food may make urban dwellers more vulnerable to shocks in the global system. Climate change, economic crises, food contamination, and increasing fuel costs could all result in price hikes that dramatically reduce the affordability and availability of food within a city.

One increasingly popular response to the many problems with the conventional, industrial food system is to source more food locally—either within or very close to cities themselves. Urban and near-urban agriculture can take on many forms, including commercial farms and markets or community and household gardens. Urban agriculture brings consumers back into close contact with the source of their food and can allow them to make more informed decisions about what they are buying into. It can enhance urban resilience by maintaining a city's capacity to feed itself and providing a number of ecosystem services such as waste recycling, wildlife habitat, economic opportunities, and microclimate improvements.

Chapter 10

Urban Ecology, Green Networks and Ecological Design

Edward Cook

Introduction

This chapter examines how the field of **urban ecology** and planning concepts of **green networks** and **ecological design** contribute to making cities more sustainable. The scientific basis and relevance of urban ecology for sustainable cities is reviewed and an overview of green networks and ecological design is provided with discussion of particular challenges and opportunities of working with this concept in urban settings. The importance of **hierarchy in ecology** and planning for green networks is illustrated through exploration of multi-scalar plans that range from continental to individual local sites. A set of examples are provided that illustrate how urban ecological design helps realize the goal of creating more sustainable cities. Finally, a discussion is provided that clarifies how linking science, policy, planning, and design can ultimately lead to making green networks and ecologically designed sites in cities a reality, providing a foundation for urban sustainability.

Urban ecology The study of the interactions of organisms and their environment in an urban context.

Green networks The system of interconnected patches and corridors that provide and sustain ecological functions and values within human-dominated landscapes.

Hierarchy in ecology The order of interactions and relationships within an environment.

Urban and Landscape Ecology

The scientific field of ecology focuses on the study of interactions between organisms and their environment. Historically, ecologists have largely tended to study organisms within natural ecosystems and undertaking less research in **human-dominated ecosystems** such as cities. With global shifts toward increasing urbanization and the understanding that impacts on the environment are seldom contained locally, urban and landscape ecology have emerged as important topics of increasing interest to ecologists, urban and regional planners and designers, landscape architects, social scientists, and others in recent decades.

Human-dominated ecosystems Ecosystems that are managed and transformed by humans.

The integration of ecology in urban planning and design took hold in the late 1960s as a result of the environmental movement and contributions of a number of influential academics and practitioners. In 1969, Ian McHarg, a professor of landscape architecture and urban and regional planning at the University of Pennsylvania, published his seminal book *Design with Nature*. In it, he issued a plea for integrating ecological thinking into planning and design decisions for cities and regional landscapes. He also articulated a method for gathering, analyzing, and synthesizing ecological information to inform planning and design decisions. Several students who studied and worked with McHarg continue to contribute to the development of ecological planning and design.

Frederick Steiner, currently Dean of the School of Architecture at the University of Texas at Austin, is making ongoing contributions to the development of ideas and applications of ecological planning and design. He is among the most prolific writers on this topic with numerous books and articles of his own (see *The Living Landscape: An Ecological Approach to Landscape Planning* 2002; *Human Ecology: Following Nature's Lead*, 2002; *Design for a Vulnerable Planet*, 2011). He also leads a growing effort to document exemplary work that demonstrates ecological planning and design as it is realized.

Michael Hough, another disciple of McHarg, and founder and former head of the landscape architecture program at the University of Toronto, developed an urban focus to his ecological planning and design work through his practice and contributions as an educator and author. He took the lessons of McHarg's approach to ecological planning to the city and helped transform sterile urban environments into living urban ecosystems. His book, *City Form and Natural Process* (1984), laid a foundation for many others who then saw the potential of the city to become a living and much more sustainable environment.

John Tillman Lyle, a professor of landscape architecture at California State Polytechnic University in Pomona (CSPU Pomona), made important contributions through the establishment of the Center for Regenerative Studies at CSPU Pomona. The Center explores and tests theories and principles of regenerative design as articulated in his books *Design for Human Ecosystems* (1985) and *Regenerative Design for Sustainable Development* (1995).

In addition to the efforts of planners and designers, ecologists have crossed into the realm of urban planning and design to provide greater levels of scientific understanding as a basis for making decisions about the future of cities. Richard T.T. Forman's books *Landscape Ecology* (1984), *Land Mosaics* (1995), *Urban Regions: Ecology and Planning* (2008) and *Urban Ecology: Science of Cities* (2008) are now used regularly by planners and designers. These references enable them to incorporate more knowledge about how ecosystems work and how human interactions with natural systems change the dynamics and, in turn, the relationships that exist. Forman, an ecologist, is a professor in the Harvard University Graduate School of Design and works with landscape architects, and urban designers and architects to inform the planning and design process with sound ecological science.

Following this model, teams of scientists, planners, and designers are being formed to work on urban ecological problems in an attempt to increase the potential for urban sustainability. New movements are emerging that are embracing nature in the city, recognizing that humans and nature must inevitably co-exist in the same space and discarding old notions that humans and nature are better separated or managed independently. Timothy Beatley's growing organization promoting biophilic cities (see Chapter 4) and other efforts that are increasing "urban wildness" by fostering urban woodlands, re-naturalizing or re-wilding rivers, and re-creating urban meadows are embracing the idea of true forms of nature occupying the same space as humans.

Cities impact **ecological functions** in many ways. Alberti (2005, p. 169) notes that urbanization "fragments, isolates and degrades natural habitat; simplifies homogeneous species composition; disrupts hydrological systems; and modifies energy flow and nutrient cycling." **Urban ecosystems** and natural ecosystems have similar interactions. Urban ecosystems, however, are a blend of natural and human-created elements and, as such, interactions are significantly affected by human intervention as well as natural processes.

Integrating ecology into cities should be an important sustainability goal as it 1) ameliorates human impacts on ecosystems, 2) enhances **ecosystem services**, 3) adds **biodiversity value** to cities, 4) ensures equitable access to nature and resources, and 5) maintains a healthy functioning planet for future generations. For instance, urban and landscape ecology are interdisciplinary sciences dealing substantially with the interaction of natural and human processes (Forman and Godron, 1984; Forman, 1995). Landscape ecology addresses how spatial variation in the landscape affects ecological processes. Theories and applications provide a rigorous scientific methodology that can be integrated into urban planning, design processes, and urban sustainability policies, providing the basis for sustainable urban planning and design decisions.

The goal of green networks and ecological design is to preserve or restore the ecological integrity of critical natural systems while allowing for compatible human activities that continue productive use of the landscape for human benefit. **Ecological integrity** is a concept that refers to the health of an ecosystem or a landscape at which the system functions viably. Forman (1995) notes that to achieve ecological integrity, near-natural levels of production, biodiversity, soil, and water characteristics must be present. He also notes "ecological integrity could be measured as the single most important or sensitive attribute of an ecological system" (p. 499). As a measure for sustainable development, the challenge becomes the quantification of the idea of "near-natural." It is quite simple to assess many areas and determine that they are not near-natural because of the evidence of excessive deterioration. But, because there are so many natural attributes that are difficult to quantify, determining which areas are near-natural may be a major challenge.

Noss (2004) characterizes ecological integrity as an "umbrella concept," embracing all that is good and right in ecosystems. It encompasses other conservation values, including **biodiversity, ecological resilience**, and "naturalness." Noss notes that although urban areas will never have the biodiversity, naturalness, and ecological resilience of pristine wilderness areas, there are reasonable standards that can be met. He outlines an index that considers structure, function, and composition of the urban ecosystem. These effectively align with the three basic tenets of landscape ecology: structure, function, and change. Utilizing these and other concepts from landscape ecology, planning and design strategies are developed within a nested hierarchy that relates to various levels of ecological functioning and correlated levels of human activity and management.

The three interrelated and fundamental concepts (landscape structure, function, and change) enable a greater understanding of urban ecosystems and how they interact. Landscape structure refers to the spatial and structural characteristics

Ecological functions The interactions that occur between organisms and their environment.

Urban ecosystems Dynamic ecosystems that have similar interactions and behaviors as natural ecosystems, but consist of a hybrid of natural and human-made elements whose interactions are affected not only by the natural environment, but also by culture, personal behavior, politics, economics and social organization.

Ecosystem services The combined outcomes of an ecosystem that are beneficial to all species and the Earth overall including processes such as the production of oxygen, decomposition of waste and the production of clean water.

Biodiversity value An intrinsic value of achieving biodiversity that is work protecting regardless of its value to humans.

Ecological integrity The healthy compositions, structure and functioning of an ecosystem consisting of near-natural levels of production, biodiversity, soil and water.

Biodiversity The variety of life forms as indicated by the number of species of plants and animals in a particular ecosystem.

Ecological resilience The measure of an ecosystem's ability to absorb changes and disturbances and still function in a productive manner.

of the landscape. Vegetation, soils, hydrology (rivers, streams, lakes, etc.), and topographic conditions, including slope and landform, are all integral to understanding landscape structure. Landscape function refers to the interactions that occur between organisms and the environment. Landscape change occurs constantly. As living dynamic systems, landscapes are in a constant state of flux.

Landscape structure includes patches, corridors, edges, and the matrix (Figure 1). Patches are the irregularly shaped elements in the landscape mosaic, whether it is developed or not, that differ from their surroundings. Corridors are linear elements that traverse the landscape mosaic and facilitate flows and connectivity between landscape elements. Connectivity, in ecological terms, refers to the interactions that occur between species and landscape structure across landscape elements. Edges are those areas that bound patches and/or corridors. They are critical zones for interaction with adjacent ecosystems. These are useful terms for describing and understanding the physical nature of landscape patterns.

As noted previously, these concepts are interrelated. By understanding how they interact, we can facilitate ecological functions and restore or strengthen sustainable urban ecosystems. In addition, there are a variety of types of patches and corridors and a range of different ways that edges accommodate interaction between ecosystems. By understanding landscape structure and the relative inherent ecological value of different types of patches and corridors, a process can be developed by which we can structure spatial arrangements on the landscape that facilitate high levels of ecological functioning and create opportunities for efficient, ecologically friendly land-use activities. Resilient urban ecosystems require effective balances of these **socio-ecological systems**.

Socio-ecological systems Complex, integrated systems consisting of natural, socioeconomic, and cultural elements and marked by interconnectedness within and between these elements.

Ecological functioning occurs and is linked at many scales. Many of the global environmental challenges we face today result from the aggregation of many actions that have occurred on small scales but taken together have a global impact. Furthermore, the organisms within ecosystems function at numerous scales, ranging from micro-biotic activity to interactions of meta-populations at broad scales. This hierarchy of systems doesn't always correlate between levels of ecological functioning and levels of government or other organizational structures intended to manage land-use and environmental objectives. Thus, it is important to find ways to allow organizations at various scales to collaborate in order to facilitate the range of ecological functions that must occur to sustain ecosystems and provide ecosystem services. One goal in sustainability is to develop new institutional objectives for governance that specifically support these socio-ecological systems.

There are several harmful effects of human landscape changes that the concepts of green networks and ecological design specifically intend to mitigate, including fragmentation, isolation, and edge effects. Fragmentation of landscapes is a pervasive and significant problem. Prior to human occupation, natural systems were connected and sustained by the flows that occurred within and between

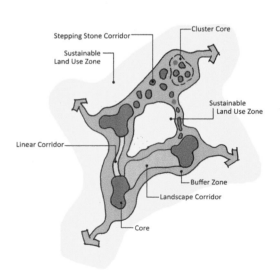

Figure 1 Conceptual Diagram of a Green Network.

various ecosystems. Examples of these flows are the movement of water through rivers, streams, and drainage corridors, the movement of air in and around various landscape elements, the migration of various organisms, and the movement of nutrients, energy, and genetic material.

For example, a river system does not just move water from one location to another. Along the way, the water carries particles of soil and other organic material. This sediment is deposited along the way in various locations. This not only changes landscape structure, but also provides nutrient-rich soil for the germination of new plants that then become food and habitat for other organisms such as fish. Fish, and other aquatic species, move up and down rivers to lakes and the sea, adding nutrients to the ecosystem. They, in turn, become food themselves for other animals, and spawn to create new populations.

Other types of flows are just as critical, but may not be as obvious. The movement of cold air from higher elevations to low-lying areas creates microclimatic variations in the landscape. Since cold air is denser than warm air, it naturally flows downward. This movement creates cooler zones in which cold hardy plant communities that differ from the surrounding areas can become established. This variation creates the opportunity for increased diversity and, as a result, a broader range of species and potential ecological value.

Human interventions such as building roads across valleys, drainage corridors, or other low-lying areas can block cold air movement, thereby changing the microclimate. This may lead to extinction of some species that require specific climatic conditions, ultimately increasing ecosystem fragmentation. Fragmentation leads to the isolation of various landscape elements and populations that inhabit them. Over time, isolated areas will decline in richness and diversity. Often the decline is not immediately observable because the landscape structure changes slowly and the organisms that inhabit these areas are not always easy to observe and document.

Because these areas are no longer connected to a larger system, the introduction of new genetic material to renew populations becomes less likely. The subsequent decline in populations and the reduced level of ecological functioning causes decline in ecological viability and resiliency of the ecosystem. Essential habitats are lost and biodiversity declines. The loss of biodiversity has a direct effect on a system's capacity to be resilient, which is the ability to respond and bounce back from disturbances and retain vital functions and structure.

Interactions between ecosystems occur naturally, but when human activities are introduced within or adjacent to natural areas, new stresses are placed on ecosystems that are not always able to adapt to the changing conditions. As a result, even though some areas may still remain as natural areas or important green areas, they may become less viable from an ecological perspective because of the incompatibility of adjacent land uses. It is important, therefore, to maintain and facilitate connections in the landscape or accommodate the flows that are vital to maintaining healthy functioning ecosystems.

From an ecological perspective, it is also important to understand the landscape morphology or history of the evolution of ecosystems with which we interact. Landscapes are dynamic and constantly evolving. Therefore, understanding how they have come to be the way that they are now helps identify the ecosystem's tendencies and provides a knowledge base for making decisions about how we might

Deep structure Rhythms, changes, long-term tendencies and latent characteristics that occur naturally in an ecological system.

best interact with them. One of the tendencies of ecological systems is to have what we call a **deep structure**. That is they have underlying attributes, such as the climate, topography, altitude, and slowly changing, dynamic characteristics, such as hydrological, geological, and other inputs that impact the natural fluctuations, distribution, abundance, and physiology of organisms within the ecosystem and enable it to thrive. Within developed areas, such as cities, it is useful to understand what the ecoystem's deep structure is so that plans can accommodate and take advantage of the rhythms and changes that occur naturally in these systems. Thus, an important dimension of urban sustainability is to allow nature to continue to thrive and essentially manage itself without the infusion of energy and resources to keep ecological systems functioning.

The problems associated with maintaining viable ecosystems in urban areas are significant. Specific challenges often surface around development, conservation, landscape restoration, edge effects, and site-scale ecological design (Van der Ryn and Cowan, 1996). Sustainable urban landscapes are a finely structured mosaic of property owners and land uses, resulting from numerous political, economic, cultural, and physical determinants, where competing interests for undeveloped land are intense. The outcome is that urban forms are an amalgamation of the most resilient human creations and ecological processes.

Nature's deep structures are the most resilient ecological processes. They are ever present in our cities and continue to provide evidence that when they are ignored in city design, nature will recapture critical components either through catastrophic natural events or through incremental change. Typically, valuable economic and labor resources are used to hold back the forces of nature or re-build urban infrastructure after reoccurring natural disruptions such as floods, soil movement, or weathering. The main goal of urban ecology is to understand these forces and work together with natural processes to achieve a sustainable future for cities.

Ecosystem Services and Socio-economic Benefits

The principal reason green networks and ecological design are gaining acceptance as urban sustainability strategies is that they provide a variety of ecosystem services (Daily 1997) and cultural benefits. Ecosystem services provide many benefits that we often take for granted. These services include breathable air, clean and plentiful water, food, pharmaceuticals, clothing, fuel, climate, waste disposal, pollination of plants, carbon sequestration, and much more. Green networks and ecological design protect ecosystems and the benefits they provide for all species and organisms, including human beings. They also provide a number of important socio-economic and cultural benefits such as increased property values, recreational opportunities, and a sense of community and identity.

A recent report by Odefey et al. (2012) quantifies the economic benefits of green infrastructure for cities. These are things that we have historically relied upon for basic human existence; however, as ecosystems have become more stressed, both locally and globally, they are declining. Following is a brief description of various ecosystem services and cultural benefits realized through the establishment of green networks and employing ecological design.

Increased biodiversity—The diversity of life on this planet is immense and it is difficult to fully comprehend, never mind document, the variety of species that exist. But this diversity is being threatened. Scientists estimate that we lose dozens of species every day. This is well beyond the natural "background" rate of one to five species daily (Center for Biological Diversity, n.d.). To protect and increase biodiversity, both habitat and conduits for species migration are among the most important ecological functions that can be accommodated. Plants and animals are dispersed through corridors and patches in natural systems. These zones serve a conduits for nutrients, energy, and gene flow. Within an urban context, some areas may not be suitable as primary habitat for all but a few species, but may be quite suitable as islands of refuge or places to forage if connected to node or primary source areas.

Hydrologic Processes—The preservation and restoration of hydrologic processes is critical. This includes lakes, rivers, streams, drainage corridors, flood plains, and groundwater recharge basins, among others. Flood containment and protection against soil erosion are also important (Cook, 2007). When in a viable state, drainage corridors serve as filters for surface runoff, helping to purify water before it returns to water supply sources. They also serve as sinks for groundwater recharge. Many cities rely on groundwater for municipal water supplies. Yet, groundwater is being extracted more rapidly than it is being naturally replenished in many locations. Ecosystems located in areas where groundwater recharge is most viable can be incorporated into green networks, allowing surface water to be filtered as it percolates to replenish groundwater resources.

Climate Amelioration—In many metropolitan areas, and particularly in hot arid climates, an **"urban heat island (UHI) effect"** has significantly increased average temperatures, reducing human comfort and causing increased energy consumption (Bowler et al., 2010). See Chapter 14 for an in-depth discussion on UHI. In urbanized areas, climate modification can be achieved by increasing vegetative cover in appropriate locations. Street trees and other green spaces help to mitigate increased temperatures through shading and evapotranspiration. McPherson (1992) calculated that there are significant energy savings that can be achieved with urban tree plantings. Negative effects of wind can also be mitigated through increased plantings.

Urban Heat Island (UHI) effect A global phenomenon in which surface land and air temperature in urban areas is higher than in bordering areas. The primary causes are related to the structural and land cover differences of these areas.

Recreation—The most common human activity that may occur in more natural areas is recreation. As concerns increase over the lack of fitness, rampant obesity, and rising health costs associated with these conditions, opportunities for urban recreation become increasingly significant. Suitable activities might include hiking, cycling, horseback riding, nature observation, picnicking, and light camping in specific locations.

Carbon Sequestration—As noted in Chapter 8, the emission of carbon into the atmosphere is creating significant environmental problems. In addition to poor air quality in urban centers, the aggregation of carbon in the global atmosphere is known to be causing changes in climate that may have serious implications for future generations. Many scientists and engineers are exploring ways to artificially sequester carbon to help offset these anticipated climate changes. Natural ecosystems, particularly wet ecosystems, can sequester carbon from the atmosphere. Maintaining viable functioning ecosystems can help preserve natural carbon sequestration processes throughout the planet.

Reduced Management and Maintenance Costs and Aesthetics—Elements of green networks that are predominantly comprised of more natural ecosystems can be self-sustaining and thus provide areas in which a range of other activities can occur without having to be maintained and managed using public resources. Although it is difficult to place specific monetary values on beautiful scenery, it is generally understood that aesthetic qualities are important. Research has shown that properties adjacent to nature areas have increased economic value. In many cases, the image of an entire district is formed because of the existence, or lack of, natural landscape characteristics. The spiritual or emotional value of beautiful natural areas should also be recognized.

Education and Human Psychology—Education and human psychological ties with nature can be reinforced by having nature accessible within cities. As society becomes more urbanized, the danger of losing touch with nature becomes real. Functioning urban nature areas can provide opportunities for city dwellers to learn first-hand about urban nature processes, and green spaces can provide sanctuary from the strains of urban life. In the long term, this may promote a stronger environmental ethic in society (see the Biophilic Cities insert by Timothy Beatley in Chapter 4).

A number of other functions could be identified, but these are some of the most relevant in urban areas. All of these ecosystem services or functions would likely not occur simultaneously throughout a green network. However, there may be several compatible functions with varying levels of priority in certain segments.

Green Networks

The concept of green networks embraces urban ecology as an essential determinant of city form and provides a guiding philosophy for sustainable new urban development. It also offers opportunities to retrofit existing urban structure to the ecological patterns nature has shaped over time. It is a human interpretation of relationships that have occurred in nature since the beginning of time, which grew out of Dutch planning. It is also a response to the detrimental effects of fragmentation and ecological degradation.

A green network is a system of interconnected or related patches and corridors that provide and sustain ecological values within a human-dominated landscape mosaic. Related terms or concepts include ecological networks (Cook and Van Lier, 1994, Jongman and Pungetti, 2004), green infrastructure (Benedict and McMahon, 2006), greenways (Hellmund and Smith, 2006), green or ecological structure (Werquin et al., 2005), and habitat and dispersal networks (Asbirk and Jensen, 1984). This emerging planning idea, when applied effectively, can create an "ecological infrastructure" for cities and contribute to long-term urban sustainability.

To be effective, green networks, must be designed as a coherent system of natural or semi-natural landscape elements configured and managed with the objective of maintaining or restoring ecological functions. This will not only conserve biodiversity, but it will also provide opportunities for sustainable use of natural resources and ecosystem services. The principal benefit is that nature is allowed to thrive and essentially manage itself without the infusion of energy and resources to keep urban ecological systems functioning. Figure 1 on page 198 is a conceptual diagram of the various components of a green network.

Core areas are relatively large natural or semi-natural open space or landscape elements that provide secure habitats for a variety of species and a broad range of ecological functions. These areas are critical as prime habitat for organisms that are not tolerant of high levels of human activity or deterioration in ecological value. Clustered cores are collection of smaller natural or semi-natural open space elements compromising a larger core area. This approach helps connect fragmented ecosystems in urban areas and achieve higher-level ecological functioning. This retrofitting enables ecosystems to **adapt**.

> **Adapt** The ability to evolve or maintain based on changing circumstances, including environmental conditions.

Each of the core areas are linked by corridors. There are three different types of corridors—landscape, linear and stepping stone corridors. Landscape corridors link wide continuous landscape elements. They allow the use of these zones in similar ways as core areas but they also facilitate flows and migration between cores. Linear corridors are narrower and function primarily as connections from one core area to the next. They typically do not provide sufficient area to be used as primary habitat zones. Stepping stone corridors allow for connectivity between core areas across less hospitable zones. These corridors operate as points of refuge and temporary habitat as organisms move along the corridor.

Surrounding all of these areas are buffer zones. These zones play a critical role in filtering contaminants, invasive species, unnatural predators, and reducing other detrimental impacts introduced by human activity such as noise and toxic materials. It provides protection for core areas and corridors, minimizes edge effects, and allows the core or corridor to function at its maximum potential. Connectivity to other green network elements enables the system to carry on throughout a larger region as an interconnected system of core areas (patches), corridors, buffer zones, and other landscape elements. Surrounding the green network elements are sustainable use areas. These zones are dominated by human activity such as agricultural areas, heavily urbanized zones, suburban development, or other types of human settlement.

Green networks vary in size with a variety of core areas, corridors, and buffer zones. These have different inherent landscape characteristics and opportunities for implementation and acquisition of various landscape elements. In some locations, it may be possible to design green networks based largely on the remaining natural ecosystems and habitats. However, in many urban settings ecological restoration is often necessary. Another approach is to design "synthetic" corridors that mimic natural systems. This may be appropriate to establish linages where no natural connections are possible. The concept of green networks is also applicable at multiple scales. It can be applied at the continental, national, city, or local level.

The Nested Hierarchy of Green Networks

Although the scales or levels may vary depending on the situation, we can identify four principal scales at which this nested hierarchy can be established. At the broadest level, or "mega-" scale, the focus is on linking green network elements of significant size such as at the continental level. At the next level, or "macro-" scale, national or regional level significance is important. Cities typically make up the next level, or "meso-" scale. And, at the finest level or "micro-" scale individual projects or sites become most relevant. Linking these scales together creates connectivity and flows that are essential for their long-term sustainability. There are several initiatives focusing on implementation of green networks across scales.

In North America, a proposal entitled "Yellowstone to Yukon Conservation Initiative" (Schultz 2005) aims to establish a conservation corridor, extending from Yellowstone National Park following the Rocky Mountains up through Canada to the Northern Territory of the Yukon. This initiative is intended to protect critical habitat for bears and other large mammals that need vast undisturbed areas to thrive. More than 100 organizations partnered in the initiative that links major national parks and wilderness areas and other critical habitat as a part of a **continental-scale conservation strategy**.

After 20 years of planning, the first complete version of the Pan European Ecological Network was implemented (Jongman et al., 2011). This was undertaken in three distinct pieces: one plan for Western Europe; another for Central Europe; and the third for South Eastern Europe, including Turkey. This extensive effort required the generation of national plans that were then integrated into a continental scale initiative. Important European bird migration routes and habitats as well as varying levels of conservation for a variety of other species and critical ecological sites were protected and conserved through this planning effort.

In Central America, the Mesoamerica Biological Corridor (MBC) project ranges from the Yucatán Peninsula of Mexico through all of Central America to the southern reaches of Panama (Miller et al., 2001). This plan was started by a number of international nonprofit conservation organizations. They worked with national governments in a cooperative effort to identify important habitat and ecological zones throughout the region. Each of the participating countries developed a plan and coordinated their efforts with adjacent national governments. They also developed strategies and incentives for formulation and implementation of local level plans. The conservation strategy is title "Paseo Pantera" (Path of the Panther) and is intended to protect biodiversity by providing protected corridors and patches (Carr et al. 1994). The plan has been broadened to incorporate numerous other areas of ecological importance.

An example of a multiple scale strategy is the case of Hacienda Baru in Costa Rica (Ewing 2005). Costa Rica participated in the cooperative international MBC planning effort. The country has a strong national-level plan identifying parks and other conservation areas that have significant ecological importance. Over 25% of the country's total land area has National Park or other protected status. Part of Costa Rica's economic development strategy is to embrace the idea of ecologically based tourism to preserve their critical natural resources and the ecological integrity of the landscape. As a part of this strategy, tax incentives,

Continental scale conservation strategy An approach to ensure the ecological integrity and connectivity of vital habitats across entire landscapes or regions and establish strategic partnerships involving mixed land uses to promote ecological restoration in the reconnection of large ecosystems.

preservation easements, education, decentralized administration, partnerships with international organizations, and land purchases were initiated to secure the most important ecological zones and encourage more sustainable land-use activities (Ewing, 2005).

Hacienda Baru converted an 820-acre parcel of land from an active cattle ranch to an eco-lodge, re-establishing critical wildlife habitat and supporting sustainable tourism. The landscape transformations were vast. In 1971, the land was completely deforested and open for cattle grazing. Several years later, the US packing company who owned the land decided to concentrate only on its production and meat packing operations in the US and sold the land to ranch employees. The new owners created a series of pastures separated by fence lines.

A common practice in this region is to create these fences using cuttings from live trees as stakes. These take root and leaf out becoming living trees while serving as fence posts. This protects the fence posts from rot and insect infestations. It also provides a lasting and living fence system. As the fence lines became established, monkeys, kinkajous, opossums, iguanas, and olingos from the upland forested areas started to use them as migration corridors to travel down to lowlands near the water's edge (Ewing 2005). In essence the new landowners created a site-scale green network that facilitated the migration of certain species in the area.

Subsequently, the landowners embraced the opportunity to convert this series of pastures into an eco-lodge complete with nature trails, rustic lodging, environmental education opportunities, and guided tours. Over time, the pasture naturally regenerated and the former barren landscape of Hacienda Baru is now completely reforested. It is now home to a wide range of plants and animals that coexist with visitors and eco-lodge owners and employees.

Ecological Design

The more encompassing concept of urban sustainability and green network planning strategies can only be realized through implementation on a site-by-site basis. This is where ecological design becomes an essential element for creating more sustainable urban futures. Nature has provided and does provide us with exceptional examples of functional, efficient ecological systems. We can use these to inform the design and restoration of urban ecosystems. The more we know about how nature functions, the better we will be at designing systems that provide the benefits we seek. Along with the natural sciences, we need to embrace technology in developing effective strategies for designing urban ecosystems. This will enable us to implement a range of ecosystem types from natural to artificial. This section provides examples of ecological design that illustrate how knowledge of natural patterns and processes can inform design and be applied in urban situations to contribute to the long-term sustainability of cities.

Re-wilding Urban Rivers and Streams

Rivers, streams, and other riparian zones are among the most biologically diverse and rich areas in most landscapes. They are also critical lifelines since they are essential for providing connectivity that facilitates flows and other ecological functions. It is important to preserve or conserve the critical corridors where these rivers and streams occur naturally. In many places, however, humans have settled near rivers and streams, altering these landscapes.

In most cities, rivers and streams and their drainage basins are usually significantly modified. For example, corridors are sometimes **channelized** or straightened in an attempt to manage flooding and maximize adjacent land for urban development. While this can remedy short-term problems immediately surrounding the watercourse, many unintended complications can arise.

Channelized Straightening of a river using furrows and/ or passageways.

Channelization and straightening rivers increases the speed of water flow, which in turn increases the water's natural force. This can erode stream banks and further instability downstream. Rapidly moving water also carries more sediment, reducing water quality. Additional problems occur downstream when rivers no longer follow their natural course. The sediment loads carried by the rapidly moving water drops once the river returns to its normal course and over time these areas fill with the sediment that restricts the passage of water, resulting in flooding beyond historical flood zones. Many rivers and streams that have been previously modified are now being restored and **re-naturalized**, which includes reintroducing the natural meander and reestablishing stream bank vegetation.

Re-naturalized A habitat or ecosystem that was brought back into conformity with nature, including reintroduction of ecological functions.

An example of a river that was formerly modified and has been re-naturalized is the Enz River in Pforzheim, Germany. Pforzheim is a city of about 120,000 people that sits at the edge of the Black Forest in Germany. The Enz, which runs through the center of the city, was channelized and straightened in the early 1900s. The land area adjacent to the river has been developed through the decades as urban housing, industrial complexes, shops, and other forms of intense urban development. The river restoration project was part of a larger scheme to create a new urban park following the course of the river for about 15 kilometers through the city. The design was undertaken in the late 1980s and construction was completed in 1992. Figure 2 shows the river in its channelized, straightened condition and Figure 3 in its re-naturalized condition.

The re-naturalization process reintroduced the natural meander course based on historical analysis of the river form. Trees and shrubs were replanted

Figure 2 Channelized Enz River in Pforzheim, Germany.

Figure 3 Re-naturalized Enz River in Pforzheim, Germany.

in locations where they would typically be found in a natural river channel structure. The landscape architects who designed the project followed ecological design principles, using lessons from nature to inform their design strategies (bio-mimicry). Some 20 years later, the river corridor is now well re-established in its natural regime, vegetation has matured providing habitat for many species, and the river corridor now looks much like it did in its original natural condition. Most importantly, it has reestablished many of the ecological functions that were lost when the river was modified and has once again become an important ecological corridor linking the Black Forest to other important ecological zones.

Preserving Urban Nature

Source and core areas are large areas of natural or near-natural landscape, although they do not need to be original landscapes if they show high levels of ecological functioning. There are many examples of landscapes that have been rebuilt after some significant disturbance or deterioration that provide many of the same benefits as the original or core areas. They are important elements in a green network as they provide stability are generally more resistant to external influences.

These areas function as prime habitat for a variety of species and contribute to shaping climate, hydrological, and other biological functions within the larger urban area. The term "source area" refers to the fact that these areas are places where significant genetic populations spread to other zones via the green network. Because these areas are usually covered by vegetation, they supply oxygen and absorb carbon helping to improve air quality. The presence of vegetation also can help mitigate UHI effects by keeping the surface cooler and through the process of evapotranspiration, which returns moisture to the air. Depending on the specific characteristics of the core or source area, they may perform important hydrological functions such as water retention and groundwater recharge, which returns and filtration of surface water. From a biological perspective, these areas usually occupy sufficient terrain to provide habitat and range for the migration for many species that may not be tolerant of frequent interaction with, or influence from, humans.

The Phoenix Mountain Preserves in Phoenix, Arizona, USA, are an example of a series of open space elements in an urban area that function as multiple core or source areas. Figure 4 shows a natural or near-natural, open space preserve that is part of the city of Phoenix Park system. The first of these preserves, South Mountain Park, was established in 1924 and occupies 16,500 acres. It is the largest city park in the US. An additional 10,500 acres were added to the preserve system in 1972 with the establishment of the North Mountain Range. The most recent addition to the preserve system came with the establishment of the Sonoran Preserve Master Plan (Figure 5) which identifies approximately 21,500 acres in North Phoenix to be added to the system.

Figure 4 Open Space Preserves in Phoenix, Arizona, USA

Image courtesy author

Grey to Green Infrastructure: Transportation and Utility Corridors

Most roads, rail lines, utility corridors and other infrastructure elements form barriers, interrupt flows, and accelerate fragmentation and isolation. Strategies such as ecological bridges and eco-ducts are being employed to overcome these barriers. It is most effective, however, to research natural systems and understand how alternative road alignments or making infrastructure corridors more compatible with ecological functions can be accomplished. In addition to mitigating barrier affects, designing or retrofitting infrastructure systems as ecological corridors

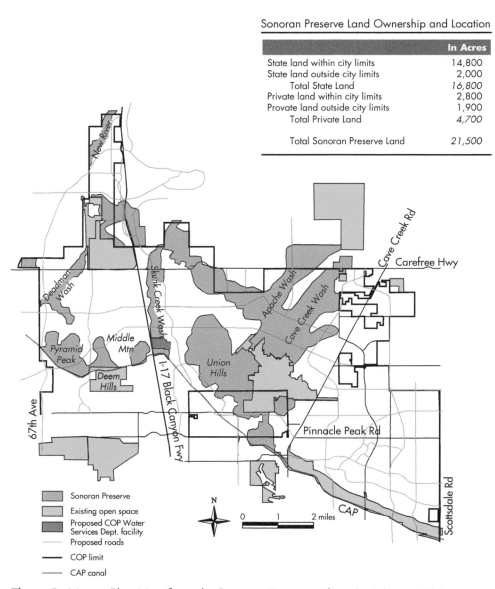

Sonoran Preserve Land Ownership and Location

	In Acres
State land within city limits	14,800
State land outside city limits	2,000
Total State Land	16,800
Private land within city limits	2,800
Provate land outside city limits	1,900
Total Private Land	4,700
Total Sonoran Preserve Land	21,500

Courtesy James P. Burke

Figure 5: Master Plan Map from the Sonoran Preserve, Phoenix, Arizona, USA

Source: From *Sonoran Preserve Master Plan* by James P. Burke. Copyright © 1998 by James P. Burke. Reprinted by Permission.

Figure 6 An Ecological Bridge along the A-50 Highway in the Netherlands

Figure 7 An Eco-Duct Passing Under a Major Road in Eindhoven, Netherlands

creates opportunities for establishing linkages within urban areas in places where there are no natural connections. For instance, most roads, rail lines, and utility corridors travel through rights of ways that are much wider than the actual infrastructure element. The residual space along the roadways and rail line, and underneath overhead utility lines or above underground utility lines can become an ecological corridor by simulating natural corridors through ecological design. These areas can also become useful as recreational corridors for hiking, running, cycling, and horseback riding.

Figures 6 and 7 show how an ecological bridge and an eco-duct can help mitigate the barrier affects of major transportation elements. These corridors and crossing mechanisms should be situated at the location of known migration corridor for animals that inhabit the region. On initial inspection, the cost might seem prohibitive to accommodate the movement of animals. However, these bridges provide greater safety for motorists as they minimize the potential for accidents involving wildlife. An eco-duct is a simple, inexpensive way to facilitate movement of small animals underneath roadways rather than limiting migration possibilities to crossing roadways risking the animal's lives. In addition to managing and protecting wildlife corridors, green infrastructure has other benefits.

Portland, Oregon has established a progressive "Green Streets" program to manage urban storm water by localizing collection. Figure 8 includes curb extensions in a typical residential neighborhood. This is a very simple adaptation to a typical street curb and gutter system. It involves extending curbs into the street, narrowing the roadway at the corners and leaving gaps along the existing curb line to allow water to flow from the street into a landscaped area at the end of each block. The water that runs off the streets usually carries contaminants from automobile exhaust, oil, and other elements and while the landscaped zone helps to filter out contaminants, the water also provides sustenance for the plants. Figure 9 is another example from Portland Green Street program.

Figure 8 Green Streets in Portland, Oregon, USA with Curb Extensions to Capture runoff Water from the Street

Image courtesy author

Figure 9 Green Streets in Portland, Oregon, USA with Curb Inlets Directing Runoff Water to Landscape Strips in Sidewalk

Courtesy author

Figure 10 Green Parking Lot in Essen, Germany with Bio-swales Between Parking Bays to Capture and Infiltrate Runoff Water from Paved Surfaces

In this case, water that runs off the street collects along the gutter and drains through inlets into the landscape zones rather than being carried below ground into the storm sewer system.

In many cities, parking lots occupy extensive areas and are sterile, hot environments. Figure 10 shows a parking lot in Essen, Germany that incorporates a bio-swale to collect water from the adjacent paved areas, allowing it to percolate into the soil. The actual parking spaces are made up of a permeable paving system that incorporates unit pavers with gaps between the stones to allow the water to percolate through rather than running off. The increased green space also helps to cool these areas and improve the aesthetic quality of parking lots.

Forgotten Space: Vacant Land, Underutilized Sites, and Brownfields

Post industrial brownfield sites Underused or vacant land previously used for industrial or commercial purposes often containing hazardous substances and contamination that are made available for redevelopment.

Interim re-vegetation Plant species used to stabilize the soil and nutrients of an idle site in an effort to provide barren land with productive use.

Natural regeneration A dynamic process by which life recolonizes when vegetation, land and/or ecology has been partially or completely destroyed.

Vacant and underutilized land, and derelict or **post-industrial brownfield sites** can play important roles in green networks. They can act as stepping stones within more developed urban structure and provide linkages between other more stable ecological zones. They represent an opportunistic resource in most cities. Through their research of 70 US cities, Pagano and Bowman (2000) determined that on average 15% of the land area in these cities was vacant. These vacant parcels often remain fallow for decades. Since they are privately owned, it can be difficult to establish programs for temporary use. However, many cities are now exploring ways to use these parcels until they are transformed to a more permanent land-use activity. Strategies such as **interim re-vegetation**, temporary urban agriculture, or in some cases, **natural regeneration** can return some ecological value to these sites.

Brownfields are sites that have been previously occupied, often in some industrial capacity, that are now abandoned and may have significant site

Figure 11 Constructed Wetlands in Westergasfabriek Park in Amsterdam, Netherlands

Figure 12 Wood Deck Pedestrian Pathway through Constructed Wetlands in Westergasfabriek Park in Amsterdam, Netherlands

contamination issues. Typically, brownfields require intensive research and investigation to determine appropriate strategies for site restoration and removal of contaminants before or in conjunction with developing suitable strategies for re-use. Ecological approaches to site remediation, including phytoremediation (Suresh and Ravishankar, 2004), bioremediation (Diaz 2008), and in-situ oxidation (Huling and Pivetz, 2006), can provide many ecological benefits beyond just cleaning up hazardous waste or other contaminants.

There are also many new parks being created on sites of former industrial facilities or other land-use activities that resulted in degradation and contamination. These projects turn formerly undesirable or unusable spaces into useful public sites that improve ecological functions in urban areas. The Westergasfabriek Culture Park in Amsterdam, Netherlands is an example of how an urban park can transform a derelict site into a vital ecologically functioning system.

The site is located in the central part of Amsterdam and occupies approximately 36 acres. It was originally developed as a coke gas production facility in 1898, which operated until 1967. The site was left with serious soil contamination. The cleanup and restoration process began in 2000 when it was determined that the site would be converted into a new urban park. The design for the park established a number of ecological elements that help manage some of the contamination problems. This included infusing the site with a wetland that provides habitat, helps filter and clean water, and incorporates a naturalistic aesthetic element in the heart of the city. The photographs in Figures 11–13 show a variety of elements in the park integrated with the old industrial structure.

Figure 13 Foundations of Old Gas Storage Tank now used in Designed Wetlands in Westergasfabriek Park in Amsterdam, Netherlands

Conclusions

While it is clear that a need exists for maintaining the viability of critical ecological systems in urban areas, it is uncertain whether this goal can be achieved with the strategies that many cities currently employ in urban planning. Ecological and urban theories have evolved in different directions and we are only now starting to understand the impacts of urbanization on ecosystems and the substantial efforts required to restore them. Numerous perspectives exist on how to conserve existing viable systems and restore those with degraded quality. Understanding urban ecology, employing the planning concept of green networks, and incorporating ecological design shows promise for making urban areas more sustainable. Linking science, policy, planning, and design is the most effective way to integrate ecology into cities and provide a foundation for a more sustainable future.

Landscape metrics can be used to assess the viability of these strategies. These tools are used to analyze landscape structure; inherent characteristics of landscape elements; and the interrelationships between natural landscapes and between urban and natural landscape elements; and how external factors affect the functioning of these landscapes (Boutequila et al., 2006; Cook, 2002; McGarigal and Marks, 1995; Dramstad et al., 1996). The use of metrics also provides the opportunity to have more objective conversations about ecological outcomes and urban planning decisions. One of the most important issues, therefore, is generating public awareness and support for sound concepts such as green networks and ecological design. If the public is knowledgeable about the potential contribution these concepts make toward actually realizing a sustainable future for cities, they are more likely to influence and/or support the efforts of the urban development community, government agencies, and politicians.

Chapter 11

Managing Water and Its Use: The Central Issue for Sustaining Human Settlements

Ray Quay

Introduction

Water is essential for the survival of life as we know it, and its abundance, or lack thereof, has been a major factor in the growth or decline of almost every human civilization. Today, adequate water supplies to meet current and future worldwide needs of humans and the environment could be one of the greatest challenges of the 21st century. This challenge is largely due to an emerging global water crisis of three interrelated dimensions: the lack of sanitary, **potable water** and **wastewater treatment** in developing countries; degradation of **freshwater** supplies and associated habitats by human action such as pollution; and a looming shortfall between freshwater supply and **water demand**. These problems are further complicated by the fact that they are not uniform for all parts of the globe.

Since the adoption of the United Nations Millennium Development Goals in 2000, the percent of people worldwide that do not have access to potable water decreased from 24% to 11%, and those who do not have access to sanitation facilities decreased from 51% to 33%. These are significant achievements but in sub-Saharan Africa, between 43%–52% of the population in 9 countries are still without access to sanitary, potable water (Water Aid, 2016) and 70% do not have access to sanitation (Cho, 2011). In 2010, it was estimated that 80% of the world's population lived in an area impacted by degradation of a river system. This impact also has global spatial patterns. Some countries are able to deploy technology to reduce these threats, while others, such as Africa and South Asia, are more limited in their ability to do so (City of Phoenix, 2011). This can create significant regional problems. For example, it was estimated that 54% of China's seven main rivers were unsafe for human consumption in 2005 (McDonald, Weber, Padowski, Boucher, & Shemie, 2016). Lastly, many developed and undeveloped countries have already highly **stressed** water supplies.

Research has shown that agriculture and development is impacting watersheds for 9 out of every 10 large cities (>750,000), increasing pollutants such as sediment (40%), phosphorus (47%), and nitrogen (119%) and increasing water treatment costs (McDonald et al, 2016). In addition, it is estimated that over 1 billion people globally live within **watersheds** where demand exceeds available

Potable water Water that is suitable for drinking.

Wastewater treatment The process of removing material from sewage that is harmful to humans or the environment. Also referred to as sanitation.

Freshwater Water that is not salty especially when considered as a natural resource.

Water demand The consumption of water to support human and environmental activities.

Stress The result of a change that threatens the functionality of a system.

Watersheds The geographic extent from which rainfall will eventually flow into a river or lake.

water supplies and that half of the world's 16 largest cities are experiencing freshwater shortages (National Research Council, 2012). The gap in **water supply** and demand will likely increase with continued growth and potential **climate change** impacts.

Discussion of these three water crisis issues requires more space than afforded by this book. Therefore, this chapter focuses on how US urban water demand and supplies are managed to maintain a sustainable water supply. We look specifically at conditions of change and **uncertainty** that will be the most critical issue for our future.

Water supply The total amount of water available to be used to meet water demands.

Climate change Long-term deviations from historical records of precipitation, temperature, and storm intensity.

Uncertainty Lack of knowledge about when or what will be a future state.

Water Supply

Earth's surface is covered by both water (71%) and land (29%). Of this water, 97% is in the oceans and 3% is freshwater. Over half of the freshwater is locked in ice caps, glaciers, and permanent snow mass. This leaves less than 1% of the earth's freshwater for use by life on the earth's landmasses (Sato, Qadir, Yamamoto, Endo, & Zahoor, 2013). Importantly, this 1% is not evenly distributed across the earth's landmasses. Some areas such as the northern regions of North America, Europe, and Asia and tropical forests have more; desert areas of the southwestern US, northern Africa, and the Middle East have less.

Groundwater Water found underground in aquifers.

Surface water The water that flows or is impounded on the surface of the earth, such as that found in rivers and lakes.

Aquifer An underground feature of porous stone or river gravel/sand that is saturated with water.

Distribution of the World's Water

Freshwater supplies can be organized into three major sources: **groundwater**, **surface water**, and **reclaimed water**. Sources vary from one region to another. Some regions have access to surface water from lakes and rivers. In other regions, groundwater from shallow aquifers is widely available while other regions must pump water from aquifers thousands of feet below the surface. In yet other regions, water supplies are scarce and the lack of surface or groundwater limits human activities.

Despite the variation, all water moves within a natural cyclic system from the oceans to the atmosphere then to land and back to the oceans. Solar radiation evaporates water from the oceans and surface water on land. This evaporated water exists in the atmosphere where it condenses into clouds and eventually precipitates as rain and snow on land and over the oceans. On land, some of the rain and melted snow becomes surface water flowing into streams and rivers and eventually back to the oceans. Some surface water is stored in lakes. Some percolates into the ground and into groundwater **aquifers**. Water in the aquifers moves underground, eventually reemerging at springs that flow back to the oceans via streams and rivers. Surface water can

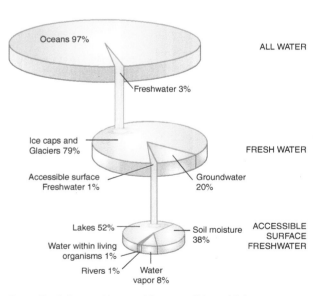

Source: Earth Forum, Houston Museum of Natural Science at http://earth.rice.edu/mtpe/hydro/hydrosphere/hot/ freshwater/0water_chart.html

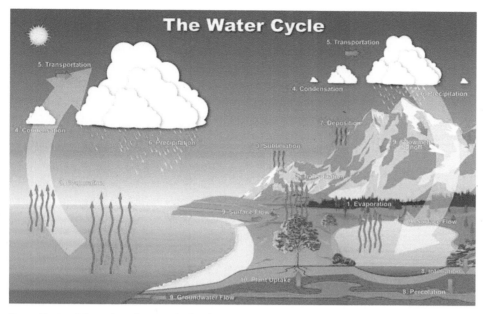

Source: National Oceanic and Atmospheric Administration (NOAA) at http://www.noaa.gov/
resource-collections/water-cycle

also take a slower path as glaciers and permanent snowpacks, storing water for decades or centuries. This **water cycle** makes freshwater a **renewable resource** that is constantly replenished as water moves from the oceans to the land.

Groundwater

Not all paths of the water cycle occur in the same time frame or in the same place. The time it takes for **recharged water** to move into and through an aquifer can vary from decades to centuries. Shallow water aquifers are more closely linked to surface waters and can often be found in association with a river. Water will percolate into these aquifers quickly. However, the volume of supply is often limited and high-volume pumping can quickly deplete these aquifers. They are also susceptible to contamination from surface sources such as septic tanks. Deepwater aquifers contain water that may be centuries old. Typically, they are within geologic formations that can span hundreds of square miles and contain large volumes of water. Although their location is independent of surface water features, they are often re-charged where rivers and streams cross over them. Given their volume and depth, recharge can take decades to centuries.

In ancient times, humans relied on three primary sources of freshwater: rainfall; surface waters that could be found in rivers, lakes, and springs; and groundwater. Infrastructure such as canals and wells were developed to convey freshwater from where it was found to where it was needed for agriculture and consumption and impoundments where it was stored. But the technology limited the depth from which groundwater could be drawn to the surface. Thus, access to groundwater was restricted to locations near rivers and lakes, which were the source of shallow

Water cycle The cyclic process that continuously moves water back and forth between the land, oceans, and the atmosphere.

Renewable resource A resource that continually renews itself through natural processes.

Recharged water The flow of water into an aquifer, typically from the land surface.

aquifers. This limitation changed in the middle of the twentieth century when advances in drilling and turbine pump technology allowed large volumes of water to be pumped from aquifers thousands of feet below the surface.

Today, groundwater is a major supply of water worldwide with technology ranging from methods not much more advanced than those used by ancient civilizations to highly advanced systems of reverse osmosis to remove salt from seawater. Many major urban areas, such as Boston, New York, San Francisco, Los Angeles, and Phoenix, move water from watersheds hundreds of miles away to meet the water demand of their region. In Arizona, the 337-mile long **Central Arizona Project** canal diverts water from the Colorado River to Phoenix and further south to Tucson mostly for agricultural irrigation and municipal water needs.

Central Arizona Project A quasi-governmental agency that manages the delivery of a portion of Arizona's allocation of the Colorado River to central Arizona primarily using a canal that stretches from the Colorado River to Tucson, Arizona.

Although the natural variability of groundwater supplies is not as high as that for surface waters, its variability is typically a function of the depth of the aquifer providing the supply. Shallow aquifers, where the time of surface water to recharge the aquifer can be measured in years, will be responsive to climate patterns of less than a decade. Thus, such groundwater is susceptible to drought conditions. Deeper aquifers, where the recharge time can be measured in decades, will only be responsive to climate patterns that last for several decades or longer. Regardless of the time until these resources are impacted, groundwater is not an inexhaustible water supply.

Withdrawal from aquifers at rates higher than they are recharged is a growing global problem. Aquifers in India, China, and the US are experiencing declining levels because of groundwater pumping. The Ogallala Aquifer, one of the largest in the US, stretches under 174,000 square miles of the eight high plain states and has been the primary source of water for this region for the last 100 years. Unfortunately, it is being pumped so fast that estimates suggest it may be pumped dry in the next 25 to 30 years. Extracting more freshwater than the earth can naturally recharge presents a significant challenge in managing it. This is in large part due to the fact that water is a finite resource.

The ~1400 million cubic kilometers (~49.4 million cubic feet) of water moving through the water cycle is the only water there is (Loftus & Ross, 1995). We cannot make more. The most we can do is change it from one form to another (solid to gas to liquid) or reclaim it (treatment and desalination). Reclaiming water is expensive and its use is often contentious; albeit, controversy is lessening as the public gains a greater understanding of what reclaimed wastewater is and how it is used (EPA, n.d.).

Surface Water

The availability of surface water is a function of the climate, topography, and dynamics of the water cycle within a region. For most regions, it comes from seasonal upland precipitation that flows into streams and rivers and moves down to lower areas. Large watersheds with high volumes of rainfall, such as the Mississippi River, sustain flows throughout the year with some seasonal fluctuation. Smaller watersheds and those with low rainfall may only flow during the wet season. Some areas have natural storage within snowpacks at higher elevations. Accumulated winter snow typically thaws slowly in the spring, gradually feeding water to streams and rivers. When these snowpacks melt quickly, flooding can occur. Across the US, dams have been built to catch and store water during the wet season so it can be released during the dry season as well as for flood control. Streamflow variability is not just seasonal, however.

Precipitation is influenced by climate cycles that vary in length from decades to centuries. For example, the southwestern US has experienced oscillations in its regional climate between wet and dry periods over the last 5,000 years. These oscillations are highly uncertain but are characterized by wet and dry periods lasting between 10 to 100 years. Recent data for these periods has been obtained from observations and monitoring stations. Older data has been obtained from **tree ring records**. The Colorado River provides an excellent example of how these measurements inform our understanding of regional climate.

Tree Ring Records A tree ring is the layer of growth a tree produces in a year. A trees age can be estimated by counting these tree rings. Each ring has a light and dark colored part. The light part is produced during its growing season. The wetter the growing season the wider this ring, thus past rainfall can be estimated from a trees rings.

There are written records of the Colorado River's streamflows for the last 100 years. Estimates for streamflows beyond the records were done using precipitation estimates based on regional tree ring records. These records show periods of drought as long as 35 years, but there is little evidence of a normal pattern of wet and dry periods. In fact, the records show high variability. For instance, written records show that the southwest recently experienced a 15-year dry period, but within this period, 2005 and 2006, respectively, were the driest and wettest years in the last 100 years.

In other regions, the oscillation between wet and dry can be very slow, sometimes centuries long, thus there may be low variability in streamflow records. For many regions, long-term streamflow records do not exist. Thus the severity of dry years, like that which affected Texas and Georgia in 2011 (Box 1), seem like abnormal events, when they may actually be part of a climate profile that is longer than we have written records. The importance of understanding long-term streamflows and precipitation estimates is that the information can be used to prepare for unforeseen events and long-term climatic changes.

Reclaimed Water

Reclaimed water is water extracted from sanitary sewage. It is classified as blackwater or grey/graywater. Blackwater is water that typically comes from domestic and business sewage, including toilets, dishwashers, and food preparation sinks. It is not sanitary and therefore, not comsumable or reusable without treatment. Greywater (graywater) is non-sewage water that comes from washing machines, bathroom sinks, bathtubs, space, and pools as well as rain and stormwater runoff. This water can be reused for outdoor use without treatment. It does, however, require treatment for human consumption.

Reclaimed Water Wastewater that has been used in homes, business, or as part of an industrial process. There are two basic types of wastewater: Blackwater and grey/graywater.

A 2013 study shows that approximately 75% of the wastewater in Canada and the US is treated. Yet, only 2.3 km^3, or 3.8%, is used annually (Sato et al., 2013). Developed countries in general treat more of their wastewater. As municipal water supplies become more stressed, reuse of this water is becoming more common. Some countries, such as Singapore, Australia, and Namibia, are using reclaimed water as a source of drinking water. In the US, reclaimed water is most often used as a source for non-potable uses such as landscaping and toilets. Until recently, US federal and state regulations restricted the use of reclaimed water as a direct source of potable water. This is beginning to change.

Box 1 Drought: A Natural Disaster Worth Paying Attention To

Most natural disaster news we see in the US is about earthquakes, tornadoes, hurricanes, floods, landslides, forest fires, and extreme winter storms. Rarely is drought described as a natural disaster on par with these others. Recently in some regions of the US such as California, Georgia, and Texas, however, drought and reduced water supplies has received quite a bit of attention in the news. Even in these cases, drought is rarely described as a natural disaster. Is this warranted or have we become complacent in thinking of drought as a potential natural disaster?

There are likely four reasons for this current view of drought: 1) a drought unfolds slowly over many months or years and does not create the same sense of emergency as other natural disasters; 2) in US cities, even during the most severe droughts, it is rare that water will not flow when the water faucet is turned on or a toilet is flushed; 3) the biggest impacts for urban dwellers are mandates for decreased outdoor water use or higher prices at the grocery stores when farm yields are impacted; and 4) deaths are attributed to heat waves not drought, even though the weather conditions that cause summer heat spells are the same as those that cause drought.

Generally, the term "drought" implies an extended period of time, usually a year or more, of abnormally low rainfall resulting in a water shortage that makes it difficult to meet the normal water demand for agriculture, urban, or natural environmental systems. In regards to water resource management, what constitutes a drought, including its scale, severity, length and impacts, varies widely with geography and climate. All watersheds are dynamic systems that undergo cycles of wet and dry conditions. The magnitude and duration of these cycles are unique for each watershed.

Communities have different watershed compositions and portfolios of water resources; some only have one source while others have several. Even the impacts of drought will vary based on a region's ability to respond. As of 2016, Arizona has been experiencing drought conditions for 17 years, yet it is just approaching a point where some of its surface water supplies may be restricted. In contrast, two years of drought in Texas in 2011 led to the rare event of the entire state being declared a national disaster area. Thus, talking about drought in Texas is different than talking about drought in Arizona.

In regions where water management and agricultural practices are less sophisticated, even short-term droughts can eliminate the ability to meet a region's water demands. In western Africa, drought conditions regularly create disastrous conditions that result in hundreds to thousands of deaths from thirst and famine (Gettleman, 2011; Lacey, 2006). US history includes cases where droughts have created disasters as well.

The drought that led to the US Dust Bowl in the 1930s devastated the farms of the Great Plains, forcing thousands of people to leave the area (see Chapter 2 for more on the Dust Bowl). In the late 1950s, another severe seven-year drought killed livestock and crops in Texas, resulting in a similar exodus of people from their ranches and farms (Burnett, 2012). In 1988, a Midwest drought destroyed half the nation's corn crop, forcing thousands of farmers out of business (Schneider, 1993).

There is a growing perception in the US that drought is occurring more frequently and more severely (Travis, Gangwer, & Klein, 2011). In the past decade, drought conditions have begun to affect more than just rural agriculture regions. In late 2007, after 18 months of drought conditions, many cities in the South reached a point where they had only enough water to last a few months. As a result, the state of Georgia and a number of cities enacted mandatory restrictions on urban outdoor water use, with some requiring a 50% reduction in water use (Goodman, 2007). Yet in 2012, following another two years of extreme drought in Georgia (WALB, 2012), there was little public discussion about actions needed to manage water (Jones, 2012).

In 2013, during one of California's worst droughts of the last 100 years, the governor of California called for cities to encourage their residents to voluntarily reduce their water use by 20%. But, California fell far short of this goal. State surveys have shown that urban water use in May actually increased by 1% over May use in previous years, mostly driven by consumption in Southern California (Boxall, 2012). In 2016 during continued drought, the Governor mandated 20% reduction in urban water use. This time around there was a much stronger response such as strictly enforced irrigation restrictions. California is estimated to have met this goal, but the effectiveness varied from city to city.

Defining whether drought is getting worse is difficult to do because even the study of drought impacts has taken a back seat to disasters like floods and tornadoes, which cause sudden fatalities, and we have limited information to assess if trends are changing (Travis et al., 2011). Our anecdotal record seems to imply that droughts are occurring more frequently, but this could be due to growth in cities putting more strain on limited water resources, which are then more easily strained by drought. Regardless, it is likely in the future that drought will get worse as the US experiences higher temperatures and changes in precipitation due to climate change. Given this, and our recent experiences, perhaps it is time to give drought more attention as a potential natural disaster, which will manifest itself in different ways for each region.

References

Boxall, B. (2012). California approves big fines for wasting water during drought. *LA Now.* http://www.latimes.com/local/lanow/la-me-ln-water-wasting-fine-20140715-story.html

Burnett, J. (2012). How One Drought Changed Texas Agriculture Forever. *NPR News US Around the Nation.* http://www.npr.org/2012/07/07/155995881/how-one-drought-changed-texas-agriculture-forever

Gettleman, J. (2011). Misery Follows as Somalis Try to Flee Hunger. *The New York Times.* http://www.nytimes.com/2011/07/16/world/africa/16somalia.html

Goodman, B. (2007). Drought-Stricken South Facing Tough Choices. *The New York Times.* http://www.nytimes.com/2007/10/16/us/16drought.html

Jones, W. C. (2012). Georgia's newest drought stirs less political interest. The Augusta Chronicle. http://chronicle.augusta.com/latest-news/2012-08-22/georgias-newest-drought-stirs-less-political-interest

Lacey, M. (2006). In Deep Drought, at 104°, Dozens of Africans Are Dying. *The New York Times.*

Schneider, K. (1993). Recalling '88 Drought's Disaster, Farmers Say Deluge Is Not as Bad. *The New York Times.* http://www.nytimes.com/1993/07/08/us/recalling-88-drought-s-disaster-farmers-say-deluge-is-not-as-bad.

Travis, W. R., Gangwer, K., & Klein, R. (2011). *Assessing Measures of Drought Impact and Vulnerability in the Intermountain West.* Western Water Assessment White Paper.

WALB. (2012). *Special Report: South Georgia's Drought.* Retrieved from http://www.walb.com/story/16975753/special-report-south-georgias-drought.

Two communities in West Texas have been permitted by the state to use treated effluent directly as a source of drinking water and California is considering standards to allow this as well. This may seem disgusting to some. But the reality is that if the source of the water you drink is a river then you are drinking treated effluent as most wastewater treatment plants discharge effluent into rivers. This effluent mixes with the river water, flows downstream, and eventually is pumped out of the river into a water treatment plant. The water is then treated and delivered to someone's faucet to be consumed.

Demand for Water

The availability of freshwater has been a key factor in the rise and fall of ancient civilizations (Water Aid, 2016). Humans use water for numerous purposes. Obviously, we need water to drink. We use it to bathe, wash and cook food, and dispose of human waste; irrigate crops and raise animals; produce raw materials and goods; and electricity generation. We also use it for less critical activities such as for watering lawns and gardens, swimming pools and fountains, and recreation. In the US, on average, the primary source of freshwater is from surface

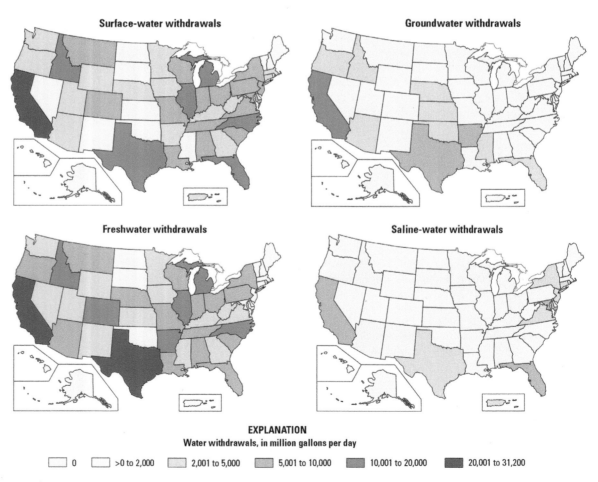

EXPLANATION
Water withdrawals, in million gallons per day

| 0 | >0 to 2,000 | 2,001 to 5,000 | 5,001 to 10,000 | 10,001 to 20,000 | 20,001 to 31,200 |

Source: https://pubs.usgs.gov/circ/1405/pdf/circ1405

waters (77%) with the rest coming from groundwater (23%). In 2010, the US withdrew ~355,000 million gallons per day (Mgal/d), or 397,000 thousand acre-feet per year (acre-ft/yr), from various sources of water for these activities. Not all water withdrawn was freshwater, as approximately 14% came from saltwater sources. The Total Withdrawals images show total surface water and groundwater as well as freshwater and saltwater withdrawal for the US by state.

In terms of surface water, thermo-electric power production used 51% of the total freshwater withdrawn and almost all of the saltwater (mining accounted for some saltwater withdrawal). Irrigation was the next largest use accounting for 29% of the total. In terms of groundwater, 65% went for irrigation and 22% for public systems. Most people in the US get their water from a public system.

Public systems vary from a single well that serve a dozen customers to large municipal systems that use multiple sources of water and serve millions of people. Typically, the water is collected from a groundwater or surface water source, treated within a water treatment plant, and delivered via pipes to individual customers. The water delivered from these systems must meet national **water quality standards**. How much water each of these public systems uses and for what purpose can vary widely between systems and regions. Several key factors account for these differences, including climate, types and extent of commerce and industry, and the age and efficiency of the systems.

For instance, hotter and dryer climates typically require more water to meet energy and cooling demands as well as outdoor water use such as irrigation. Southern parts of the US, particularly Southern California, Arizona, and Florida, have more than one growing season as well. Therefore, their annual agriculture water use is much higher. Cities with older water systems, such as Boston, have high rates of water that is unaccounted for. This is essentially water that leaks from main water lines. In some cases, the rates are as high as 30% to 40%. Newer cities have less water unaccounted for. For example, Phoenix, Arizona has an unaccounted water rate of about 6%. A common standard for measuring the efficiency of public system water use is total gallons used per capita per day (GPCD). GPCD varies between and within regions, ranging from as low as 50 to as high as 400. When compared to other countries, the US has some of the highest GPCDs in the world. Domestic use represents the largest volume of water use on average, accounting for 60% to 70% of total GPCD.

Domestic water use can be separated into two types of uses, indoor and outdoor. Residential outdoor water use varies from 15% to 60% depending on the climate. Irrigating grass, gardens, and trees being the largest. Most of suburban America is landscaped with turf, which requires the use of significant amounts of water. Cities in northern latitudes with high amounts of annual rainfall, such as Seattle, Washington, have natural rainfall that adequately supports most landscapes and thus use less water for irrigation. Cities in dryer regions, such as Phoenix, have high rates of evapotranspiration and less rainfall requiring more water to maintain landscaping. Outdoor water use also has a seasonal factor, with more being used in the summer months than winter months. Thus, cooling towers and pools in warmer climates represent a second major use.

Indoor water use can also vary. Las Vegas, Nevada and Seattle provide examples of this variation, as Las Vegas' indoor water use is 54 GPCD while Seattle's is 71 GPCD. It can also vary from one home to the next. Large and higher income families generally use more water than small and lower income families. Typically,

Water quality standards Quantitative standards set by regulatory institutions for the purposes of protecting human and environmental health

Gallons used per capita per day (GPCD) - a measure of water use efficiency for a community.

over half of indoor water use occurs in the bathroom with most of that coming from the toilet. Another 25% is used in the laundry room by the clothes washer. But these ratios are changing. Federal standards for water efficiency have greatly increased the efficiency of new toilets, shower heads, and faucets, and most manufacturers now produce more efficient clothes washers and dishwashers. As a result, newer homes typically have more efficient appliances and thus lower average water use. Much of the decline in per capita water use over the last 20 years can be attributed to these efficiency improvements.

Most urban water-resource planning focuses on human water use, but water is also critical for natural environments and wildlife habitats. Habitats in upland areas are dependent on seasonal rainfall and snowmelt. **Riparian areas** and wildlife habitats adjacent to streams and rivers often depend on shallow groundwater replenished by the river or stream. Some habitats require year round streamflows, while others have adapted to intermittent streamflows. For example, many species of fish require specific levels of streamflow to breed. Changes that reduce streamflow can result in changes to these habitats and impact wildlife. Such changes arise from several causes. Damming rivers and streams, overpumping or altering the natural recharge cycle of aquifers that are a source of water for streams, and reduced precipitation due to climate change can all lead to reduced streamflows. Such changes are one factor in the steady loss of riparian habitat in the US.

Riparian areas Natural areas of flora and fauna that are dependent on shallow groundwater that is recharged by an adjacent river or stream.

Water Treatment

Groundwater typically requires little, if any, treatment as the soil filters it as it percolates down to the aquifer. In contrast, surface water requires some level of treatment to be usable for potable purposes. How much treatment is required varies with the nature of the watershed and water source. Streams fed by mountain runoff such as those that provide water supplies to Colorado Springs, Colorado and Portland, Oregon require minimal treatment because the water has low suspended solids and organic material. On the other hand, Memphis, Tennessee, which uses the Mississippi River as a water source, must provide extensive treatment to remove suspended solids and organic material like bacteria that come from stormwater runoff and other sources.

Reclaimed water requires the most extensive treatment. There are typically four processes, involving both natural and engineered mechanisms, that can be applied—preliminary, primary, secondary, and tertiary or advanced. Preliminary steps include screening out large solid materials and removing grit. Primary treatment aims to remove contaminants and sediment that settle at the bottom of collection tanks as well as those that float to the surface. Secondary and advanced treatment processes are the most involved. Secondary processes are typically chemical or bio-based mechanisms (i.e. bacteria) that remove total suspended solids, dissolved organic matter (measured as biochemical oxygen demand), and, with increasing frequency, nutrients such as nitrogen and phosphorus, among others (NRC, 2012; p. 67). Advanced treatment, or tertiary treatment, usually involves microfiltration and reverse osmosis to remove any remaining solids as

An aerial view of a water treatment facility

well as bacteria, viruses, and pharmaceuticals. The treated water is then released back into groundwater or surface water reservoirs (Cho, 2011).

Urban Water Sustainability

Water is required to sustain human life, but it also plays an important role in lifestyle and culture. The landscapes that we find desirable, our preferences for recreation, our hygiene and health habits, and many of our technologies depend on adequate water. For instance, riparian areas, wetlands, lakes, and rivers require sufficient water to provide critical habitat and ecosystems services. Agriculture and industrial economies rely heavily on water to produce and ship goods. Many tourism- and recreation-based economies depend on water and snow. Thus, water is a critical foundation for social, environmental, and economic sustainability. Management of these resources is done under highly uncertain conditions such as drought and climate change, requiring tradeoffs. The next generation of community leaders will need to understand these systems, the tradeoffs inherent in managing water for them, and how uncertainties of drought and climate change can be accommodated in decision-making processes. It will also necessitate new approaches to planning.

Water sustainability can be defined in the terms of the 1987 Brundtland Commission report as the ability of institutions to manage water in a manner that meets the needs of present social and environmental systems without impairing the ability of future generations to do the same. There is no silver bullet to accomplishing this goal. As this remainder of this chapter will present, water sustainability is a complicated issue fraught with high uncertainty about the future state of social and environmental systems. Four conditions of water resource sustainability should be considered:

1. Availability of freshwater supplies under conditions of uncertainty, including long-term drought, climate change, and regional growth.
2. Ability to provide infrastructure needed to store, withdraw, treat, recycle, and deliver water for power, agricultural, and domestic use.
3. Ability to manage power, domestic, industrial, and agricultural demands for water.
4. Ability of private and public water management institutions to anticipate and adapt to changes within social and environmental systems.

Challenges to Water Resources Sustainability

The concept of water resources sustainability has been around a long time and has its roots in the concept of sustainable yield for renewable resources. Sustainable yield is the balance between the rate at which a resource is renewed and the rate at which it is harvested and consumed. At first glance, this seems to be a simple concept to implement; that is, identifying the rate of water renewal and managing demand so it does not exceed this renewal rate. If the social and environmental systems that generate demand for water and the natural cycle of water renewal were simple and stable, such an approach to water sustainability could be easily achieved. Unfortunately, this is not the case. Social and environmental systems are highly **complex adaptive systems** that are constantly changing in response to a number of internal and external forces. Our ability to understand how these forces impact the complex relationships within these systems and predict future forces and their impacts is very limited.

This is in stark contrast to the idea of stationarity that has been at the heart of water management for the last 100 years. "Stationarity—the idea that natural systems fluctuate within an unchanging envelope of variability—is a foundational

Water sustainability Based on the Brundtland Commission definition of sustainable development, the ability of institutions to manage water in a manner that meets the needs of present social and environmental systems without impairing the ability of future generations to do the same.

Complex adaptive systems Systems that have complex internal dynamics and functionality that are costly changing in response to internal and external stresses.

concept that permeates training and practice in water-resource engineering." It suggested that future yield can be predicted and that a plan to allocate it based on this prediction could be developed. An example of a plan based on this premise is the 1926 **Colorado River Compact**. In 1926, the seven basin states of the Colorado River watershed (Upper Basin: Colorado, New Mexico, Utah, and Wyoming; Lower Basin: Arizona, California, and Nevada) agreed to allocate a share of the Colorado River to each state. This allocation was based on the assumption that the previous 20 years (1906 to 1926) of the river's flow, identified via instrument gauging, was representative of the normal pattern of wet-dry cycles for the river. This stationarity-based assumption yielded an average flow for the Colorado of 17.5 million-acre feet per year.

Colorado River Compact An agreement signed in 1926 that allocated to Wyoming, Colorado, Utah, Nevada, New Mexico, Arizona, California, and Mexico, different portions of the Colorado River's normal and stored river flow.

Today, we have been able to reconstruct the flows of the Colorado River back almost 500 years using both instrumentation and tree ring records. The analysis shows that the river experiences very high and unpredictable levels of fluctuation between extreme highs and lows. The 20-year record used for the Compact was among the wettest periods over the last 500 years. The result was an over allocation of the river's resources, which are now estimated to be closer to an average of 15.5 million acre feet per year. A long-term impact could be that flow periods less than original annual estimate of 17.5 million acre feet will result in shortages. These would particularly impact the lower-basin states, as they would experience shortages sooner and more often than originally estimated. Since Arizona and Nevada water rights are lower than California's, these states would experience such shortages first. Fortunately, the impacts have been minor to date. The upper-basin states are not using their full allotment, so river flows have been adequate to store sufficient water in the reservoirs to meet the lower-basin state allocations.

Climate change introduces another challenge and a new layer of uncertainty for water supplies. Current estimates of temperature resulting from climate change generally agree it will increase globally with only the magnitude of the change in question. However, exactly what impact this will have on water supplies and demand is less certain. The relationships between snowpack volumes, snowpack melting, and stream flows are quite complex. Rising temperatures may cause more rain and less snow, reducing the volume of water stored in snow packs and causing snowpacks to melt sooner. For areas that rely on snowpack for storage of water such as Boston, California, Denver, Phoenix, Portland, and Seattle, this could create problems for existing systems of water storage and flood control. Temperature increases will also result in higher evapotranspiration rates, which may increase water demand for landscape irrigation, fountains, and pools. More air conditioning may be required, increasing demand for power and water for cooling towers.

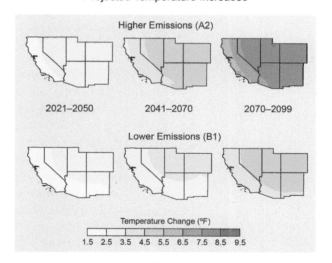

Projected temperature increased for the Southwestern US based on global climate models (GCMs) using high and low GHG emission estimates. Adapted from: Regional Climate Trends and Scenarios for the U.S. National Climate Assessment: Part 5. Climate of the Midwest U.S. NOAA Technical Report NESDIS 142-5.

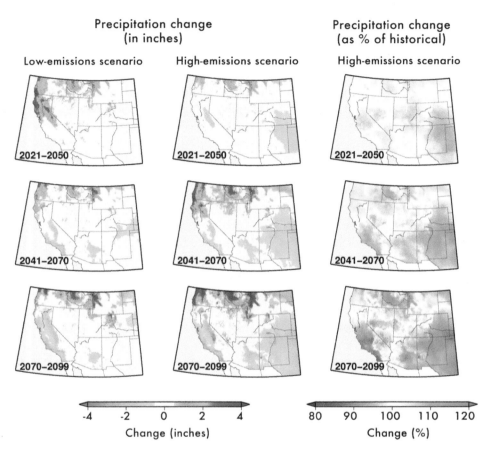

Potential precipitation changes in inches for low and high GHG emission scenarios (left and middle) as well as the percentage of change for the high GHG emission scenario.
Source: Southwest Climate Change Assessment Report (2013).

Global climate models Models that simulate the complex functions of global climate systems over many years.

Current estimates of the impact of climate change on location and magnitude of precipitation are not as certain as temperature. Temperature increases will affect how much snow falls in higher elevations and when it will melt. Generally, most of the 26 **global climate models** (GCM) included in the Intergovernmental Panel on Climate Change Fourth Assessment Report indicated higher levels of precipitation in the northern parts of the US and decreases in precipitation in the southwest. Yet, some models indicate the opposite—less precipitation in the north and more in the southwest. Thus, our ability to use GCMs to understand possible changes in extreme events and seasonality of precipitation is still limited. It is possible that a region may experience a decline in overall average precipitation but experience an increase in the severity of individual precipitation events.

Additional challenges include the uncertainty associated with social systems, financial resources, demand, and drought. Social systems are subject to high uncertainty regarding their future. For the last two decades, regional growth has been considered a certainty. The recent downturn in the economy and the collapse of the housing market has caused planners to pause and reconsider the future of growth in their regions. Even during the past periods of growth, there was a lot of uncertainty associated with what type and where growth would occur. These uncertainties make it difficult to estimate the future demand for water. Financial resources are needed to fund water supply projects, but financial markets undergo cycles of bear and bull. Predicting the timing and magnitude of these cycles remains limited. Thus, the question of whether financial resources will be available when needed adds additional uncertainty to water resource management.

The factors that affect demand are complex, and our understanding of them is limited. For example, how people will behave in response to increases in the price of water is not well established. We generally know that increases in price result in declines in demand, but how much reduction and how fast it might happen are unclear because few regions have experienced rapid and significant changes in the cost of water. How technology will affect water use is also unclear. Is there a limit on how little water can be used to flush a toilet or wash a load of clothes? Temperature impacts on outdoor water use is unclear as well. Although we have experienced variations in temperature from one year to the next and can document the impact such changes have had on outdoor water use, these variations have been minor and our experience is over a small range of change. Is the affect linear or will it decrease or increase over a larger range of change that may occur because of climate change?

Our experience with **drought response** to reduce demand is also limited. Drought response can be voluntary or mandated, but requires people to reduce water use that is likely to impact their quality of life or the economics of their business, hopefully for only a short period. Such actions include removing landscaping (grass, trees, and gardens) or allowing it to die, not filling a pool, taking fewer showers, not planting water-intensive crops, raising or lowering the thermostat, using water at off-peak periods, or reducing production of water-intensive products. Drought response is different than normal water conservation activities, which focus on reducing people's water use without negatively impacting their quality of life. Conservation actions include improving the efficiency of water-using appliances such as low-flow toilets and water-efficient clothes washers, changing habits to use water more wisely, and low-water use landscape design and maintenance. These actions can be implemented with little, if any, impact on a family's lifestyle.

Drought response government calls or mandates for reducing water use. These are typically short-term mechanisms that are designed to be suspended once the drought has resolved.

Some utilities have had experience with managing drought response using conservation mechanisms for short periods of a year or two. But little is known about how the public will respond and the impacts of drought response over prolonged periods, i.e. two or more decades. We cannot know if people or businesses will leave the region for places with more water or if they will stay and adjust their water use habits. Nor can we know if these changes will be prolonged but temporary or permanent. So, how do we achieve sustainability of water resources when there is so much uncertainty about the future? The answer does not lie in efforts to reduce uncertainty by better understanding social and environmental systems. Past experience has shown that the more we understand about these systems, the more inherent uncertainty exists within them. The answer lies within the complex adaptive systems themselves.

Solutions

Resiliency The capacity of a system to absorb change or reorganize and retain essential functions.

Adaptability The capacity of the actors in a social system to manage the system to successfully adapt to change.

Adaptation Reorganization of a system in response to internal or external change, in a manner that still retains the essential functions of the system.

Resiliency has emerged as a key concept for long-term sustainability of both social and natural systems. When systems are stressed by external forces, they change, or adapt, to maintain essential functionality. Thus, **adaptability** is one of the major components of resiliency. History shows many examples of where social systems were subject to some social or environmental forces such as drought and war. In some cases, the social systems were unable to adapt and the civilization collapsed. Others were able to anticipate change and effectuate a response, successfully adapting. These are two key factors of successful **adaptation**, anticipation and response for social-ecological systems (SES). For a full discussion on SES systems, see Chapter 2 in this text.

Unfortunately, even with advancements in the science of systems modeling, our ability to forecast the future is still significantly limited and our social-political systems often seem lethargic in all but the most critical of situations. Therefore to achieve water-resource sustainability in an uncertain environment, we must create tools and methods that allow us to anticipate multiple scenarios for the future and develop flexible strategies that can be quickly implemented as the future unfolds.

Anticipatory Governance

Anticipatory governance The process of using foresight to anticipate a wide range of possible futures, plan adaptation strategies for these futures, monitor changes over time, and act to adapt to change as anticipated.

A new model of planning and implementation called **anticipatory governance** is being used by water utilities to anticipate a range of futures, prepare a range of flexible strategies to respond to these futures, and implement these strategies over time. It consists of three basic steps:

1. Futures analysis: Development of an ensemble of possible futures that represent the full range of futures that can be currently foreseen for a particular issue. This can be based on expert opinion or can be developed using a model that forecasts future conditions across a range of values for one or more factors. Once the possible futures are identified, strategic concepts and trends are distilled across the entire ensemble to explore the sensitivity or risk of various factors and impacts.
2. Anticipate adaptation: Using the futures analysis, possible actions to adapt (react to change or effect change) to individual or groups of possible futures are developed. Such actions may be directed at preserving future options or responding to specific changes that may occur.
3. Monitor and adapt: Monitoring the present situation on a regular basis is necessary to identify changes that may indicate the realization or exclusion of one or more of the anticipated futures. As changes are realized, adaptive actions are implemented as anticipated or refined as needed. The City of Phoenix is one of the communities that has embraced this anticipatory governance model.

Case Study: Phoenix Water-Sustainability Planning Model

The City of Phoenix is located in central Arizona's Sonoran Desert at the southern edge of the arid southwestern US. The region's average precipitation is eight inches per year. In 2015, Phoenix's population was 1.5 million people located in the center of a region of 4.5 million people. Phoenix water utilities delivered an average of 269 million gallons per day, or 175 gallons of water per capita per day, to its customers. The City has a robust water supply portfolio consisting of surface water, groundwater, and reclaimed water. It is one of a few US cities that reuse almost 100% of its effluent, using 30% for agriculture and turf irrigation, 30% for power production, and 40% for environmental uses.

It currently relies primarily on its surface-water supplies from the Colorado River, delivered via the Central Arizona Project (CAP) canal, and the Salt and Verde rivers, delivered via the Salt River Project (SRP) reservoirs and canals, although it has substantial groundwater supply (650 billion gallons, or 2 million acre feet of Colorado River credits). Phoenix estimates it has enough water supplies to meet its needs today and growth for the next 100 years under normal conditions. But water supplies in the southwest are anything but normal. Given its extremely variable conditions, Arizona's surface water management over the last 100 years has been based on storage of water during wetter years and delivery during dryer years. Colorado River reservoirs provide storage for California, Nevada, and parts of Arizona, while Salt and Verde River reservoirs provide storage for central Arizona.

During Phoenix's growth boom in the 1960s and 1970s, agricultural and urban uses were increasing their reliance on groundwater, which threatened to drain central Arizona's aquifers and create a **deficit**. Surface water delivered via the Central Arizona Project canal in the 1970s provided an opportunity to reverse this trend. In 1980, the state adopted the Groundwater Management Act (GMA), creating **Active Management Areas (AMAs)** in the central part of the state where groundwater withdrawal was occurring. A goal of the GMA and AMAs was to achieve sustainable yield by 2025. It was anticipated that the primary method for accomplishing this would be to switch groundwater use to surface-water use. The AMA also required all new subdivisions to demonstrate that they have adequate water supplies (surface and groundwater) to meet the subdivision's water needs for 100 years before building was approved. However, the GMA assumed that surface water supplies were stationary and did not account for climate variability or long-term droughts.

In 2002, as the City of Phoenix began the update of its Water Resource Plan, it wanted to move away from the stationarity model of predicting and planning embedded in Arizona's standard requirements. As a result, it moved to a model of anticipatory governance. This effort included elements of **foresight** and flexible **adaptive strategies**. Since then, the City has updated its Water Resource Plan twice. The 2005 Water Resource Plan Update included a complete assessment of Phoenix's Water supply. The 2011 Water Resource Plan reassessed the water supply, looked specifically at long-term drought and climate variability, and incorporated near- and

Deficit When demand exceeds supply.

Active Management Area A regulatory term used in Arizona to identify regions of the state that must comply with the 1980 Groundwater Management Act goal of sustainable yield for the regions' ground-water resources.

Foresight The process of anticipating a possible future.

Adaptive strategies The unique way in which each culture uses it's particular physical environment; Those aspects of culture that serve to provide the necessities of life - Food, clothing, shelter, and defense.

long-term deficit management strategies. Finally, it integrated the City's water resource, conservation, and drought management plans to "reduce the risk of future deficits resulting from surface water shortfalls" (City of Phoenix, 2011). See the full plan at https://www.phoenix.gov/waterservicessite/Documents/wsd2011wrp.pdf.

Foresight

City of Phoenix planners, officials, and stakeholders began by identifying three key factors that held the most potential significance for water-resource planning as well as high degrees of future uncertainty. They included 1) delivery of surface-water supplies (Salt River Project and Colorado River), 2) growth and development patterns, and 3) water conservation levels. They defined ranges of possible future conditions for each of these factors. Relying on tree-ring stream reconstructions to provide a wider range in variability in wet and dry periods for the Colorado and Salt/Verde river systems, the team estimated allocations from these systems under normal, moderate, and severe drought conditions.

Using a trend-analysis approach, they developed several spatial growth scenarios, including an accelerated growth rate, changes in the regional economic base, higher density in peripheral areas, higher densities in the core area, and a transit-influenced growth pattern. These scenarios were used to estimate demand based on residential and employment growth. Three levels of customer water use were anticipated based on past trends and possible future trends. These factors were combined to generate 144 scenarios of water supply and demand.

In developing its 2011 plan, Phoenix incorporated climate change and drought as part of its foresight planning. Phoenix participated in a partnership of local, regional, and federal agencies (City of Phoenix, Salt River Project, Central Arizona Project, Bureau of Reclamation) together with state universities (University of Arizona and Arizona State University) to regionally downscale scenarios of precipitation and temperature estimates from several global climate change models, estimate the impacts of these scenarios on the streamflow of the Salt/Verde river system, analyze impacts on reservoir management, and estimate impacts on local municipal water allocations. These new scenarios were used to create new scenarios of surface water supplies under average and extended drought conditions and inform water management decisions.

Flexible Adaptive Strategies

As part of its 2005 Water Resource Plan, Phoenix identified two types of flexible strategies: robust short-term strategies and a worst-case infrastructure timeline for drought response. Robust strategies are those that can be implemented immediately and work well across a wide range of scenarios. For example, an analysis of all the scenarios showed that if no more growth or only little growth were to occur, adequate water supplies were available to meet existing demand even under the most severe surface water shortage scenarios with moderate drought response to decrease water demand. However, if growth were to occur even at moderate rates, this growth would require infrastructure to deploy existing water supplies and, under conditions of water shortage, acquisition of supplemental supplies would be required. In response, the City placed the burden of financing new infrastructure and supplies for new growth under normal and shortage conditions on growth itself by increasing its water-acquisition impact fee, a fee required to be paid with the building permit of each new home or building. This strategy works well under normal and shortage conditions, and under conditions of slow or fast growth.

Water supplies are essential to a community, and the failure to meet demand can result in the community's economic failure. Various strategies can be used to aggressively reduce demand or enhance supply in critical water shortage conditions. If these strategies fail to meet a community's essential water needs, sustainability will be lost. One approach to minimize such failure is to anticipate a worst-case scenario and develop flexible strategies to implement incrementally as a worst-case situation unfolds. The basis for Phoenix's worst-case scenario was to assume that current trends were proceeding to the most severe water-shortage scenario that could be reasonably anticipated, a 35-year dry period. Phoenix was in its 10th consecutive dry year in 2005. In its 2011 plan, a timeline was developed that estimated the magnitude of water shortages and when they would occur over a 25-year time frame.

Using a worst-case projection of demand based on aggressive growth, a timeline of trigger points is developed that identify when new water resources or drought-demand reductions might have to be deployed to meet basic community water needs. Phoenix utilities, like most utilities in the US, have been experiencing substantial stress on its financial resources that are limiting its ability to adapt to long-term drought and climate change in the near term. Using its infrastructure timeline as a just-in-time **decision tree** has been determined to be essential in minimizing the investment needed to adapt over time.

Decision tree For a specific problem, a diagram of questions (nodes) and answers (branches) that guide the decision maker to a recommended solution to the problem.

Using an infrastructure timeline enables each potential strategy to be assessed by water volume, certainty of availability, cost, and time to deploy. This assessment can then be used to create a decision tree that indicates the latest point in time that action would be required to preserve future options, plan future options, or fund and deploy future options. Using the decision tree, short-term plans for appropriate near-term actions can be created. Indicators of climate, drought, and certainty of supplies can be monitored, and the decision tree and short-term plans reassessed and updated as conditions change.

Summary

The social and environmental systems that support water resources and water management are complex adaptive systems that are constantly changing in response to internal and external forces. Our limited ability to predict the future of these forces and their impact on these systems creates a high level of uncertainty about the future of water resources. Using adaptive governance, foresight, and adaptive strategies to anticipate a range of possible futures and plan flexible responses increases the resiliency of our institutions, allowing us to successfully adapt as changes in environmental and social systems occur. Our view of sustainable water resources must embrace a wide range of viewpoints about water sustainability and the values they represent. Flexible, adaptive strategies allow us to balance a wide range of water needs by understanding the value-based tradeoffs needed to reduce demand, enhance supplies, and finance and build the infrastructure needed to treat and deliver freshwater under a variety of possible future conditions.

Chapter 12

Sustainable Transportation

Aaron Golub and Jason Kelley

Introduction

Some of the world's most pressing problems result from how its urban systems operate. These systems consume huge amounts of energy and materials and create intense local "hotspots" for emissions, solid waste, water pollution, congestion, safety, and other challenges to livability. Of course, well-managed urban systems can be fairly efficient and effective at providing sustainable livelihoods for large numbers. Urban transportation systems, in this vein, can be both an asset and a liability to the development of sustainable cities. In the US, a large share of energy consumption, carbon emissions, and preventable death and injury result from urban transportation systems. For example, because of enormous growth in automotive use, about half of Americans currently live in counties that fail the now four-decades-old National Ambient Air Quality standards, even after spending billions of dollars on technology to reduce emissions from automobiles.

In this chapter, we explore the sustainability challenges and solutions of urban transportation systems in four steps. First, we define the sustainability problem as a special class of urgent and harmful multi-scale and multi-sector problems. We can then clarify sustainability problems that US urban transportation systems create.

Second, to better understand our contemporary transportation issues, we must consider the linkages between transportation and land use. The history of US cities illustrates these connections. By examining the past, we can recognize how innovations in transportation allowed for today's unsustainable patterns of low-density development or what is commonly termed suburban "sprawl."

Third, to move from defining a problem to solving it, we must understand the problem's drivers or causes. For example, looking at car exhaust is just not enough to understand urban air pollution. We need to investigate all forces behind this pollution and its impacts. If we could snap our fingers and eliminate air pollution, we would have solved the problem a long time ago. We need to recognize that air pollution is an outcome of a complex system of institutions with their own web of rewards and feedbacks and that much of society benefits from the activities that generate it.

Fourth, we can then move toward solutions. Thus, understanding the causes and impacts allows us to uncover solutions. We can also develop intervention strategies, such as policies and practices, to focus efforts for change. This includes "low-hanging fruit," the easy changes in practice that yield large benefits, as well as the deeper underlying forces and values that may take decades to change.

This chapter will first introduce and define the special class of problems that are sustainability problems, and then explore the various problems that stem from urban transportation systems in the US. Next, it will examine the way in which transportation and land use have historically interacted in US cities, resulting in our present-day patterns of automobile-dependent suburban sprawl and its associated problems. Then, it will explore the various drivers of those problems, which form barriers to moving forward toward solutions. Finally, examples and cases from both the US and internationally that show some promise toward solving these urgent and messy problems will be explored.

Sustainability Problems and Solutions

Human-ecological systems Humans are dependent on natural systems, such as the water cycle or other nutrient cycles. On the other hand, humans alter these cycles by their effects on the natural environment. This combination of dependence and effects are characteristic of human-ecological systems.

Complexity A characteristic of a system based on webs of interrelated and interdependent institutions and subcomponents. Complexity means that changes in certain inputs yield unforeseeable and unintended consequences.

Multiple sectors Urban systems are built on various sectors, such as housing, transportation, and the various systems that supply them with the resources they need, such as fuels, electricity, and other materials.

Not all problems we face are sustainability problems. Sustainability problems are a special class of problems that pose a particularly urgent threat to **human-ecological systems**. Their urgency is compounded by their **complexity** and their involvement of **multiple sectors** and actors across multiple scales. This complexity means that they are best tackled using interdisciplinary approaches as diverse as the many systems that affect the problem or by applying new methods to these problems.

An example of a sustainability problem is urban air pollution. It poses significant harm to the health of many urban residents and causes billions of dollars in damage to infrastructure and crops, lost worker productivity, and additional burdens on the healthcare system. It is caused by a complex array of factors: emissions from electric power generation, exhaust from trucks and automobiles, and emissions from construction and industrial sites, among others. Cultural and psychological factors underlie household and society practices that contribute to air pollution and need exploration as well. To tackle the broader air pollution problem requires understanding such diverse issues as household energy use, freight logistics, the demand for automobile and air travel, and the technological and regulatory factors governing automobile, freight, industrial, and construction site emissions. It gets even more complex when governments set air pollution standards, implement strategies to meet these standards, and monitor them. As we can see, something as "simple" as clean air is not really a simple matter, but requires a team approach involving efforts from across many different disciplines, such as urban planning, engineering, business, and public policy.

This chapter focuses on sustainable urban transportation. We will explore the underlying sustainability problems current transportation practices pose, uncover the sources of problems in urban transportation systems, and explore sustainability solutions.

US Urban Transportation and its Sustainability Problems

Following from such a definition of a sustainability problem, we can see the complexity of urban sustainability problems and how urban transportation contributes both directly and indirectly to these problems. This relationship means that focusing on urban transportation is an effective approach to solving some of urban areas most pressing problems. Here, we introduce some of these urgent problems.

Most urban travel in the US is by automobile. In 2000, 88 percent of US workers drove or were driven to work, and fewer than 5 percent took public transit (Ardila-Gomez, 2004). For all trips, not just those to work, only about 6 percent are by human power (biking or walking). Compare this statistic with other industrialized countries like England, where 16 percent walk and bike, or Germany, where more than 34 percent bike or walk (Ardila-Gomez, 2004). Thus, the significant problems arising from our transportation system are largely the result of using private automobiles for mass transportation. Here, we review briefly some of these problems, grouped along social, economic, and environmental realms.

Social Problems

Social Disruption from Traffic Fatalities and Injuries

National Highway Traffic Safety Administration (NHTSA) statistics show that 35,092 people died in auto-related accidents in 2015. This is a significant increase (7.2%) from 2014 (32,744 deaths). It is also the largest increase in nearly 50 years (NHSTA, 2016). On top of these fatalities are about 200,000 injuries from traffic crashes monthly, resulting in thousands of permanent disabilities and days, weeks, or months of physical therapy and recovery, and countless days lost from work or school.

Social Inequality, Exclusion, and Isolation

Planning a transportation system around the need to own and operate a personal vehicle means that the system will be poorly configured for those who are unable to do so. In most metropolitan areas, around 25 percent of the population is too

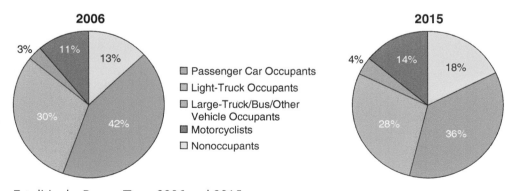

Fatalities by Person Type, 2006 and 2015

Source: NHTSA. (2017). Summary of Motor Vehicle Crashes (Early Edition). Traffic Safety Facts 2015 Data. https://crashstats.nhtsa.dot.gov/Api/Public/ViewPublication/812376

old to drive, too young to drive, or not able to afford an automobile. The dispersion and suburbanization of jobs and housing and the resulting automobile dependence means that this population has a harder time finding work. As a result, they can become isolated and excluded from the mainstream of society. In many central cities where there is often a lack of large grocery chains, automobiles, and decent transportation, "food deserts" can emerge. This forces the low-income populations that live in urban cores to rely on fast food, convenience stores, or small grocers with limited selections of healthy foods. The reliance on cheaper but less healthy food options has been shown to create health problems, especially in inner-city neighborhoods. The "Food Desert" is discussed in Chapter 9.

Sedentary Lifestyles and Detrimental Health Impacts

Studies have shown that transportation has a significant impact on how active people are, and in turn, on their health as well. The lack of "walkability" in many metropolitan areas leads to low rates of cycling and walking. This inactivity is linked to higher body mass indexes and poorer health. Obesity and diabetes are at alarming rates in many segments of the population, including children.

Ethical Dilemmas of Petroleum Dependence

The strict reliance on petroleum for the operation of the economy is called "petroleum dependence." Today, US demand for petroleum overwhelms the country's own domestic supply: more than half its petroleum needs are imported from other countries (See Box 1). Ethical problems arise as dependence on oil imports constrains US political decisions. More fundamentally, however, dependence on

Box 1 Oil and United States—A Short History

Oil was initially used in the US to produce kerosene for illumination. It was discovered in large quantities first in Pennsylvania in the 1850s, though it would not be until the 1890s when demand for it would rise that these sources would be exploited. Oil was discovered in California and Texas around 1900, during which time a single company dominated the system for distributing and refining it—Rockefeller's Standard Oil Company. After 1900, the demand for oil products exploded—and so did Standard's wealth, along with threats from new competitors and government displeasure with Rockefeller's ruthless anti-competitive behavior. In 1911, Standard was broken up into more than a dozen independent companies, many of which still exist today, such as Chevron, Exxon, Mobil, Amoco, ARCO, and Conoco, among others. The international system of oil production and distribution also began during the last years of the 1800s, with tight competition between Standard Oil and Royal Dutch Petroleum Company (now Shell) in Indonesia. Soon after, Mexico, Venezuela, Iraq, and Iran became locations of significant oil extraction.

While in 1900 only about 25 percent of the few thousand existing automobiles ran on gasoline—most were steam or electric—in a short time, gasoline would emerge as the dominant fueling technology. Significantly as well, military prowess became linked to petroleum as World War I

(1914 to 1918) showed for the first time how important planes and tanks would become to the future of warfare. After 1920, the demand for gasoline in the US would explode with the number of automobiles reaching 27 million in 1939 (Philip, 1994; 36). Thus, the die was set—our economy would become increasingly dependent on petroleum, and those firms who could supply it would become increasingly powerful. On the eve of World War II, the US contained over 80 percent of the world's automobiles and consumed 65 percent of the world's oil produced (Philip, 1994; 36–37).

Between the late 1930s and 1950, large oil deposits were discovered in the Gulf countries of Saudi Arabia, Iran, Iraq, and Kuwait, completely rearranging the world oil map. Initial dominance of the British in the Middle East followed from their colonial administration of it after the fall of the Turkish Empire during World War I. During and after World War II, diplomacy was increasingly used to support oil production by US-based companies in the Middle East and edge out British competition. By the 1950s, five US oil companies, together with two European firms, controlled nearly all of the Middle Eastern oil resources. These "Seven Sisters" included British Petroleum, Royal Dutch–Shell, Gulf Oil, Chevron, Exxon, Mobil, and Texaco.

This close relationship between the US government, private oil companies, and the governments in the Middle East would become problematic. Rapidly rising demand for oil in the US following the full shift to Fordist consumption after World War II would put pressure on the US government to preserve these special relationships in the name of oil supply stability. (The US demand for oil would outstrip its own internal production in the 1960s.) The power struggle over oil, while complicated by the tension caused by the presence of the Soviet Union in the Middle East, forced the US to support many anti-democratic regimes over the years.

For example, the relationship between the US and Saudi Arabia is particularly close—it provides the US with access to some of the largest known oil reserves on the planet in exchange for guaranteed revenues and protection of the Saudi government against external or internal aggressors. In another example, the US supported a coup against Iran's Mossadeq government after it nationalized (converted from private to public ownership) the Anglo-Iranian Oil Company in 1951 (ironically, the Anglo-Iranian Oil Company was itself a national company of the British government). This coup installed the pro-US Shah as ruler. A similar coup was supported in Iraq when it too threatened to nationalize its oil system, leading eventually to the rise of Saddam Hussein. His desires to restrict oil production to raise prices were, in part, responsible for the 1991 US invasion. While Hussein had always cooperated with the US, his new activism did not follow US plans. US strategy in the region also forced the US to overlook Iraq's attack on the Kurds in 1973 and forced it to arm groups such as the Mujahedeen in Afghanistan.

The US and Europe became so dependent on Middle East oil that changes in oil production policies among Middle East oil producers had profound effects on their economies. In late 1973, responding to US and European support for Israel during its war with Egypt and Syria, several countries in the region restricted or completely halted oil production. This quickly quadrupled the world price and caused shortages throughout the US. World prices would soon stabilize, but spike again during the 1979 Iranian revolution over the US-backed

(Continued)

Box 1 Oil and United States—A short history (*Continued*)

Shah. To this day, oil prices continue to rise and fall with changes in production and policy in the oil producing countries. The US has reduced some of its dependence on Middle East oil by improving its energy efficiency and moving to alternative sources for oil, such as Mexico, Venezuela, and Nigeria. Still, it remains in a highly vulnerable position in the world system. In 2008, it imported 57 percent of its petroleum needs, about one-third of that from the Persian Gulf.

For more information on the history of oil and the development of US policy, see The Political Economy of International Oil by George Philip (1994) and The Prize by Daniel Yergin (2008).

Bird killed in Gulf of Mexico oil spill.

© Nate A/Shutterstock, Inc.

oil means we are forced to use it, even if we do not want to. A lack of alternatives or high costs associated with available alternatives limits peoples' choices. Petroleum dependence poses a significant ethical dilemma for those residents hoping to choose how their lives affect the larger world.

Economic Problems

Costs of Traffic Fatalities and Injuries, Traffic Congestion, and Petroleum Dependence

Traffic fatalities and injuries impose large financial costs on our society. Some of these costs are borne by car insurance holders, others fall on society at large. These "externalized" costs are estimated to be between $46 and $161 billion per year (Cervero, Golub, & Nee, 2007). In good traffic conditions, driving is normally the fastest way to travel in US cities. However, during rush hour, the average traveler can suffer from long delays. Petroleum dependence also imposes other financial costs on the US economy.

Significant costs, estimated to total between $7 and $30 billion per year (Etkin, 2001), result from lack of flexibility in the economy to respond to changes in price. Additional costs result from the non-competitive structure of the oil industry, resulting in prices that are higher than what a competitive market would charge. The sum of these costs since 1970 is estimated to be over $8 trillion. Finally, the US military incurs costs for its presence in locations of strategic importance to the oil industry, with estimates of between $6 and $60 billion (Ewing, Bartholomew, Winkelman, Walters, & Chen, 2010).

Local Air Pollution

The Clean Air Act, enacted in 1970 and enforced by the US Environmental Protection Agency, has had a major impact on regulating and reducing pollution emissions from automobiles for more than 40 years. Most pollution is reduced to just a small percent of what it was before regulation. But, large increases in driving and worsening congestion in metropolitan areas means that although each vehicle is cleaner, local air pollution remains a national problem. More than 120 million Americans live in counties that fail at least one of the National Ambient Air Quality Standards that define what levels of pollution in the air are safe to breathe (Ewing, Bartholomew, Winkelman, Walters, & Chen, 2010).

Infrastructure Barrier Effects

Infrastructure for automobiles, such as freeways and arterial roads, are large, intrusive, and can separate neighborhoods from each other and cause barriers to mobility. Studies show that these "barrier effects" exacerbate automobile dependence because they can deter residents from walking or cycling for even short trips.

A lot of used cars in the junkyard.

Greenhouse Gas Emissions

Greenhouse gases (GHGs) in the atmosphere manage the planet's greenhouse process, whereby temperatures are regulated. Most GHGs are created from the burning of fuels to generate energy either in electric power plants, factories, or in vehicles which burn fossil fuels for energy. Overwhelming evidence suggests that the significant GHG emissions created by human activity rival the amount of gases produced by natural ecological systems and influence the planet's normal climate. Reducing GHG emissions is essential to avoid the worst effects of climate change. Unfortunately, because human-induced climate change has already begun, we are already too late in avoiding some significant impacts.

Transportation systems burn fossil fuels, which create GHG emissions such as carbon dioxide and methane, either in the vehicle's own engine or in power plants that make electricity for electric vehicles' batteries. Transportation is responsible for about one-third of our country's GHG emissions. About 70 percent of that is for cars, light trucks, and SUVs (Fitch, Thigpen, Cruz, & Handy, 2016; Freund & Martin, 1993).

Production and Disposal of Vehicles

Cars and lights trucks use a large amount of non-renewable steel, glass, rubber, and other materials. Data from 1990 showed that automobile production in the US consumed 13 percent of the total national consumption of steel, 16 percent of its aluminum, 69 percent of its lead, 36 percent of its iron, 36 percent of its platinum, and 58 percent of its rubber. Around 10 million automobiles are disposed of every year (Golub & Henderson, 2011).

Environmental Impacts of Petroleum Extraction, Transport, and Refining

Negative environmental effects occur throughout the supply chain—from spills at the local sites of oil extraction to toxic pollution emissions at ports and refineries to service stations where fuels can cause groundwater contamination. Oil spills, large and small, are particularly

Gulf shores Alabama.

problematic. Roughly 10 million gallons are spilled into US waters every year (Harry, 2008). This does not include the large spills such as the 2010 BP oil spill of around 170 million gallons or the 1989 Exxon Valdez spill of 11 million gallons. Worldwide, more than 3 billion gallons have been spilled into waters since 1970. Oil is toxic to animals and oil spills disrupt natural ecosystems and kill thousands of species. Inhaling or ingesting oil is fatal. During an oil spill, food sources are often coated with oil and animals consume this tainted food. It also coats water and land surfaces, suffocating fish and small invertebrates, and animals' fur and feathers, preventing their skin from breathing and their bodies from maintaining ideal body conditions, resulting in death (ORR, n.d.)

The Transportation and Land-Use Connection

The unsustainable problems resulting from our contemporary urban transportation systems may seem like a daunting list. Taking time to understand what contributed to them is an important first step. Thus, before any attempt is made to address these problems, we must first ask how we got to this point, were our cities always like this, and what can we learn from the past.

A critical point to remember is that transportation does not work in isolation. If a new road is built in an area that was once inaccessible, that land may suddenly become attractive for new development. The resulting construction creates additional travel demands, placing stress on existing transportation infrastructure. To meet this new demand, roadways are often expanded, speeding up travel time and increasing accessibility. This, in turn, makes the area even more attractive to new development. More homes and businesses are built, more demand is created, and the cycle continues to escalate. This illustrates the explicit link between transportation and land use. A change in one creates additional demands on the other. This section examines these linkages by looking back in time to illustrate how urban transportation and land use have been mutually interdependent throughout the course of US urban history.

The Early Walkable US City

During the early 1800s, US cities were "walkable" places. Cities were compact out of necessity as the primary mode of transportation was by foot; therefore, residents needed to be within a reasonable distance from jobs located in the urban core. With the introduction of the steam locomotive, wealthy residents could move farther from the urban center and "commute" to their downtown jobs. Working-class residents lacked such an alternative. Streets were typically unpaved and turned to mud when it rained, further making cross-town travel difficult at best. The omnibus, a sort of urban stagecoach, provided protection from the elements, but traveled as slowly as a person on foot.

The Streetcar Shapes US Cities

The early horse-drawn streetcars of the 1850s allowed for what could be considered the first "suburban" development in US cities. Though they traveled just slightly faster than the omnibus, the horsecars ran on rails avoiding the problems created by muddy streets. As tracks were laid and horsecar service expanded, a thin band of once inaccessible land on the urban fringe was now within reach. These street-cars relied on horses, however, which required food, water, and rest. Horses were also subject to disease and other debilitating conditions. In fact, horses often got sick and died in the middle of the street. As a result, city services were organized to remove the thousands of horses that succumbed every year.

The electric streetcar, introduced in the late 1880s solved these problems. Powered by overhead electric lines, the electric streetcar quickly replaced horsecars and became the primary form of mass transportation in cities around the country. The electric streetcar also allowed for rapid suburban development. Often financed by local land owners, the lines expanded the range of land development further from the city center. Rail lines extending from downtown provided accessibility to undeveloped, but now accessible land, which was sold and subdivided for residential development. Residents required easy access to the rail lines, however, so neighborhoods were built for walkability. To accommodate this, numerous houses were built on narrow lots that lined the streets; thereby, shortening walking distance. Land that was too far from a rail line and outside of the walkable range remained undeveloped. As streetcar lines extended beyond the urban periphery, many cities took on a distinctive "star-shaped" pattern when viewed from above, with suburban growth radiating from the city center along these routes.

The Automobile Takes Hold

The electric streetcar dominated US cities until the turn of the century. This was not to say that the automobile was absent from the scene, however. First patented by Karl Benz in the 1880s, the automobile was to remain exclusively a toy of the wealthy for the next few decades. High production and consumer costs kept the automobile out of reach for the average worker. With the advent of the Model T in 1908 and the introduction of assembly-line production in 1913, automobile prices began dropping to affordable levels. By the 1920s, what was once a toy of the rich was now viewed as the wave of the future for everyone.

As automobiles began to clog city streets originally designed for pedestrians, new urban problems emerged. Traffic congestion, accidents, noise, and automotive air pollution further encouraged people to look to the suburbs for relief. Unrestrained by rail lines, the automobile could go were no streetcar could go, only faster. Lands once inaccessible by rail were now open to development. The only thing needed was roadway infrastructure. While the New Deal funding of the Great Depression allowed for further expansion of our national road system, it was the interstate highways of post-war America that allowed for the suburban building boom.

The Freeway and Suburban Sprawl

Sold to the public for its defensive and economic benefits, the Interstate Highway System had a significant impact on urban America. Extending out of downtown areas, the freeways allowed commuters to now live and work much farther apart. Lured by the promise of the "American Dream" and the suburban lifestyle, the freeways provided high-speed mobility and opened vast expanses of undeveloped land for suburban sprawl and tract-home development. Employers, lured by cheap land and a growing available labor force relocated to suburban locations, while retailers abandoned downtown areas in favor of climate-controlled shopping malls strategically located for freeway access. Edge cities, new suburban "downtowns," emerged near the interchanges of major freeways. These edge cities were designed with the automobile in mind, containing large-scale developments isolated from one another by ample surface parking lots and wide access roads.

Meanwhile in the central city, freeway construction was targeted for designated "slum" areas, typically home to low-income and minority residents. Providing a "path of least resistance," these politically and economically marginalized communities were powerless to oppose the urban freeway construction through their neighborhoods. At the same time, powerful downtown land interests viewed the freeways as a means of easing downtown traffic congestion as well as attracting shoppers back to the central city. In the end, however, the urban freeways simply sped the exodus of residents, jobs, and retailing to suburban locations, often destroying entire urban neighborhoods that stood in their way.

From Accessibility to Automobility

In little over 200 years, the American city was transformed from a compact, walkable, urban fabric built with accessibility in mind to a sprawling metropolis geared exclusively for automobility. The automobility paradigm is centered almost entirely on providing mobility for those with cars. The results of this transformation have already been discussed, with a long list of contemporary unsustainable transportation issues now facing US cities. Before examining the potential solutions, however, it is necessary to examine the key players shaping our contemporary system.

Agents of Automobile Dependence

Urban transportation systems in the US are driven by a complex set of factors resulting from the combination of numerous histories, cultural norms, expectations, and practices. Urban transportation is shaped and reshaped, produced, and consumed across several groups of actors. Entering into a discussion about fundamentally changing urban transportation systems in the US means one must consider the needs of these various actors, how they interact with each other, and how they respond to demands for change. We must understand that a web of actors *benefits* from the current system. In this section, we briefly consider these different actors and how they interact. This discussion will then lead us into our final section, where we discuss strategies for change.

The Individual and the Household

The individual and household sit at the most micro level of activity, making decisions about how to travel, home location, and vehicle ownership. These decisions are made mostly on minimizing travel times and maximizing convenience. As work and home became more decentralized, the automobile was a clear choice for travel as it delivers significantly higher performance than public transit systems. This is due primarily to the government's much greater attention and resources placed on guaranteeing the performance of driving than ensuring convenient and accessible public transportation. It is not a result of any technical or "natural" advantage that cars have over public transit.

Moreover, individuals and households engage with larger cultural forces. Automobile ownership is often seen as a symbol of status, patriotism, and of belonging to and supporting mainstream society. For instance, car ownership in general and specific vehicles in particular are often powerful representations of an individual's status. In automotive manufacturing hubs, ownership of vehicles manufactured locally demonstrates both patriotism and support for the local economy. These cultural contexts can become significant shapers of decisions regarding automobiles.

Planners and Developers

Urban planning emerged as an important force in the process of urban development in the US. Early in the 20th century, planners felt that suburban areas offered a better quality of life compared to the crowded and dirty industrial cities of the time. To this day, much urban planning practice merely reproduces suburban, automobile-oriented models.

Developers reproduce the suburban model, not necessarily out of a particular preference, but because it represents the least risk. Banks are more likely to lend construction loans to build traditional suburban developments, and developers find it easier to develop fresh "greenfield" sites on the edge of cities. Whereas, urban infill runs the risk of potential neighborhood conflict and higher or unpredictable construction costs.

Modern suburban neighborhood.

Greenfield Urban growth that happens on previously undeveloped "green" land, such as agricultural or forested areas. This is in contrast with urban growth, which happens by reusing previously developed land within the city's boundaries.

The State and Federal Government

State governments have a special role in urban transportation systems because they are tasked with overseeing the design and construction of the interstate highway system as well as public transportation. Most states also collect their own gasoline taxes, which is primarily used for investment in roads and freeways. The federal government plays an important role in supporting automobile use and urban development around automobile dependence as well as regulating it. Federal funds have long been used to support automobile use. They were first used to build roads during the 1920s and the 1930s New Deal stimulus package. The 1956 Interstate Highway Act solidified support into a set of financing systems, based on the national gasoline tax and federal planning support, to build the national network of interstate highways we have today. As noted previously, US foreign policy is heavily tied to the stability of oil supply in order to keep gasoline prices low and predictable.

Federal policies have also been important in managing automobile use. These include regulations to control safety, pollution from automobiles, and fuel economy standards that all automobile companies must follow. Federal funds also support public transportation systems, though in small amounts compared to the monies spent for roads.

Industrial Structures: Oil and Automobiles

The oil and automobile sectors are some of most heavily concentrated in the entire US economy—relatively few companies account for nearly all of their industry's production. This concentration means that they can easily coordinate their concerns, influence public policy, and shape consumer demands through organized action. Thus, urban transportation systems, and dependence on petroleum and automobiles, are tied directly to the needs of the oil and automotive industries.

The Post-war 1956 Interstate Highway Act, passed by a federal commission with ample automobile-related representatives, solidified the course toward automobile dependence, guaranteeing financing and planning support for freeways. Though homebuilding in the suburbs pre-dated the Highway Act, the pace of suburbanization exploded after its passage. Automobile manufacturers rode the wave of public investment in freeways and suburbia in the post-war economy, overcoming competition from transportation alternatives, such as streetcars, in most cities.

Surrounding larger oil and automotive industrial sectors sit countless small companies—automobile parts suppliers, independently owned car dealers, service stations, repair shops, along with other related sectors such as automobile insurance companies, drive-through restaurants, and suburban homebuilders. In the 1970s, one study found that all together automobile-related industries contributed one-seventh of the total US economy.

Sustainability Solutions: Toward Sustainable Transportation in the United States

Thus far, the chapter has explored that sustainability problems that urban transportation systems produce, the historic linkages between transportation and land use in US cities, and some of the most powerful drivers of the current transportation system. Now, we envision sustainability solutions to particular problems by appreciating the drivers of those problems. Understanding the drivers helps us formulate specific strategies. Since we have emphasized the social nature of these problems, we will emphasize a social approach to their solution by looking at examples of social change and social movements that work towards advancing sustainable transportation. This approach generally involves challenging and reducing automobile dependence, which, in turn, reduces driving, fatalities, emissions, and costs, among other things. Challenging the broader role of driving in society yields compounding benefits across different problem areas.

In this section, we explore several solutions that build on the social context of transportation systems across various scales and institutions. The first is an urban planning approach that reorients the city away from traditional automobile-based planning. We look at the cases of Curitiba, Brazil and Portland, Oregon as inspiring examples where social change and social movements led to the rejection of the automobile-dependent model. Next, we examine the growing movement for managing automobile demand, using incentives and disincentives to influence traveler behavior. Then, we look at the international **car-sharing** movement and how technology has been used to facilitate automobile sharing. Rather than

Car-sharing A system that allows members to use cars on a short-term rental basis. The car-sharing system owns and maintains the cars, and allows members to access and use them at any time of the day, for as little as 30 minutes, removing the hassle of going to a rental car facility. The cars are placed in public parking facilities or on streets to facilitate easy access. As of 2012, there are an estimated 500,000 car-sharing members in North America.

redesigning the automobile, these systems redesign how people use the automobile, replacing ownership with short-term usage. Finally, we consider the social movement around bicycling in the US, looking at the specific example of San Francisco.

Proactive Urban Planning Paradigms

Research shows that urban planning, and variations in the mix of land use and transportation systems in a region, can have profound effects on automobile dependence and the accompanying problems it produces. Urban travelers have the choice of different modes of travel. They can choose between walking, bicycling, public transportation, and driving. The relative convenience and costs of the different options affect how travelers decide to travel. Some modes become more or less convenient depending on the arrangement of land uses, (i.e., close proximity to work and shopping), and prices associated with each mode, (i.e., gasoline, parking, and bus or rail fares). Strategies to reduce automobile use then rely on a range of land use and price changes.

Transportation systems in many US cities were designed to facilitate **mobility**. This paradigm of mobility planning refers to the dedication of urban resources and space to moving people and goods between different destinations including residences, workplaces, and shopping areas. Mobility is expensive, however, requiring substantial financial resources, including fees (e.g. tolls or parking), fixed costs (e.g. costs of automobile ownership or infrastructures), time costs, and health or environmental damages, as described earlier. It is only convenient and inexpensive when individual and government investments are coordinated (Harry, 2008). Without significant public investments in traffic engineering, road construction, parking systems, and emergency systems, automotive mobility would be very inconvenient and few people would be expected to choose it for their travel.

Related to mobility is the idea of **accessibility**, which considers more explicitly the objective of travel. Ultimately, the value of mobility results from the value derived from the completed trip. Accessibility is the attainment of that value from the trip. Ultimately, accessibility is the aim of any mobility system. Thus, in urban areas where origins, say residences, and destinations, say workplaces, are far apart, accessibility results from being mobile. But, accessibility doesn't have to be provided by mobility (Harry, 2008).

Locating destinations close to origins or placing them close to a coordinated public transit network can improve access. This is referred to as accessibility planning. For instance, the density of jobs or houses in an area dictates how proximate origins and destinations are in space. When jobs, houses, and other uses are close together, it makes walking or bicycling relatively more convenient and a more likely option for a greater number of travelers. If land uses are spread out, then the average distances between places is much farther, making travelers rely on faster modes such as driving or public transportation. Making land uses closer together is a key strategy to reduce the reliance on automobiles. It can also improve the walkability of a place.

Transit-oriented development (TOD) integrates the location of public transportation and land uses such as job and housing centers. TOD combines density with the convenience of public transportation such as light rail or **Bus Rapid**

Accessibility The measure of ease with which something can be reached in an environment.

Transit oriented development (TOD) The practice of mixing land uses, such as retail and housing, and increasing the density of urban development near public transit stations. This enables more residents and workers to use public transit for their travel.

Bus Rapid Transit (BRT) Using buses to offer rail-like public transit services. BRT systems are based on using normal buses in exclusive, dedicated lanes that allow them to avoid traffic. BRT also relies on passengers prepaying their fares at kiosks or stations like rail systems to speed up passenger boarding.

Transit (BRT) stations. Evidence shows that compact development approaches such as TOD reduce the need for driving by around 20 to 35 percent, depending on the specific design (Harry, 2008). In fact, residents in one Atlanta TOD area drive only one-third as much as the average Atlanta resident (Harry, 2008). Combining land use strategies such as TOD with measures such as increasing the supply of public transportation, reducing the rate of highway construction, and increasing fuel prices (whether by raising taxes or through the natural increase in petroleum prices) have been estimated to reduce total driving by about 38 percent (Harry, 2008).

It is clear that the arrangement of land use and transportation in a region dictate how people travel. Knowing this does not get us any closer to solutions, however. Implementing alternative land use and transportation systems is not easy, considering the array of institutions listed earlier in this chapter. We proceed with three cases, two in the US and one in Brazil, showing how citizens and elected officials worked to change their land use and transportation planning in an effort to reduce automobile dependence.

Portland, Oregon

During the 1950s and 1960s when many US cities were being retrofitted for automobile use, communities in Portland, Oregon rejected a significant part of the freeway plans developed for it by the State of Oregon's Department of Transportation. The Mt. Hood freeway was slated to connect downtown Portland to the southeast, running through established neighborhoods as it made its way to connect to an outer interstate running north-south. The Mt. Hood, as a designated interstate, was to be funded by the federal gas tax—only 8 percent of the funds would need to come from local sources.

Despite this, neighborhoods in the area organized themselves to challenge the freeway using new regulations in the National Environmental Protection Act of 1969 and the Clean Air Act of 1970. These acts required environmental impact analyses for infrastructure projects—tests never required before for freeways. Using these procedures and growing support from local political bodies, the community and city government got the freeway cancelled in 1974. New rules in the 1973 renewal of the 1956 Interstate Highway Act allowed the use of federal highway funds for mass transit and the $180 million planned for the Mt. Hood freeway was redirected to build a light rail.

Dozens of other communities across the US successfully fought freeway plans in their communities. The most famous cases include San Francisco's cancelling of the planned Embarcadero Freeway connection to the Golden Gate Bridge, and Boston-area residents' stoppage of the "inner beltway" around Boston. For more on the history of freeway revolts, see "Stop the Road: Freeway Revolts in American Cities," written by Raymond Mohl in 2004 (Oak Ridge National Laboratory, 2010).

Phoenix, Arizona

While Phoenix, Arizona may not initially conjure up images of sustainable urban living, the city has been making great strides towards accessibility planning over the past decade. In fact, streetcars played an important role in the development of the city and by the late 1800s streetcars provided a significant part of the city's transportation needs (Office of Response and Restoration, n.d). From the 1950s, however, the Phoenix metro area experienced rapid exurban and suburban growth based around the mobility planning paradigm. In 1999, several cities in Maricopa County created a proposal for the Central Phoenix/East Valley Light Rail project. The 20-mile, 28-station line opened in December of 2008 and passes through central and east Phoenix, connecting neighboring cities of Tempe and Mesa to the east. Annual ridership averages around 16,322,830 as of 2016, up substantially from 11,348,343 in 2009, its first year of operation (Pisarski, 2006).

Native American Connections manages the Devine Legacy complex, an example of Transit Oriented Development in Phoenix, Arizona.

Supporters argued that it would stimulate and re-center growth and revitalize downtown Phoenix and the surrounding neighborhoods. To try and jumpstart development, the cities of Tempe and Phoenix developed special land use regulations such as station area plans and TOD zoning. The City of Phoenix planning department then partnered with Arizona State University and the Saint Luke's Health Initiative on a federally funded project called "Reinvent Phoenix" to assist neighborhoods located along the light rail to create long-term visions of station area development (Pucher & Dijkstra, 2000).

The project synthesized land use, economic, transportation, housing and health assessments using participatory planning processes to create long-range plans in contextually sensitive ways aimed at revitalizing and preserving the existing stable neighborhoods. For example, housing market analyses revealed a lack of afford-able housing in some neighborhoods along the Light Rail, while **Health Impact Assessments** carried out with community members showed that some neigh-borhoods are "food deserts," lacking easy access to healthy food (see Chapter 9). Adaptation to climate change was also a focus of the project, and plans were created to reduce temperatures through tree, shade, and park investments. Community members active during the planning process were asked to create a steering committee for their neighborhoods to steward the plans forward into the future.

Health Impact Assessments Detailed assessments, often led by community members, of the various health impacts of a project or planning process used to highlight important concerns or risks.

An example of the kind of TOD proposed in the Reinvent Phoenix plans can already be found in Phoenix. A local non-profit housing developer called Native American Connections builds and manages apartment complexes for low-income families, one of which is located right along the light rail (San Francisco Municipal Transportation Agency [SFMTA], 2009). The proximity to rail allows families access to many destinations without the burden of owning an automobile. Many essential services such as schools and medical care are also close by.

Curitiba, Brazil

Curitiba is generally considered an example of best practices worldwide in terms of urban transport planning. The city's history shows the importance of the relationship between investment in public transport improvements and urban development. An important element of its success was a master plan adopted in

1966 to direct the city's growth and development (Lublow, 2007). It was decided at this early stage that the city's growth would be directed along certain corridors and that these corridors would be focused around public transit, not automobiles. This was, in effect, a rejection of the US model of automobile-based urban development being exported around the world, especially in Brazil, during the post-war period. In fact, at around the same time Curitiba was growing, Brazilian planners had just completed Brasilia, the capital city, based on a decentralized, automobile-oriented model. It has since been retrofitted with heavy and light rail systems to improve its functionality.

In Curitiba, the relationship between land use planning and public transport was solidified when higher densities were enforced along the bus corridors through a strict zoning code. This made public transit very attractive to most residents of the city, since most live close to the various bus routes and most destinations are also close to the bus system. In addition, these plans were made before there were residents—the proactive planning channeled development with explicit goals rather than having to respond to ad hoc development after problems arose.

Jaime Lerner, mayor of Curitiba during this period, favored a bus-based system over the often-proposed rail system because of its relatively lower cost and ability for flexible and fast deployment. In 1974, a hierarchical bus system was developed with the introduction of express and local buses and featuring a single ticket. A "trinary road system" was developed wherein an exclusive bus way (lanes only for buses) and slower travel lanes make up the main axis of the system. Buses in the bus way, sometimes called Bus Rapid Transit, function like a light rail. They use mini-stations with platforms that help speed up boarding. The exclusive bus lane speeds buses ahead of crowded local streets. The rail-like boarding system makes boarding much faster since passengers have already paid to get into the bus stop, thus eliminating the need to pay the driver.

Lerner's proposals were not always embraced by the car-owning public in Curitiba. He engaged in several political battles with proponents of the automobile model, and was forced at times to use drastic measures to protect the public spaces he was trying to create. In one instance, car drivers threatened to drive through a pedestrian plaza he had created by closing several downtown streets. In response, Lerner brought hundreds of schoolchildren to play in the plaza (Mohl, 2004). Curitiba's comprehensive and proactive planning, in contrast to more typical piecemeal and reactive planning in US cities, makes it an inspiring example. Curitiba's thinking has gone on to influence urban planning and sustainability thinking around the world. For example, Bogota, Colombia recently implemented a citywide BRT system to inexpensively add significant capacity to its public transit systems.

A recent national study showed planning approaches such as these that link land use and public transportation will be essential for the US to reach carbon dioxide reduction targets (60–80 percent below 1990 levels) required for climate stabilization (National Highway Traffic Safety Administration, 2017). Though nothing as sweeping as Curitiba's approach can be found in the US, there is an increasing awareness of the interaction between transportation, land use, energy, and emissions. Dozens of new TODs are appearing in areas across the country.

New Urbanist principles are being increasingly used in what would have been more conventional suburban master planned communities. Moreover, Curitiba's BRT innovations are now being built in dozens of cities across the country with funding and planning support from the Federal Transit Administration.

Demand Management Strategies

Proactive urban planning can provide travelers with alternative options to the automobile. In urban areas where existing infrastructure has already literally been set in concrete and asphalt, however, alternative strategies may be needed. One such alternative, demand management strategies attempt to provide drivers with incentives or disincentives designed to change their travel behavior.

HOV and HOT Lanes

One common demand-management strategy, high-occupancy vehicle (HOV) lanes, uses existing roadway space to incentivize carpooling by providing a designated lane during busy rush-hour periods. Drivers are encouraged to share a ride, leaving their vehicle at home. Carpooling, of course, requires coordination between individuals and a certain degree of flexibility in scheduling. As such, it may not be a viable option for all travelers. The result is that some HOV lanes are underutilized, and therefore, less effective than intended. With advances in technology, however, underutilized HOV lanes can be converted into more advanced high-occupancy toll (HOT) lanes.

HOT lanes provide drivers traveling alone the option of using the higher speed lanes. They can continue to drive in the free congested regular roadway lanes, or they can pay to drive in the HOT lanes with carpoolers. In theory, pricing HOT lanes according to the amount of congestion on the roadway will enable the priced lanes to remain congestion-free and provide an incentive for those willing and able to pay for the benefit of free-flowing traffic. Those who need to get somewhere fast, and are willing to pay, can select the HOT lanes.

Congestion Pricing

In some cases, entire sections of a city can be gripped with heavy congestion. Downtown areas of European cities have been facing this problem for some time. Some have turned to the demand management strategy of congestion pricing. The idea is to charge drivers for their travel behavior. When a person decides to drive their automobile during congested times of day or in congested areas, they are adding to that congestion and hence delaying other drivers. In addition, congestion results in more air pollution emitted. Society at large, not the driver alone, pays for this pollution through environment and health-related impacts. A congestion charge requires drivers to pay a fee for driving during peak times. The fee is typically collected automatically using newly emerging technology.

Congestion pricing comes in two forms. In one system called *cordon pricing*, drivers during busy times of day pay a fee when they cross a "cordon" boundary that surrounds the congested area. Only drivers crossing this cordon pay the fee, while trips made exclusively within the cordon do not. Stockholm, Singapore, and various cities in Norway have implemented this strategy. In the second system called *area pricing,* drivers traveling anywhere within a designated area are charged a fee. In this case, travelers crossing into the area as well as those making trips made entirely within the zone are charged. London uses this system. Cameras installed throughout the pricing area automatically image and charge drivers. Though no congestion pricing systems are currently operating in the US, both New York City and San Francisco have studied the potential for cordon pricing to reduce their traffic congestion problems.

Traffic Calming and Car-Free Zones

In some cases, existing roadways need to be "calmed" to traffic levels or speeds that are more appropriate for their given context. Residential streets may suffer from vehicles traveling at unsafe speeds. Neighborhood connector streets may become cut-through routes for frustrated commuters. Major commercial streets may become clogged with high-traffic volumes, undermining pedestrian access to the area. In these cases, traffic-calming strategies may be a potential strategy.

These strategies are aimed at slowing traffic, reducing traffic volumes, or both. Some strategies, such a speed humps and roundabouts, required drivers to slow down before passing over, around, or between obstacles on a street. Other strategies, such as half-street closures and diagonal diverters, discourage cut-through traffic by eliminating direct routes through neighborhoods. Another approach, called a "road diet," removes through lanes and converts them to on-street parking, turn lanes, or bicycle lanes, thereby slowing traffic and discouraging cut-through use.

Alternatively, streets may be "pedestrianized" and turned into car-free zones. Though successful in many cases around the world, the US experience with pedestrian-only streets has been mixed at best. Attempts in the 1960s and 1970s to pedestrianize commercial streets in US downtowns and compete with suburban shopping malls were unsuccessful in many cases. Success stories do exist, however. Denver's 16th Street Mall, Church Street in Burlington, VT, and Third Street Promenade in Santa Monica, CA are prime examples of successful pedestrian-only conversions.

The demand management strategies mentioned here work on the assumption that travel demand can be influenced through pricing and physical changes to the roadway infrastructure. Other strategies attempt to go one step further and provide drivers with an alternative to the automobile. The following section discusses some of these strategies for reducing vehicle demand.

Rethinking Automobile Ownership

At first glance, trading the convenience of one's personal, private car for the occasional use of a shared car, owned and maintained by others and located somewhere out in the public realm seems supremely countercultural in the US. It appears that there are places all over the nation, however, where this idea makes sense and is increasing in popularity. Car sharing is a system that allows members to use cars on a short-term rental basis—for as short as 30 minutes in some systems. The cars are placed in public areas in cities, rather than in car rental agencies. Though

Car-share in Berkeley, California.

Car-Share in San Francisco, California.

no car-sharing programs existed before 1994, in the US by mid-2009, there were roughly 280,000 car-share members sharing about 5,800 vehicles (SFMTA, 2009), with these numbers growing roughly 20 percent per year.

Car sharing dates back to the 1940s in Northern Europe, and most notably, to the electric car-sharing system in central Amsterdam during the 1970s and 1980s (SF Municipal Transit Authority, 2009). San Francisco saw an early experiment in car-sharing in its Short-Term Auto Rental program, though it only lasted from 1983 to 1985. Eventually, with improvements in communications technologies, modern car sharing took off with systems introduced in Europe and Canada in the early 1990s. Portland, Oregon was the site of the first car-sharing system in the US, with its CarSharing-PDX that opened in 1998.

Car sharing takes place when a member makes a reservation online or by phone through a voice-operated menu some time before he or she needs it (though the reservation can be made instantly, so long as a car is available). Based on the person's location, the system show where vehicles are available for the reservation period requested. The user can specify the kind of vehicle they want or do searches anywhere in the system—even in other cities for members of a multi-city network like Zipcar. The car is available to the user once the reservation is made. The member's cardkey activates the car. Reservations can be extended on-the-go as long as the car is still available.

Numerous studies have been made of the transportation impacts of car sharing. It can have effects on several aspects of transportation systems, such as household car ownership, parking demand, car use, and the demand for "alternative" transportation such as public transportation, cycling, and walking (Shaheen, Cohen, & Chung, 2009). Research across North America shows extremely significant effects. After joining car-sharing groups, households went from owning an average of 0.47 vehicles (already somewhat lower than typical North American

The windshield-located card reader on a car-share vehicle in Berkeley, California.

households) to 0.24 vehicles. For instance, in the group of about 6,000 surveyed households that joined car sharing, almost 1,400 vehicles were "shed"—equal to almost half of the vehicles the group owned. Even more vehicles were reduced because car-sharing households avoided planned purchases of vehicles.

The Rise of Bicycle Activism in the United States

Bicycling makes up a very small share of daily travel in the US, constituting only about 1 percent of all trips. But, with increased gasoline prices and traffic congestion, growing concern about climate change and interests in physical activity, bicycling has experienced a boom in many US cities (Switzky, 2002). Chicago, New York, Portland, Seattle, and many smaller university cities have experienced significant increases in utilitarian bicycling. In San Francisco, it is estimated that 5 percent of adults use bicycles as their main mode of transportation (up from 2 percent in 2001), and 16 percent ride a bike at least twice a week.

Vehicle miles traveled
A measure of total travel by all vehicles. If 100 vehicles travel each 100 miles, the total vehicle miles traveled is 10,000.

Bicycling is poised to be a substitute for many short-range automobile trips and has enormous potential to contribute to reductions in **vehicle miles traveled** (a measure of the total distance in vehicle travel). Nationally, roughly 72 percent of all trips less than three miles in length are by car, a spatial range that an average cyclist can cover easily. Bicycles do not require expensive, long-term capital investment or operating costs like that of transit and so can be deployed quickly. And, in many respects, bicycling is among the most equitable forms of transportation because it is affordable and accessible to almost everyone. **Bicycle space,** or an interconnected, coordinated, multifaceted set of safe bicycle lanes, paths, parking racks, and accompanying laws and regulations to protect and promote cycling, has been extremely difficult to implement in the US, however.

Bicycle space Coined by Jason Henderson, the well-connected set of bicycle-related infrastructure, such as bike lanes and paths, as well as bicycle storage facilities like bike racks and larger storage facilities at public transit stations.

Lack of political will to develop bicycle space has been a major barrier. There is no strong national bicycle policy with dedicated funding programs as there is for automobiles. Advocacy for bicycling has been a largely local, fragmented, and isolated effort. Therefore, the few cities, such as San Francisco, that have established a political will to promote bicycling—and that have seen significant increases in bicycling—are worth emulating.

San Francisco, California

In San Francisco, an 11,000-member bicycle organization has lobbied hard for the production of bicycle space and the city has experienced a rapid upsurge in bicycling. Between 2005 and 2009, bicycling increased 53 percent, accounting for 6 percent of all trips in 2009 or 128,000 daily trips. In some inner city neighborhoods bicycling accounts for over 10 percent of all trips. This is despite the fact that much of the city terrain is quite hilly.

Critical Mass A number or amount large enough to produce a particular result; in this case, the number of bicyclists that can recapture urban space from the automobile, enabling the mass to progress through streets unimpeded, forcing motorists to have to wait for its passage. As such, critical mass is an act of civil disobedience meant to illustrate what a city might be like without automobiles and if street space were used for bicycle travel.

Through the early and mid-1990s, despite a growing and vocal San Francisco Bicycle Coalition (SFBC), the City of San Francisco's unspoken priority was to ensure that bike lanes did not impact car space. The frustration over the lack of political will to create bicycle space led bicyclists to create their own spaces. These were the spaces created by the **Critical Mass** bicycle rides which, beginning in 1992, occurred on the last Friday of every month in downtown San Francisco. Similar

Critical Mass rides eventually spread to New York City, Chicago, and globally to cities like Rome and Vancouver. The name Critical Mass is as implied: a critical mass of cyclists that once reached, can recapture urban space from the automobile, enabling the mass to progress through streets unimpeded and forcing motorists to have to wait for its passage. Critical Mass helped reframe the questions about urban sustainability, pushed open the debate about the use of street space, and showed the possibilities of a bicycle-friendly city.

Bicycle demonstration in Budapest.

Eventually, struggles between Critical Mass and the City's mayor led to a particularly violent clash in the fall of 1997. In response, the mayor directed his traffic department to hold hearings around the city in 1997 and 1998. Hundreds of cyclists attended these meetings and the SFBC took a more aggressive position in its lobbying efforts. Despite taking made pains to differentiate itself from Critical Mass, the fall out from Critical Mass campaigns strengthened the SFBC By 1998, it had 1,700 members and was gaining allies with some members of the city's board of supervisors. As a result, a number of key streets were made more welcoming to cyclists.

When bicycle lanes were added to Valencia Street in 1999, bicycling increased by 144 percent. The success of the Valencia Street project further emboldened activists and proved that, with adequate infrastructure, more people would choose to bicycle. By 2004, 16 percent of all trips on the street were by bicycle. By the mid-2000s, and with almost 5,000 members, the SFBC had almost every local elected official concerned about the "bicycle vote" and very few elected officials spoke against bicycling. Despite the support, ambiguity about how to implement bicycle space was widespread among many politicians. In 2009, San Francisco adopted the San Francisco Bicycle Plan, designating near-term, long-term, and other improvements to the city's 208 miles of existing bikeways.

Conclusions

Taking a social view of transportation-related problems leads us to a social view of their solutions. The influences of industry and government, along with individual choices for convenience and status, create a complex web built around automobile dependence. Challenging this process requires profound and difficult social changes. The examples assembled in this chapter give us a flavor of what social change can look like.

The public-transit oriented example from Curitiba shows a clear departure from standard models at the time. This model was developed and supported by a team of active citizens, planners, architects, and elected officials. Finding similar examples of such sweeping efforts in the US is difficult, but Portland's efforts are inspiring.

Similarly, car sharing illustrates that many communities are willing to sacrifice convenience for other goals. Of course, for many, car sharing is a rational alternative to high parking and automobile ownership costs. We must wait and see how large of a dent car sharing can make.

In a similar vein, we find bicycle activists challenging the dominant urban management of roadways for automobiles, asserting an alternative vision for urban streets filled with bicycles rather than cars. Producing bicycles is a minor part of the national economy, yet it can provide a significant amount of our mobility if that possibility were made real. Urban streets were used for thousands of years by humans, horses, bicycles, and streetcars before automobiles came along. There is nothing intrinsic about any one of their roles in the streets—it is purely a social construct. Thus, we can see more clearly how solutions will require social changes involving a constellation of actors across multiple scales and approaches.

Several important lessons are found in the efforts of cities and citizens attempting to reduce their automobile dependence. The larger lesson is that urban practices, such as automobile dependence, water or energy use, and pollution, are the result of institutional webs consisting of citizens and neighborhoods, city and state governments, and federal policies. Achieving sustainability begins with understanding these institutions and how they respond to and resist change.

As demonstrated in this chapter, effective cases of reducing automobile dependency can be found across all of these scales and institutions. We examined how at the regional scale, proactive citizens, neighborhoods, and city governments led movements against the expected city planning paradigm. They emphasized goals of walkability and public transportation over the typical reliance on roads and automobiles.

Other citizens have implemented visions for reduced reliance on automobiles by proposing reasonable alternatives. Groups around the world have implemented car-sharing services that reduce the need to own an automobile. The impacts on urban travel have been shown to be profound. Likewise, groups have used collective action (i.e., Critical Mass) to illustrate what bike-friendly cities might look like. These actions have successfully translated into real policy changes at both the city and regional scales.

This chapter emphasized the role of citizens and activists together to show that there is no one "right answer"—no magic bullet. The importance is in how citizens and governments implement solutions. That when we join with others with similar visions, we can create the social change needed to challenge the dominant urban planning paradigm, we can practice automobile independence.

Chapter 13

Energy in the Sustainable City

Martin J. Pasqualetti and Meagan Ehlenz

As of 2014, four billion people in the world (54%) lived in cities. By 2050, the United Nations (UN) projects 66% of the world's population will be urban residents. Nearly 90% of the increase is expected to occur in Asia and Africa (United Nations, 2015). When combined with population growth projections, cities will need to accommodate an additional 2.5 billion people over the next three decades. As a result, cities will continue to be responsible for the vast amount of energy the world uses. And they are multiplying.

Presently in China, for example, new cities are appearing even before there are people to occupy them in anticipation that by 2020, one in every eight people in the world will live in a Chinese city. According to the UN, China is expected to add one more megacity (10 million+) and six more large cities (5 to 10 million) by 2030 (United Nations, 2015). This growth will require massive amounts of energy and expanded infrastructure. This goes for existing cities as well, many of which are growing quickly both in population and area. This is exerting additional pressure on energy supplies and infrastructure. The problem is that, with no exceptions, cities, particularly large ones, are unsustainable in their current configuration and operation. This fact is recognized by many organizations and funding agencies, although progress to modify these trends is difficult to identify and harder to implement.

When we look at our cities, we can see the many manifestations of the energy they use. Skyscrapers need to be heated and cooled, highways form concrete rivers loaded with cars burning refined oil, and street lights and neon signs transform night into day. Big cities mean big energy and their existence depends on a ceaseless supply of it in every form imaginable, including gasoline, electricity, natural gas, ethanol, wood, even dung.

So great is the energy demand of cities that they can be considered terrestrial 'black holes,' sucking in all of the nearby energy and never letting any of it escape. Hyperbole aside, the limitless energy appetites of cities require vast supply networks, stretching hundreds, even thousands, of miles in all directions. In this chapter, we will examine how energy is used in cities, the consequences of that energy use, and what city planners can contribute to help cities operate more efficiently and improve the urban quality of life. We will also explore energy alternatives and strategies that can be addressed through urban planning and policy changes that will enable cities to be more sustainable.

City Size and Energy

From the standpoint of energy, scale is important when sustainability is the goal. Although small cities may not be able to provide all the services of a large city, they do have several advantages from an energy perspective. For example, smaller cities are generally easier to manage than larger ones. Their energy needs are smaller, economies and politics are often simpler, supply lines less complicated, commuting times quicker, food supplies closer, and they are unlikely to stew in the heat of their own heat islands.

Almost all cities started out small with very few growing to substantial size before the Industrial Revolution. Examples include the pre-industrial city of Teotihuacan in Central Mexico, which might have exceeded 100,000 people 1,000 years ago. However, its size was possible only because of the control of slave labor and the advantages of a mild climate. Rome had one million residents by 133 BCE, but it was unique. Few other cities held one million residents until the early-1800s. For instance, census data show that Peking, China reached 1 million in 1800 (United Nations, 1980), London in 1801 (Emsley, Hitchcock, and Shoemaker, 2015), and New York City around 1820 (Gibson, 1998).

Today, nearly 500 cities in the world exceed one million residents. In 2014, 20% of the world's urban population lived in a medium city (1 to 5 million

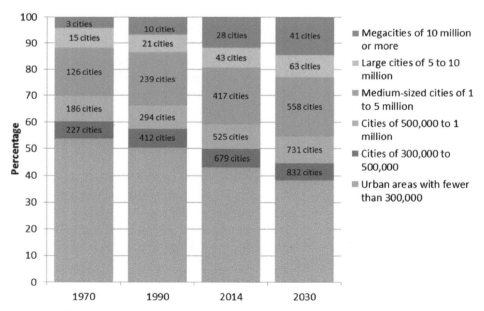

Figure 1 Distribution of the world's urban population (percentage) by size class of urban settlement and number of cities, 1970, 1990, 2014 and 2030.

Source: United Nations, 2015: p. 17.

From World Urbanization Prospects: The 2014 Revision, by the United Nations Department of Economic and Social Affairs, © 2015 United Nations. Reprinted with the permission of the United Nations.

inhabitants). The population in these cities is projected to grow 36% to 1.1 billion by 2030. By contrast, 400 million people are anticipated to be living in large cities (5 to 10 million residents) by 2030. As indicated in Figure 1, this number is expected to grow to over 650 by 2030 (United Nations, 2015). These cities will require extensive amounts of energy to sustain their populations.

Organically Grown Cities

What facilitates large concentrations of people and the growth of cities? Glasgow and Manchester in the UK, Cleveland and Pittsburgh in the US, Essen and Frankfurt in Germany, and similar cities grew because they had access to two things: concentrated forms of energy to fuel industry and transportation options such as waterways and railroads. Yet, even as technology and innovation have enabled city builders to create bigger and better urban forms, we face several new challenges. For instance, São Paolo, Beijing, New Delhi, Cairo, and Mexico City, among others, struggle to provide food, shelter, and transportation for millions of people. These cities and myriad others around the world also struggle with overcrowding, pollution, and congestion to such a degree that it seems beyond hope that these problems can ever be resolved. So, the question becomes can big cities ever be sustainable? And if yes, what role does energy play?

The problem faced by those who would hope for sustainable cities is that sustainability has seldom been *planned for* from the beginning. Rather, most cities have grown organically. They have expanded incrementally, a freeway here, a building there, expanding up hillsides and across farmlands. For developed cities, this presents challenges that relate to retrofitting the built environment in more sustainable ways. For cities yet to be built, this presents a different set of challenges. More important, it offers unprecedented opportunities, as urban decision makers and city planners are identifying ways to change the city building model and proactively plan for integrated, sustainable development. These opportunities extend to energy systems and how they are integrated into the urban fabric in terms of energy-efficient architecture, building materials, and transportation designs.

Decision makers and city planners are exploring myriad energy alternatives that will enable them to move towards sustainability. These include, but are not limited to, hydroelectric, solar, wind biomass, geothermal, and even tidal energy. These energy sources are often referred to as clean or renewable energy as their emissions are inert, or at least minimally impact the Earth's systems, or their sources are infinite or self-renewing. Decision makers and planners weigh factors such as operating costs, fuel or waste costs, reliability, and whether or not the resource is available locally as well as the infrastructure needed for generation and distribution for these alternatives to fossil fuels.

Given these divergent challenges for cities of the future, is there anything that can be done to improve urban sustainability? While the answer is 'yes,' it will not be easy or quick for one simple reason—cities are complex, in both form and function. The built environment, and the policies that guide it, are intertwined with the social, economic, and political pressures of a place. Within these places, there are vested interests that will argue against upsetting the status quo that serves them well and few policies that can remedy conditions in the short term. Like a massive aircraft carrier under full power, it is not easy to turn this ship around.

Cities as Products of Energy Innovation

Cities have always offered a number of economic, social, and physical advantages. They offer mutual security and economies of scale. They are transportation hubs and nodes for the trade of goods, services, and ideas. They also support a diversity of skills, networks, and lifestyles, and can facilitate innovation and creativity in ways less dense and smaller settlements cannot.

The earliest pre-industrial cities were located along rivers and coasts as waterways provided for irrigation, simple transportation, and, later, energy generation. The Industrial Revolution and the availability of coal changed all that. These advances led to the growth of cities as well as their untethering from the banks of rivers and coastal areas. As coal began replacing wood and other biofuels, agrarian economies started shrinking in favor of those based on industry.

North Sea A marginal sea of the Atlantic Ocean located between Great Britain, Scandinavia, Germany, the Netherlands, Belgium, and France. It connects to the ocean through the English Channel in the south and the Norwegian Sea in the north. It is more than 970 kilometers (600 mi) long and 580 kilometers (360 mi) wide, with an area of around 750,000 square kilometers (290,000 sq. mi).

Fossil fuels Carbon-based fuels in the forms of coal, oil, and natural gas that derive from animals and mostly plants. They are created over millions of years under conditions of heat and pressure.

Industrialized London, which today receives its energy primarily through wires and pipelines, first relied on coal carried up the Thames in ocean-going caravels that plied the waters of the **North Sea** from mines near Newcastle. Other cities on the British Isles such as Cardiff and Edinburgh would also prosper because of coal both by burning it as an energy resource as well as from mining and exporting it to other cities. Over time, the development of energy resources would facilitate the emergence of many other cities including Baku, Azerbaijan; Dhahran, Saudi Arabia; Valdez, Alaska; Harcourt, Nigeria; Charleston, West Virginia; Kogalym, Russia; and the city-state of Singapore to name a few.

In the early days of the US, cities tended to be near energy resources such as water and wood. For instance, Pittsburgh, Pennsylvania was not located on the coast but was in close proximity to coal supplies and easy river transport. The growth of large industrial cities was only possible when **fossil fuels** became more abundant and cheaply available. Once canals and railroads were available to reliably transport coal from Appalachia, more distant coastal cities such as Philadelphia and Baltimore began developing their own industrial centers, leveraging their ready access to the world's ocean.

As urban economies became less dependent on proximity to raw materials and industrial clustering, particularly across the US and Europe, people had greater choices about where they could live. With the introduction of larger scale power generation in the 1870s and the development of the modern electrical grid in the early 20th century, the growth of cities anywhere became more feasible. During the same period, households gained more control over their mobility with the rise of the automobile. Together, these trends facilitated more scattered and less dense settlement patterns.

This change in the built environment stimulated a corresponding expansion of a more elaborate, expensive, and complex energy infrastructure. Energy supplies no longer converge on a few big urban demand nodes, such as those of the industrialized centers of the northeastern US. Rather, energy is now needed wherever anyone prefers to live. In the US, this helps explain the population growth and settlement patterns in warmer and less crowded southern and western states. For instance, places like Atlanta, Phoenix, and Denver are home to millions of people despite their physical distance from the great energy reserves necessary to support them.

People living in these cities, especially those who migrated in the last few decades, often make a calculated decision. They have left behind the environmental and social constraints of older industrial cities or rural areas in favor of, what they hope, is a better quality of life. However, there is an unrecognized cost—their new homes are in places that are entirely dependent upon a steady supply from distant energy stores. Thus, these cities are vulnerable to energy supply interruptions whether natural or intentional in origin, which can reduce urban resiliency.

Energy and Transportation in Large Cities

Owning a car, owners will tell you, brings convenience, status, independence, privacy, and flexibility to transportation. While all this may be true, there are responsibilities attached to ownership that extend beyond personal expense including huge societal and environmental costs. It starts with the alarming inefficiency of the car. Consider the difference in energy necessary to transport a 180-pound person inside a 4,000-pound car to that required to transport the same person on a bicycle an equal distance. Research shows that a bicycle can be up to 98.6% efficient in terms of converting energy at the pedals into forward motion (Glaskin, 2013: 9) while cars use only about 14-30% of their energy to actually move the vehicle forward (US Department of Energy, n.d.). Now, imagine the aggregate cost of thousands of people choosing to drive instead of pedaling. This individual choice is happening in real time in China, as households have traded in the practicality of riding bicycles for the prestige of driving cars.

Fifteen years ago, there were few private cars in China and the principal commute choice was the bicycle. Known as the "kingdom of bicycles," people pedaled their way around the city (Energypost.eu, 2015) with approximately 40 percent of Chinese cycling to work and school (EU SME Centre and the China-Britain Business Council, 2015). Since then, the 'car revolution' has stimulated a rapid rise in the demand for transportation fuels. Politically, China's transition from bicycles to cars was intentional. As a mechanism to stimulate the economy and provide incentives to individual Chinese, the government encouraged automotive ownership, even subsidizing their purchase in 2009. Over 1.6 million cars were acquired in just two years and car ownership continues to grow (Figure 2).

China's explosive expansion in car ownership in combination with its dependence on coal and oil for energy generation has resulted in some of the worst air pollution on the planet (Figure 3). In 2011, the country was the second-largest oil consumer (US was first) and in 2012, coal accounted for 66% of China's energy consumption (US EIA, 2015). This has led to serious health concerns at both the national and global levels (Zhang et al., 2010). **Smog**, however, is not the only consequence of rapid urbanization and a car-centric transportation network.

Smog A photochemical haze caused by the interaction of solar ultraviolet radiation with volatile organic compounds, such as hydrocarbons, and nitrogen oxides. Smog-forming pollutants come from many sources, such as automobile exhausts, power plants, factories, and many consumer products, including paints, hair spray, charcoal starter fluid, solvents, and even plastic popcorn packaging. In typical urban areas, at least half of the smog precursors come from cars, buses, trucks, and boats.

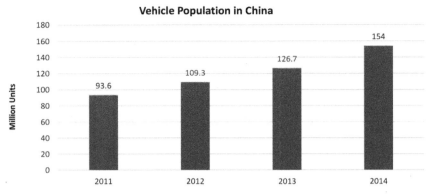

Figure 2 Vehicle growth comparison in China between 2011 and 2014.
Source: EU SME Centre, *The Automotive Market in China*, p. 7.

©axz700 / Shutterstock, Inc.

Figure 3 Notorious Beijing air pollution blankets Tiananmen Square.

In Beijing, congestion has increased significantly (Figure 4), as the city's 19 million residents and 5.3 million cars have turned streets and highways into parking lots. As these consequences become more visible, some people are already calling for a return to the good old days when cars were not so coveted and prevalent (Denmark. dk, n.d). In Beijing, government officials have begun considering policies to return to the use of bicycles. Ironically, this shift is happening at the same time as Beijing suburbs are becoming too distant for the convenient use of bicycles. See Chapter 16 in this text to learn more about what China is doing to become more sustainable.

Despite Beijing's heavy traffic and dangerous air pollution, the Chinese capital is by no means the sole example of a congested city nor is it the most polluted. This distinction belongs to Delhi, which, ironically, was originally planned to reduce travel distances. Nevertheless, it has become crushingly reliant upon automobiles (Guerro & Cervero, 2011). In Beijing and Delhi, not to mention places like Los Angeles and San Francisco, traffic is commonly at a standstill, and there is no easy solution in sight given the combined challenges of resolving air pollution and congestion simultaneously. Even if everyone in these cities converted to zero-emission vehicles overnight, congestion would remain, as would the need for more roads, parking spaces, and all the other accoutrements that accompany an automobile economy. Cars may bring status and the luxuries of independence, but these privileges come at a high cost when it comes to sustainability.

One of the principal reasons that bicycles were an effective mode of personal transportation in Beijing was the high population density within the city. For example, the density around Beijing's historic Forbidden City is about 25,000 persons per km². Such density also works to the advantage of mass transit schemes, allowing for an effective system. The Beijing Subway, begun in 1969, was expected to have 30 lines, 450 stations, and 1,050 km in length by 2012. Presently it has 18 lines including an airport express line. It is first in the world in annual ridership with 3.21 billion rides delivered in 2013 (Gibson, 1998).

The California cities of San Francisco and Los Angeles offer further examples of the effects of population density on mass transit options. San Francisco, a city constrained by the San Francisco Bay and Pacific Ocean, is about 121 km² (47 mi²) with a 2015 population of approximately 841,000. It is the second most-densely settled city in the US (after New York) at nearly 18,000 people per mi². As a result of this density, living in San Francisco without owning an automobile is not only feasible but increasingly popular (Glaskin, 2013). By contrast, Los Angeles has four times as many people (approximately 3.9 million people in 2015), but is ten times larger at 503 mi² and has few natural barriers to expansion (Figure 5A-B). It is also a **horizontal city**—as are Houston, Dallas, and Phoenix—where owning a car is almost mandatory as the low population density makes mass transit costly on a person-mile basis. Recent research suggests that light rail systems require approximately 30 people per gross acre around stations and heavy rail need 50% higher densities to be cost-effective (Guerro & Cervero, 2011). For a more detailed exploration of transportation and energy see Chapter 12 in this text.

City form is not only influenced by natural settings, but also by historical roots. For example, population densities tend to be higher in European cities and Eastern US cities than in Western US cities because the older cities took shape before electricity and automobiles afforded increased spatial flexibility. Once electricity and cars became ubiquitous, preferences shifted away from high-density settlement patterns to the suburbs that offered greater individual space and privacy. This has come at a cost, however, as low-density cities require greater energy distribution and infrastructure. Thus, these trends do not bode well for a sustainable future

©TonyV3112 / Shutterstock, Inc.

Figure 4 Bicycling and traffic congestion in Beijing.

Horizontal cities are cities that primarily go outward rather than upward. They generally have low population densities with typically fewer than 10 people per acre. This is significantly lower then the 15 to 20 people per acre in vertical cities (Rybczynski, 2010).

Figure 5A-B The population density of Los Angeles (Left) is 8,225 people per mi^2 (with downtown at 4770 per mi^2.), while the population density of San Francisco (Right) is more than twice that at 17,938 per mi^2. San Francisco's greater population density facilitates robust mass transit systems, which save fuel and shorten commuting distances.

or a reduction in the enormous concentration of energy supply infrastructure. In the following section, we examine some of the alternatives available to reduce the negative impacts of energy-intense cities.

The Energy-Sustainable City

Energy efficiency Using the least amount of energy for maximum benefit while producing the least waste. A typical internal-combustion car, for example, has low-energy efficiency, resulting in substantial wasted energy and greater pollution per mile driven. When vehicle energy efficiency is increased, waste goes down and mileage goes up.

Recognizing the rising energy use in cities, several organizations are addressing what can be done to reduce it. A prime mechanism is **energy efficiency**, as it is one of the easiest and least expensive techniques. In the US, for example, the American Council for an Energy-Efficient Economy (ACEEE) has developed a City Energy Efficiency Scorecard, ranking 51 large urban cities (Hawaii State Energy Office, 2016). In their 2015 report (http://aceee.org/research-report/u1502), Boston topped the list as the most energy efficient city, followed by New York, Washington DC, and San Francisco. Birmingham and Oklahoma City ranked at the bottom, meaning that they are particularly wasteful in their use of energy. See the full list at http://aceee.org/research-report/u1502. The ACEEE developed their Scorecard by evaluating energy efficiency in each of the following categories: Local Government Operations, Community-Wide Initiatives, Buildings Policies, Energy and Water Utility Policies & Public Benefits Program, and Transportation Policies. You can develop a scorecard for your own city by accessing the ACEEE web site: http://aceee.org/local-policy/scoring-tool.

A second organization, the European Institute for Energy Research (EIFER), specifically addresses the role of energy in all future decisions concerning urban development, spatial use, and settlements. In order to understand the role of energy in urban change, develop innovative strategies, and implement them, EIFER targets their research in the following fields (MTA.info, n.d):

- Analyzing urban development, dynamics, and impacts on energy demand using approaches to assess and understand urban transitions
- Analyzing local energy and climate change policies
- Understanding urban governance and its interaction with energy planning, efficiency, and climate change policies

- Assessing innovative urban mobility by evaluating urban policy, technological trends, spatial simulation, and its interaction with the built environment
- Integrating concepts for sustainable energy use, taking into account acceptability studies, analyses of lifestyle, models for the diffusion of new technologies, analyses of the legal framework and of stakeholders, studies on governance, and the interactions between the different influence factors

Both the ACEEE Scorecard and EIFER programmatic assessments are based on the presence of various policies dealing with operations, energy use, and transportation to enhance sustainable solutions. We recommend that cities supplement these approaches by adopting one or more strategies in each of the following categories:

Category One: Measures towards greater sustainability already taking place

- Increase energy efficiency
- Increase the use of renewable energy

Category Two: Measures towards sustainability making slow progress

- Introduce more efficient designs in cities
- Reduce the dependency on automobiles
- Rethink the concept of urban spaces

Category One: Measures Toward More Sustainable Cities Already Taking Place

Increase Energy Efficiency. Amory Lovins has been the most visible champion of energy efficiency, calling it the "**soft energy path**." (REVE, 2016). In contrast to always relying on technical solutions for energy demand, he argues there are better alternatives. Soft energy is tantamount to energy efficiency; that is, getting the job done while using less energy. His approach is called **Demand-Side Management** (DSM) and is the opposite of **Supply-Side Management** (SSM). That is, instead of generating more electricity to meet rising demand, he showed that we can achieve the same personal and communal benefits by cutting our use of electricity and other energy through greater efficiency.

As Lovins often says, it is always cheaper to save a kilowatt of power than to produce a kilowatt of power. Here's an example: Say we provide $1 billion to build a 1,000 megawatt power plant to serve one million people. That would be a supply-side approach. However, if we spent the $1 billion to make every house and building in the city more efficient, we could have the same result but with broader impact (more consumers impacted) and longer-lasting benefits (continuing consumer savings). That would be a demand-side approach. Along the way, we would also avoid everything associated with having a new power plant, including fuel development and refining, air and water pollution, and product transportation for the entire life of the power plant, usually 30-40 years.

Despite the seeming logic of a DSM approach, it has taken several decades for the concept to find wide acceptance in the US, although we do find rising evidence of its acceptance if we look hard enough. Anyone who has traveled to Switzerland, Norway, or Germany, for example, has noticed that doors and windows seal

"Soft energy pathways" are flexible, resilient, sustainable and benign. They

- rely on renewable energy flows that are always there whether we use them or not, such as sun and wind and vegetation, on energy income, not on depletable energy capital.
- are diverse, so that energy supply is an aggregate of very many individually modest contributions, each designed for maximum effectiveness in particular circumstances.
- are flexible and relatively low-technology— which does not mean unsophisticated, but rather, easy to understand and use without esoteric skills, accessible rather than arcane.
- are matched in scale and in geographic distribution to end-use needs, taking advantage of the free distribution of most natural energy flows.
- are matched in energy quality to end-use needs: a key feature that deserves immediate explanation (Lovins, 1976).

Demand-Side Management (DSM). An approach toward energy efficiency often used by electric utility companies that meets customer needs by lowering the demand for energy rather than increasing the supply. This can be accomplished by using more energy efficient appliances, better insulation materials, tinted or reflective coatings or films, and reducing lighting and energy losses, among other measures.

Supply-Side Management (SSM). The opposite of DSM, SSM tends to satisfy demand by increasing the supply of energy. For example, building a new power plant instead of building more efficient buildings.

tightly, lights switch off when rooms are empty, air conditioners shut down when windows are opened, and incandescent lights are an old and rejected technology. The US and a number of other countries are trying to catch up to these models.

Large energy consumers were the first to get on board the efficiency train. Chief Financial Officers and Executives quickly understood how much money could be saved with simple changes such as lighting, the installation of motion sensors, and the use of window screens and other coverings. Payback was often quick and impressive. For example, 20 years ago at Arizona State University (ASU), their first significant experiment with DSM was to retrofit the lighting fixtures and install motion sensors in every room of six campus buildings. The university was soon saving over $200,000 per year with a return on investment of only a few years. ASU continues to implement additional energy efficiency measures to this day, with savings approaching $1 million per year from its 2007 baseline. Gradually, such sensible changes are finding their way into private homes.

Whereas Arizona and other states are incrementally catching on to the multiple benefits of DSM, energy efficiency has been aggressively pursued elsewhere in the US, particularly in California. Today, California has one of the lowest rates of per-capita energy consumption in the US, a rate that is on par with that of the mild tropical islands of Hawaii (Rybczynski, 2010). The success in California results from one of the most robust and varied catalogs of energy efficiency measures and policies anywhere (United Nations, Department of Economic and Social Affairs, Population Division, 2015). This progress towards urban sustainability has been accompanied by the acceptance of a number of internationally recognized green building certification system programs, including BREEAM and the United States Green Building Council (USGBC) LEED (Leadership in Energy and Environmental Design) (United Nations, Department of International Economic and Social Affairs, 1980).

BREEAM is a suite of environmental assessment tools launched in 1990 by the United Kingdom group Building Research Establishment Ltd (BRE). It focuses on sustainable value and energy efficiency as outlined in its International Code for a Sustainable Built Environment (BREEAM, n.d.). USGBC LEED is a set of rating system certifications for the design, construction, operation, and maintenance of green buildings, homes, and neighborhoods. LEED-certified buildings generally operate with greater energy efficiency and, therefore, less environmental impact (USGBC, 2016) (United States Green Building Council [USGBC], 2016). Learn more about BREEAM at www.breeam.com, BREEAM USA at www.breeamusa .com, and LEED at www.usgbc.org/leed. Chapter 5 in this text provides more details on how these organizations are informing urban design.

Increase the Use of Renewable Energy. The impacts of urban energy use are not contained within city boundaries but extend beyond these areas along supply paths of railroads, highways, pipelines, shipping routes, and transmission lines. In other words, energy consumption and supply is not only an issue for cities. The more we can rely on renewable energy resources, the greater the decline of environmental and human costs of supplying urban energy demands will be—for urban and non-urban areas alike. Movement in this direction began in the mid-1980s, accelerating from the late 20th century until today, when we are witnessing an unexpectedly quick growth in the installation of wind and solar power worldwide (Figures 6 and 7).

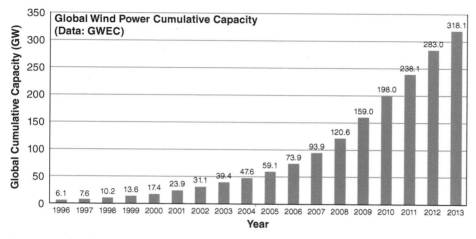

Figure 6 Global wind capacity in total gigawatts and annual gigawatt additions.
Source: REN21, Renewables 2016 Global Status Report (Paris: REN21 Secretariat).

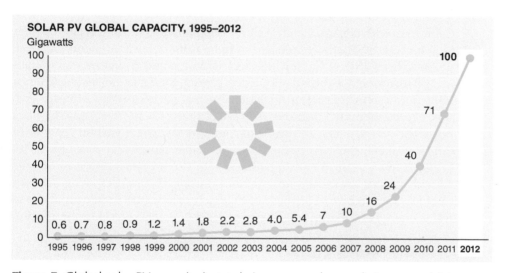

Figure 7 Global solar PV capacity in total gigawatts and annual gigawatt additions.
Source: REN21, Renewables 2016 Global Status Report (Paris: REN21 Secretariat).

According to the US Energy Information Administration (EIA), the US leads the world in wind energy production. However, it accounts for only 4.4% of total US energy generation (2016a). It trails Denmark at 39%, Portugal at 24%, Spain at 20%, Ireland at 18%, and Germany at 10% (EIA, 2016). The Spanish REVE, or Wind Energy and Electric Vehicle Magazine, in its March 1, 2016 article *U.S. number one in the world in wind energy production* found that "US wind energy equals approximately 190 million megawatt-hours (MWh), which is enough electricity for about 17.5 million typical U.S. homes. China was close behind the U.S. at 185.1 million MWh, followed by third-place Germany at 84.6 MWh" (REVE, 2016).

Solar photovoltaics (PV) produce comparatively less energy than wind, but have been quickly catching up. In 2015, 50 gigawatts (GW) of new solar capacity was installed worldwide and the cumulative capacity is now 227 GW. In terms of generation, this is equivalent to thirty-three 1-GW coal-fired power stations. A June 2015 post on energypost.eu noted "solar power now covers more than 1% of global electricity demand. In three European countries—Italy, Germany and Greece, solar PV supplies more than 7% of electricity demand" (energypost.eu, 2015). Operating these installations produces no air emissions and leaves behind no waste, and they supplant generation from energy sources that do. For more on renewable energy efforts around the world see http://www.ren21.net/gsr-online/.

In the US, rising demand for electricity in our large cities continues to drive the quick emergence of renewable energy as does the widespread adoption of state targets for renewable energy use called **Renewable Portfolio Standards**. Figure 8 outlines US RPS targets ranging from 2% (South Carolina) to 50% (California and Oregon). The two newest RPS were initiated by Hawaii and Vermont. Hawaii's Governor David Ige signed one of the most ambitious into law in 2015. The mandate raises the state's target to 100% by 2045. It is the first in the nation to do so. With renewable energy already at 23.4%, it is almost a quarter of the way there (Hawaii State Energy Office, 2016). Vermont's 2015 RPS mandates a 75% renewable energy portfolio by 2032.

California, with approximately 38.5 million people as of 2015, is aggressively reaching for 50% of its utilities retail sales to be from renewable energy by 2030. Today, more than 29% of California's electricity demand is met by renewables, including hydropower. As urban population growth continues, renewable energy, including the adoption of rooftop solar installations, is expected to keep pace and, perhaps, exceed increased demand. According to the California Public Utilities Commission (CPUC), its California Solar Initiative (CSI) is well on the way to meeting its goal of installing 1,940 megawatts of solar capacity by the end of 2016.

Progress toward greater sustainability is part of a pattern that can be witnessed with greater clarity and optimism in European countries using **Feed-In Tariffs** (FITs) as incentives. FITs have been largely responsible for the rapid rise of renewable energy installations in several countries, such as Germany and Spain.

Renewable portfolio standards A regulation that requires the increased production of energy from renewable energy sources, such as wind, solar, biomass, and geothermal. Other common names for the same concept include Renewable Electricity Standard (RES) at the United States federal level and Renewables Obligation in the UK.

Feed-In Tariffs (FITs) A policy mechanism designed to accelerate investment in renewable energy technologies. It achieves this by offering long-term contracts to renewable energy producers, typically based on the cost of generation of each technology.

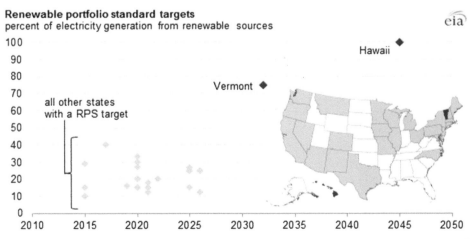

Renewable portfolio standard targets
percent of electricity generation from renewable sources

Figure 8 Overview of US states with renewable portfolio standards percentages and target dates. Source: EIA, 2015.

In Denmark, consumers are partially responsible for financing FITS. The country expects that this investment will not cost consumers more but rather "the rising efficiency of renewable energy means, that the cost to consumers of 33 % renewable energy in 2020 will be lower than the 11 % provided in 2002" (Denmark.dk, n.d; US EIA, 2015). China has had the fastest growth in the development of solar, wind, and hydropower in recent years. Moreover, they have also been installing solar water heating throughout the country, something that is common in other places such as Israel and Palestine, although not in the US.

The expansion of renewable energy is one way to reduce some of the more pernicious environmental impacts of cities and render them more sustainable. However, there are barriers. As with any 'new' technology, the success of rooftop solar hinges, in part, less on the availability of sunlight and more on planning policies that are in place to inhibit or encourage it. At the most general level, zoning or building codes that have not been updated to address rooftop solar leave the property owner open to uncertainty. From a site design perspective, development standards do not protect, nor actively encourage, solar orientation that would preserve access to sunlight.

Homeowners themselves may be subject to NIMBY (not-in-my-backyard) objections. For instance, some homeowner associations (HOAs) prohibit the installation of solar panels for aesthetic reasons, citing concerns that the panels will detract from neighborhood property values. Even without HOA restrictions, individual homeowners are subject to NIMBY complaints as they go through the local permitting process. At a larger scale, solar and wind farms are frequently subject to contestation by communities interested in preserving rural vistas and citing health and nuisance concerns, such as noise and vibration (from wind turbines) or glare (from solar panels).

Category Two: Measures Toward Sustainable Cities Making Slow Progress

Introduce more efficient designs in cities. Naturally, redesigning cities for greater sustainability is no easy task. The basic elements of such revisionist design include greater walkability, designated bike lanes, recycling, mass transit, urban farming, and—most important of all—the readiness to both think about sustainability in a holistic way and integrate it into day-to-day decision making in the first place. Some of the cities that have demonstrated such a willingness include Melbourne, Australia; Ottawa, Canada; Masdar City, UAE; and Älvstaden, Sweden. While revamping urban spaces to make them more energy sustainable is a large task, it is being done here and there with some notable success, such as in Freiburg, Germany, as described in Chapter 4.

Reduce the dependency on automobiles. Efforts to reduce the use of automobiles have produced little net benefit in most cities, especially in cities with low population densities or no history of mass transit. There are some exceptions, however. The Bay Area Rapid Transit (BART) light-rail system in the San Francisco Bay Area is very popular with 126 million trips in 2015 and an average weekday ridership of 423,100. By comparison, New York City's Metropolitan Transportation

© T Photography/Shutterstock, Inc.

Figure 9 Bus Rapid Transit is a flexible alternative to fixed-rail systems.

Authority (MTA) and Tokyo's 2015 total ridership was 1.763 billion and 3.411 billion, respectively (MTA.info, n.d.). In both instances, these transit systems have a long history with mass transit and reflect population density, so people are more accustomed to leaving their cars at home. However, when you consider population density, BART is relatively successful with 145 rides per capita in 2015 compared to MTA's 207 rides per capita in New York City. Some cities have opted for bus rapid-transit instead of fixed rails and have had some success—for example cities in Brazil, Indonesia, Australia, and Europe (Figure 9).

The reality, however, is that people who can afford cars tend to prefer them, even though the cars themselves rarely carry more than just the driver. As a consequence, every person who drives a car contributes much more than their share of greenhouse gases (GHG) than they would if they were to walk, car pool, use mass transit, or ride a bicycle. It is estimated that each gallon of gasoline used in US passenger vehicles results in about 18.95 pounds of GHG (US EIA, 2016). For a car traveling 12,000 miles per year, this means that it emits over 8,000 pounds of carbon dioxide, or more than twice its own weight. This is yet another example of why it is difficult to make cities sustainable (US Energy Information Administration [EIA], 2016a). See Chapter 12 on Sustainable Transportation for a more in-depth discussion.

Rethink the concept of urban spaces. One of the opportunities for sustainability lies in cities that have not yet been built. Here, city builders can reconsider the relationship between the natural and built environment. Undoubtedly, one of the most radical experiments in restructuring the form and function of urban spaces has been Arcosanti, an '**arcology**' developed by Paolo Soleri in Arizona to describe his designs for ecologically sound human habitats (Cosanti Foundation, n.d; US Energy Information Administration [EIA], 2016b). An urban laboratory in the desert, Arcosanti is a demonstration of how urban architecture and ecology can

Arcology (architecture + ecology) Paolo Soleri's theory of compact city design, promoting alternatives to urban sprawl.

Costanti Foundation-Yuki Yanagimoto

Figure 10 Foundry Apse and housing at Arcosanti in Mayer, Arizona. Source: Oscar Lopez, "Paolo Soleri's Arcosanti: The City in the Image of Man." ArchDaily, Sept 3, 2011.

work in tandem to create more efficient and highly livable cities. Built on 25 acres of land, the project currently includes 13 buildings and consumes approximately 20% of the energy required by conventional architecture. Among the many elements included is the use of **passive energy** mechanisms such as the apse effect, the greenhouse effect, the chimney effect, and the heat sink effect as well as miniaturization, elimination of the automobile, sustainable agriculture, and emphasis on building upward rather than outward (Figure 10).

Like many utopian communities, Arcosanti's full potential has yet to be reached. Soleri originally envisioned a dense, compact city with 5,000 residents built on its acreage. To date, the community hosts between 50 and 150 residents, both short and long term, and the projected development is approximately 10% complete. Read more about Arcosanti design principles, concepts, and history as well as plans for the future of the community at https://arcosanti.org/.

Planning for the Sustainable City

As this chapter suggests, the complex nature of cities presents special challenges for sustainability. City planning offers multiple responses. It directly engages with the built environment, urban infrastructure—from roads to utilities—and the ways people use space. As a result, planning policies and practices are positioned at the intersection of energy consumption and demand—or they can be.

Passive energy Capturing naturally generated energy for energy consumption. This is in contrast to active energy that relies on external systems, i.e., energy generating plants, to generate electricity for consumption.

These terms are most often used in reference to solar energy systems. In this context, passive systems operate without external devices. For example, greenhouses use glass windows to absorb and retain heat. Active solar energy systems include collectors to capture, store and convert solar energy into electricity.

Each of the strategies identified in this chapter can be used to achieve a more sustainable city; however, they will be effective in different ways. For developed cities, the emphasis must address policies and programs that retrofit the existing built environment, upgrade existing infrastructure, and encourage densification of the current urban form. In contrast, developing cities, and those that have not yet been built, will require comprehensive policies and investments that integrate the built environment with smart technologies and efficient energy distribution. In other words, land use and development decisions must be made intentionally, with an eye towards transportation, walkability, and energy efficiency and supplies. This section reviews some of the planning policies that can be used to create sustainable cities.

Density and, conversely, urban sprawl are one of the key considerations for the sustainable city. The density of a city dictates the form of its transportation network. If residential neighborhoods are separated from grocery stores and other commercial services, for example, it limits the ability of residents to choose walking or biking as a transportation option. Similarly, if jobs are separated from the places where people live, it directly impacts commuter traffic patterns. If you have ever been stuck in morning or evening rush hour, this is something you understand well! Sprawling development patterns also encourage reliance on cars by making mass transit less efficient. With fewer passengers and destinations concentrated along a route, the transit system becomes prohibitively expensive to operate and less able to deliver passengers where they would like to go.

How do we overcome the problems of sprawl and increase density in cities? The primary tool at the planner's disposal is zoning. Zoning codes are laws, adopted by a local government, intended to regulate the placement and density of development as well as its permitted or prohibited uses. The first comprehensive zoning code was adopted by New York City in 1916. Its purpose was to regulate the form of the city's expanding skyline, protecting access to light and fresh air at the street level. Subsequently, zoning codes were adopted across the country and were used to separate incompatible land uses. Whereas industrial uses and residential areas, particularly working class and poor neighborhoods, were comingled in the early 20th century, zoning codes were used to organize land uses and keep noxious uses away from the places where people lived. Referred to as *Euclidean zoning*, this single-use approach to zoning is still predominant in the US and, in many respects, discourages the kind of mixed-use density that would support a walkable and bikeable urban form.

Alternatively, some communities are now adapting their zoning codes to achieve a denser, more sustainable built environment. Form-based codes are one strategy that unites density with zoning laws. Instead of regulating primarily on land use, the form-based code prioritizes the relationship between the development and the public realm (i.e., streets and sidewalks), the massing of buildings relative to one another, and the scale of blocks and neighborhoods. In this context, it encourages a mix of uses—both within buildings and between parcels. Beyond adopting a new style of zoning code, what else can cities do to adapt their existing codes to a more sustainable city?

One strategy is to reconfigure codes that require a *minimum* amount of parking to a code that establishes a parking *maximum*. Parking maximums encourage shared parking between uses and/or parcels, and sets minimum bicycle parking standards. Similarly, *minimum* lot size requirements can be modified to require or incentivize density in target areas. Earlier in the chapter, we compared San Francisco, a city with natural constraints, to Los Angeles, a city without topographical boundaries and the ability to sprawl. Here too, planning policy can help mitigate sprawling development patterns by imposing regulatory growth constraints.

Urban growth boundaries (UGB) are regulations that direct development to areas within the boundary while significantly constraining development beyond the urban boundary. By restricting growth outside of the UGB, the region benefits from both increased density and protection of open space, agricultural land, and environmental resources. UGBs have been used with success in several places, including in Hong Kong, South Korea, and the US. However, this approach is not free of criticism. Development options are constrained within UGBs and increased demand places pressure on housing and commercial prices. In addition, UGBs are not a failsafe for urban sprawl, as development pressures can "leap frog" beyond the protected rural zone.

At a building level, city planning can also contribute to DSM strategies. For new construction, cities can adopt stronger site and building efficiency standards within their zoning and building codes, including regulations that facilitate (and do not inhibit) passive and renewable energy options (e.g., solar PV, passive solar design). Similarly, local governments can either require or incentivize green development practices, such as BREEAM and USGBC's LEED standards. Cities can also impact the efficiency of existing buildings, though it presents a greater challenge. Weatherization programs and incentives offer one instance of the ways local governments are striving to improve the performance of existing buildings. Weatherization assistance programs, for example, not only seek to improve the building efficiency of, often, older housing stock, but also address the financial and quality of life burdens borne by low-income households with the least access to the latest technologies and home energy upgrades.

Final Thoughts

Absent existential threats like permanent traffic jams, deadly pollution, and rampant street violence, cities will be slow to overcome the inertia of their own unsustainable ways, both in the manner we construct them and in the way we operate them. We cannot expect conditions to improve in the future as long as we continue the pattern of rural-to-urban migration, love of suburbs, the dominance of the private automobile, or the use of nonrenewable energy to fuel transport systems, heat and cool buildings, and power factories.

All urban considerations we have just discussed—including cars, highways, congestion, building design, personal comfort, landscaping, pollution, and growth trends—are spokes on a wheel fixed together at the hub of energy. Cities cannot move toward a sustainable future without considering the central importance that energy plays in how they operate and what shape they take. Further more, they will not succeed without pursuing a comprehensive approach towards sustainability, from the urban form to the inclusion of renewable energy sources. This chapter demonstrates that cites are currently unsustainable because they are not designed or operated with energy efficiency as a primary consideration and because they rely on unsustainable forms of energy to operate. We need to reduce consumption of fossil fuels, bolster the development of renewables, and increase our attention to energy efficiency in every decision we make about our urban spaces. If we do this, we will accelerate our pace toward the sustainable cities of the future.

Chapter 14

The Urban Heat Island Effect and Sustainability Science: Causes, Impacts, and Solutions

Darren Ruddell, Anthony Brazel, Winston Chow, Ariane Middel

Introduction

As Chapter 3 described, **urbanization** began approximately 10,000 years ago when people first started organizing into small permanent settlements. While people initially used local and organic materials to meet residential and community needs, advances in science, technology, and transportation systems support urban centers that rely on distant resources to produce engineered surfaces and synthetic materials. This process of urbanization, which manifests in both population and spatial extent, has increased over the course of human history. For instance, according to the 2014 US Census, the global population has rapidly increased from 1 billion people in 1804 to 7.1 billion in 2014.

During the same period, the global population living in urban centers grew from 3% to over 52% (US Census, 2014). In 1950, there were 86 cities in the world with a population of more than 1 million. This number has grown to 512 cities in 2016 with a projected 662 cities by 2030 (UN, 2016). **Megacities** (urban agglomerations with populations greater than 10 million) have also become commonplace throughout the world. In 2016, the UN determined that there are 31 megacities globally and estimate that this number will increase to 41 by 2030. The highest rates of urbanization and most megacities are in the developing world, particularly in South America (i.e., Sao Paulo, Buenos Aires, Rio de Janeiro), Asia (i.e., Jakarta, Delhi, Mumbai, Shanghai, Dhaka), and Africa (Cairo, Lagos). Megacities in developed countries include Tokyo, New York, Los Angeles, London, and Paris.

One common thread found in all cities, regardless of population size and urban extent, is that urbanization inadvertently alters the local climate system. Although impacts vary at different **spatial scales**—such as at the scale of an individual building, neighborhood, city, or urban region, clear differences occur in climate between urban and rural areas, as witnessed in patterns of precipitation (rainfall and snowfall), humidity, particulate matter, and wind speed (Landsberg, 1981).

Urbanization The process whereby native landscapes are converted to urban land uses, such as commercial and residential development. Urbanization is also defined as rural migration to urban centers.

Megacities Urban agglomerations, including all of the contiguous urban area, or built-up area.

Spatial scales The extent or coverage of any given area, such as a county, state, or nation.

This chapter examines one of the best-known urban-induced climate alterations—the **Urban Heat Island** (UHI) effect. By the end of this chapter, you will understand:

- The **morphology** of UHI as well as different UHI types
- Factors causing or modifying **UHI intensity**
- UHI tendencies among various climate regimes
- The impacts of the UHI on residents
- Linkages between the UHI effect and urban sustainability
- Various **mitigation** and **adaptation** strategies to reduce the UHI effect

The UHI Effect

The UHI effect was first recognized by the English meteorologist Luke Howard who noted that temperatures within the City of London were different from the surrounding countryside (Howard, 1833).

> The temperature of the city is not to be considered as that of the climate; it partakes too much of an artificial warmth, induced by its structure, by a crowded population, and the consumption of great quantities of fuel in fires (p. 2).

He hypothesized that the conversion of the Earth's natural surfaces into urban surfaces (e.g., asphalt, concrete, glass) for human socio-economic activities, now referred to as **land use and land cover change** (LULCC), and the generation of heat from factories and/or transportation systems, among other sources, now referred to as **anthropogenic waste heat**, were key factors in causing the UHI. It turns out Howard was not far off!

Cities are comprised of structures of different heights, providing a "rough" surface as opposed to relatively cohesive heights in outlying areas. They also tend to be warmer due to the impervious nature of the materials used (i.e., concrete, asphalt, metal, etc.), which inhibit moisture absorption and limits the cooling effect, and anthropogenic waste heat, which stems from human activities. For example, many manufacturing processes use heat (i.e., coal and nuclear plants, cement kilns, or steel furnaces). Waste heat is heat that escapes or is released from these processes. Household appliances, lawn maintenance equipment, and combustion engines, among other machines release heat into the environment when used as well. These topographical and structural differences as well as the concentration of human activities combine to increase the temperature in urban regions; thereby, contributing to the UHI effect.

The UHI effect quantifies the temperature gradient between urban and nearby rural areas, which often depicts a given urban area as an island of heat amid the native landscape. This pattern is similar to the geomorphic land/sea interface of an island, with a notable "cliff" separating the urban and rural temperature fields. This is illustrated in Figure 1, which depicts the UHI effect in Phoenix, Arizona.

There are four distinct **UHI subtypes**: 1) subsurface, 2) surface, 3) urban canopy layer, and 4) urban boundary layer (Oke, 1995). Observing and measuring temperatures within the different UHI subtypes requires distinct methodological platforms. For example, urban canopy layer studies use data either obtained from

Urban Heat Island (UHI) A global phenomenon in which surface land and air temperature in urban areas is higher than in bordering areas. The primary causes are related to the structural and land cover differences of these areas.

Morphology The study of the form, structure, process, and transformation of an organism or phenomenon and/or any of its parts.

UHI Intensity Measured as the differences in land surface temperature (LST) between urban areas and surrounding boundary area. This variance is related to meteorological elements such as cloudiness, wind speed, temperature, and absolute humidity.

Mitigation Strategies used to reduce ing climate of the countryside:ties. rature (LST) between urban areas and surrounding greenhouse gas emissions (GHG) and increase carbon sequestration in order to slow the rate of climate change.

Adaptation Adjustment strategies to increase a system's ability to adjust and reduce vulnerability to the effects of climate change.

Land Use and land cover (LULC) Land use refers to the syndromes of human activities (e.g., agriculture, urbanization) on lands. Land cover refers to the physical and biological cover of the surface of land (i.e., bare soil, vegetation, water). Land Use and land cover change (LULCC) refers to the changes forced by human activities on both land use and land cover.

Anthropogenic waste heat Heat energy released into the atmosphere from human activities such as industry and transportation.

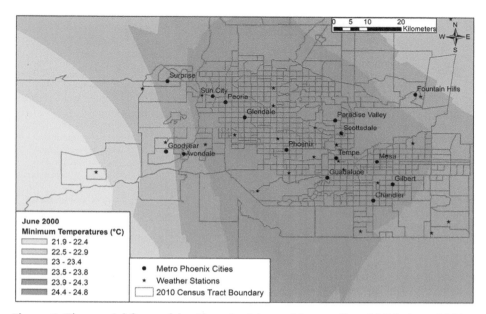

Figure 1 The spatial form of the Phoenix, Arizona Metropolitan UHI in June 2000, as measured by minimum temperatures (T_{min}) taken from several weather stations
Source: Author, Winston Chow

point measurements taken at weather stations or during mobile surveys (Brazel et al., 2000). Surface and subsurface UHI studies often use data from handheld, infrared thermometers or from remote-sensing instruments mounted on airplanes, helicopters, or satellites. The urban boundary layer UHI can be investigated through **radiosonde** data collected by weather balloons that travel up to the top of the lower atmosphere.

Stewart and Oke (2012) introduced a more inclusive approach that uses the concept of Local Climate Zones (LCZs). The LCZ system classifies study sites based on the urban structure, land cover, materials, and anthropogenic activity. The LCZ scheme relies on 17 standardized classes to assess differences between urban and rural landscapes within the given study area (Figure 2).

Parks, lakes, and open areas within cities generally observe lower temperatures than the surrounding urban area and can disrupt urban temperature peaks; these are sometimes called **park cool islands (PCI)** (Figure 3). Dense urban areas can also be cooler during the day than sprawled urban areas due to reduced incoming solar radiation and more shading (Middel et al., 2014). If daytime temperatures in the urban area are lower than in the surrounding areas, this is called an **oasis effect**. The oasis effect is mainly driven by moisture availability and shading. Knowledge about the cooling influence of PCIs within cities is important in discussing ways to sustainably mitigate the UHI, as we shall see later in this chapter.

Typically, scientists measure the UHI effect based on weather station data either by the increase of minimum temperatures (T_{min}) as its surrounding environment "urbanizes" over time, or by the difference in temperatures between urban and rural areas (ΔT_{u-r}) (Landsberg, 1981). Maximum magnitudes of ΔT_{u-r} are

Urban heat island (UHI) subtypes:
1. Subsurface: below the surface of the earth
2. Surface: the skin of the surface of the earth
3. Urban canopy layer: the atmospheric air layer between the ground surface and average building roof height
4. Urban boundary layer: the lowest portion of a planetary atmospheric boundary layer that is directly influenced by an underlying urban area

Radiosonde A miniature radio transmitter with instruments for sensing and broadcasting atmospheric conditions

Park cool island (PCI) Zones in the urban landscape reporting relatively cooler temperature conditions, often associated with parks, lakes, or open areas

Oasis effect Evaporative cooling effect due to heat advection when a source of water exists in an otherwise arid area.

Built types	Definition	Land cover types	Definition
1. Compact high-rise	Dense mix of tall buildings to tens of stories. Few or no trees. Land cover mostly paved. Concrete, steel, stone, and glass construction materials.	A. Dense trees	Heavily wooded landscape of deciduous and/or evergreen trees. Land cover mostly pervious (low plants). Zone function is natural forest, tree cultivation, or urban park.
2. Compact midrise	Dense mix of midrise buildings (3–9 stories). Few or no trees. Land cover mostly paved. Stone, brick, tile, and concrete construction materials.	B. Scattered trees	Lightly wooded landscape of deciduous and/or evergreen trees. Land cover mostly pervious (low plants). Zone function is natural forest, tree cultivation, or urban park.
3. Compact low-rise	Dense mix of low-rise buildings (1–3 stories). Few or no trees. Land cover mostly paved. Stone, brick, tile, and concrete construction materials.	C. Bush, scrub	Open arrangement of bushes, shrubs, and short, woody trees. Land cover mostly pervious (bare soil or sand). Zone function is natural scrubland or agriculture.
4. Open high-rise	Open arrangement of tall buildings to tens of stories. Abundance of pervious land cover (low plants, scattered trees). Concrete, steel, stone, and glass construction materials.	D. Low plants	Featureless landscape of grass or herbaceous plants/crops. Few or no trees. Zone function is natural grassland, agriculture, or urban park.
5. Open midrise	Open arrangement of midrise buildings (3–9 stories). Abundance of pervious land cover (low plants, scattered trees). Concrete, steel, stone, and glass construction materials.	E. Bare rock or paved	Featureless landscape of rock or paved cover. Few or no trees or plants. Zone function is natural desert (rock) or urban transportation.
6. Open low-rise	Open arrangement of low-rise buildings (1–3 stories). Abundance of pervious land cover (low plants, scattered trees). Wood, brick, stone, tile, and concrete construction materials.	F. Bare soil or sand	Featureless landscape of soil or sand cover. Few or no trees or plants. Zone function is natural desert or agriculture.
7. Lightweight low-rise	Dense mix of single-story buildings. Few or no trees. Land cover mostly hard-packed. Lightweight construction materials (e.g., wood, thatch, corrugated metal).	G. Water	Large, open water bodies such as seas and lakes, or small bodies such as rivers, reservoirs, and lagoons.

8. Large low-rise	Open arrangement of large low-rise buildings (1–3 stories). Few or no trees. Land cover mostly paved. Steel, concrete, metal, and stone construction materials.

VARIABLE LAND COVER PROPERTIES

Variable or ephemeral land cover properties that change significantly with synoptic weather patterns, agricultural practices, and/or seasonal cycles.

9. Sparsely built	Sparse arrangement of small or medium-sized buildings in a natural setting. Abundance of pervious land cover (low plants, scattered trees).		
		b. bare trees	Leafless deciduous trees (e.g., winter). Increased sky view factor. Reduced albedo.
		s. snow cover	Snow cover >10 cm in depth. Low admittance. High albedo.
10. Heavy industry	Low-rise and midrise industrial structures (towers, tanks, stacks). Few or no trees. Land cover mostly paved or hard-packed. Metal, steel, and concrete construction materials.	d. dry ground	Parched soil. Low admittance. Large Bowen ratio. Increased albedo.
		w. wet ground	Waterlogged soil. High admittance. Small Bowen ratio. Reduced albedo.

Figure 2 Illustrations and Definitions of the Local Climate Zones
Source: Stewart and Oke (p. 7; 2012)

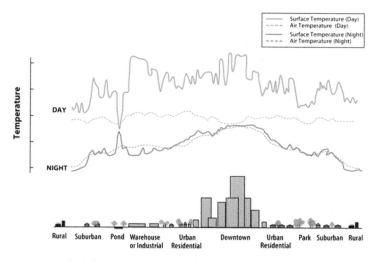

Figure 3 Daytime and nighttime surface and air temperature over various land use types

Source: Environmental Protection Agency https://www.epa.gov/heat-islands/learn-about-heat-islands

generally found at the urban core (e.g., downtown). Plateaus of elevated temperatures can also be found in other areas that have distinct urban land-use categories (e.g., commercial, residential, and industrial). UHI is also a dynamic phenomenon that arises from different urban and rural temperature cooling rates, however.

Under clear and calm weather conditions for a hypothetical city, ΔT_{u-r} usually varies in a consistent manner during a 24-hour period (Figure 4). Canopy layer temperatures generally start cooling in the late afternoon and evening, with urban areas cooling at slower rates than rural areas. This difference results in the growth of ΔT_{u-r} reaching a maximum about three to five hours after sunset (Oke, 1982).

It is important to understand the physical factors that cause UHI (and modify its maximum intensity) when modeling its occurrence and development within different cities and at different scales. Atmospheric models can range from relatively simple statistic-based regression models based on observed data to highly complex, physics-based numerical models that require enormous computational power (Masson, 2006). Results from both observation and modeled studies have great potential in urban planning and policy toward alleviating the negative impacts of the UHI. See Simulating Alternative Planning Scenarios Box below.

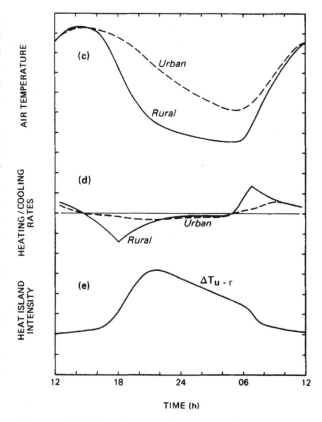

Figure 4 UHI development for a typical temperate city as seen through air temperature changes (a), heating/cooling rates (b), and heat island intensity (c) (Oke, 1987).

Simulating Alternative Planning Scenarios

Scientists use climate models to assess proposed UHI mitigation measures at different temporal and spatial scales and analyze outcomes of various "what-if" scenarios. These techniques offer insights into planning and/or redesigning more efficient living environments. A recent study by ASU scientists investigated the impact of urban form on afternoon temperatures during the summer season among neighborhoods throughout metropolitan Phoenix. They used the microclimate model ENVI-met to simulate temperatures in 5 typical Phoenix neighborhoods. (Middel et al., 2014).

Air temperature simulations at a spatial resolution of 1 meter were conducted by classifying the data models into five LCZs (Figure 5). The study found that neighborhoods comprised of mesic landscaping (e.g., grass, trees, and shrubs) were cooler than residential areas with xeric landscaping (e.g., drought-tolerant), but at the micro-scale, urban form had a more discernable impact on temperatures.

Dense urban forms were more beneficial in terms of daytime cooling than sprawled or sparsely built urban forms, mainly because of reduced incoming solar radiation and shading. This study supports the finding that dense urban areas can have an oasis effect and create a local cool island in mid-afternoon. A follow-up study using the same model found that for every 1 percent increase in tree canopy cover in residential neighborhoods, 2-m air temperatures are reduced by 0.14°C in mid-afternoon (Middel et al., 2015).

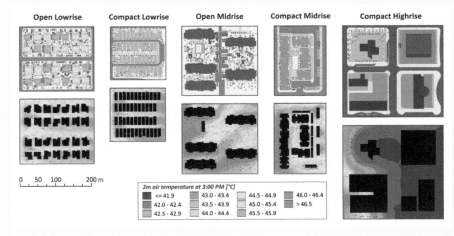

Figure 5 2-m air temperature simulation scenarios for five "typical" residential neighborhoods in Phoenix, Arizona. Blue represents cooler temperatures while reds and purples represent hotter temperatures.

Source: Middel et al. 2014.

UHI Variations and Climate Regime

While the UHI effect is a near-universal phenomenon in cities, impacts and UHI intensity vary throughout the world. For instance, the UHI effect in Mexico City (dry, high elevation) is considerably different from the UHI effect in Singapore (tropical, low elevation). Thus, environmental characteristics such as the local **climate regime** are extremely important in understanding UHI drivers and impacts.

There exist broad patterns of climate regimes across the Earth due to the general circulation of the Earth's atmosphere. In 1884, German climatologist Wladimir Köppen introduced a global climate classification system that partitions the Earth's climates based on averages, variations, and the effects of weather elements (e.g., temperature, precipitation) on the land surface environment. Köppen's system produced five major climate groups, which are summarized in Table 1 and shown in Figure 4.

Climate regime State of the climate system that occurs more frequently than nearby states due to either more persistence or more frequent recurrence. In other words, a cluster in climate state space associated with a local maximum in the probability density function (Stocker, 2013).

Table 1 Summary of Major Climate Groups of Köppen Classification System

Köppen Climate Classification Group	Climate Characteristics	Linkages with UHI	Example Cities
Group A: Tropical/megathermal	12 months of the year with average temperatures above 18°C or higher with somewhat significant precipitation	Humid, tropical climates; heavy precipitation; modest high temperatures at day and night; year-round warm temperatures; high anthropogenic waste heat	Rio de Janeiro, Lagos, Kinshasa, Abidjan, Mumbai, Dhaka, Jakarta, Bangkok, Singapore, Manila
Group B: Dry (arid and semiarid)	Little precipitation	Exposure to very high temperatures; extended summer season; high anthropogenic waste heat	Karachi, Lahore, Cairo, Lima, Tehran, Baghdad, Phoenix
Group C: Temperate/mesothermal	The coldest month average is between 0°C and 18°C and at least one month averaging above 10°C	Hot and dry or warm and wet summers; cool winters	Shanghai, Guangzhou, Sao Paulo, Delhi, Istanbul, Tokyo, Mexico City, London, New York, Bogota, Los Angeles, Buenos Aires, Paris
Group D: Continental/microthermal	Have at least one month averaging below 0°C and at least one month averaging above 10°C	Warm summer temperatures; severe winters	Beijing, Moscow, Seoul,
Group E: Polar	12 months of the year with average temperatures below 10°C	Severe winters; cool summers	NA

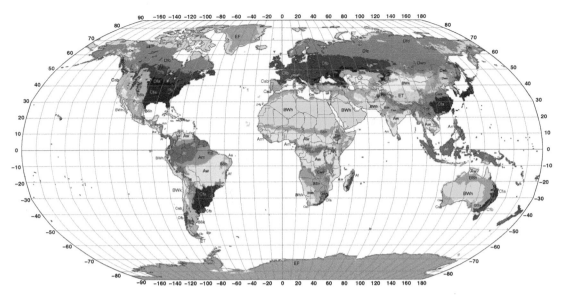

Figure 6 Updated World Map of Köppen-Geiger Climate Classification
Source: Kottick et al. (2006) http://koeppen-geiger.vu-wien.ac.at/pdf/kottek_et_al_2006_A4.pdf

What Factors Cause the UHI?

The key to understanding how UHI is caused is to examine alterations to the urban **surface-energy balance (SEB)**. The SEB describes how the available energy in the Earth's climate system is balanced and partitioned into energy **fluxes** that move towards and away from the Earth's surface. We distinguish between two major flux types: radiative fluxes and turbulent fluxes.

Radiative fluxes refer to the shortwave radiation K, incoming from the sun and reflected by Earth's surfaces, and longwave radiation L, emitted by the atmosphere and the Earth's surfaces. Turbulent fluxes are driven by wind and associated with phase changes of water (latent heat flux Q_E) and heating of surfaces (sensible heat flux Q_H). The amount of incoming and outgoing energy, i.e. the net flux, must be equal. Therefore, the net all-wave radiative flux density Q^* is defined as the sum of total radiation arriving at ($\downarrow$) or exiting from ($\uparrow$) a surface. Q^* is the sum of all incoming ($\downarrow$) and outgoing ($\uparrow$) shortwave (K) and longwave (L) radiation and can be mathematically expressed as:

$$Q^* = K\downarrow - K\uparrow + L\downarrow - L\uparrow = K^* + L^* \qquad \text{(eq.1)}$$

At the same time, the available energy Q^* can be expressed using the turbulent heat fluxes Q_H and Q_E:

$$Q^* + Q_F = Q_H + Q_E + \Delta Q_S \qquad \text{(eq.2)}$$

This representation requires two additional variables: the energy stored or withdrawn from the surface substrate, also called heat storage (ΔQ_S), and anthropogenic waste heat (Q_F). Examples of Q_F include heat emissions from vehicles, industrial plants, and air conditioners, among other sources. In most

Surface-energy balance Describes the balance between radiation, conduction and convective heat flow.

Flux Rate of flow of some quantity.

Table 2 Suggested causes of UHI through alterations of the urban energy balance, together with (1) typical urban features associated with the change, and (2) direct urban climate effects on UHI formation (based on Oke, 1982).

Surface energy balance equation symbol	Surface energy balance variable	Typical associated urban features	Direct urban climate effects
Increased K*	Shortwave radiation	Canyon geometry	Increased surface area and multiple reflection
Decreased L*	Longwave radiation	Air pollution	Greater absorption and re-emission
Increased L↓	Longwave radiation	Canyon geometry	Reduced sky view factor
Addition of Q_F	Anthropogenic waste heat	Buildings and traffic	Direct addition of heat
Increased ΔQ_S	Storage heat flux	Construction materials	Increased thermal admittance
Decreased Q_E	Latent heat flux	Construction materials	Increased water-proofing
Decreased $(Q_H + Q_E)$	Sensible and latent heat fluxes	Canyon geometry	Reduced wind speed

cities, however, anthropogenic waste heat is a relatively small component of the urban-energy balance when compared to Q_H or ΔQ_S.

Table 2 demonstrates how various factors influence the urban surface-energy budget, and therefore, UHI. For example, LULCC such as the amount and extent of asphalt have the tendency to increase heat and will affect the amount of storage heat flux. Understanding typical city features and how they vary within urban areas is critical in explaining how UHI is caused. Increases in the use of certain construction materials (such as concrete, glass, asphalt, and steel) reduce the potential for evapotranspiration and increase **thermal admittance**.

Many urban materials have properties that store and release heat. Those that retain and radiate high levels of heat to the environment over a longer period are considered to have high thermal admittance. Materials such as concrete and asphalt store, and later radiate, more heat than vegetated surfaces such as urban parks. Urban air pollution also contributes to UHI through the creation of smog. For instance, smog reduces the rate of surface radiative loss creating more surface heat and prolonging the dissipation of that heat.

There are two other urban features that influence UHI—**urban canyons** and the sky view factor (Figure 7). Urban canyons are characterized by tall buildings on either side of long, narrow streets typical of most **central business districts**. These artificial canyons increase the total urban surface area for energy storage and reduce urban wind speed. They also reduce the sky view factor, or how much of the sky is seen from the surface. This reduces surface radiative loss, especially at night (Unger, 2004; Middel et al., 2017).

Thermal admittance A property that quantifies the rate at which a material stores or releases heat. It is the square root of the product of thermal conductivity and heat capacity.

Urban canyons An artificial "canyon" formed by two vertical walls of buildings with a horizontal street or pavement surface.

Central business district The commercial and often geographic center of a city.

Figure 7 A dense urban canyon (i.e. two walls + street) of downtown Singapore that is typical of most central business districts (left), and a fish-eye lens picture that represents the low sky view factor (SVF) seen at the surface of the street (right) (*Source*: Author – Winston Chow)

Additional Factors Influencing UHI Intensity

Difference in temperatures between urban and rural areas (ΔT_{u-r}) is strongly affected by the prevailing regional climate over a city. Higher wind speeds, increased cloud cover, and lower cloud heights have been shown to reduce magnitudes of ΔT_{u-r} (Arnfield, 2003). These meteorological factors disrupt surface cooling in different ways. Higher wind speeds increase near-surface turbulent mixing, which decreases the efficiency of surface radiative cooling. Increased cloud cover limits surface longwave radiation loss, with low-level clouds (e.g., stratus and cumulus) having greater impact compared to high-level clouds (e.g., cirrus and cirrostratus). Maximum UHI intensities are thus generally associated with periods of clear, calm weather.

The size and geographic location of cities also affects the maximum magnitude of ΔT_{u-r}. Generally, the bigger the urban population, the larger the maximum UHI intensity (Oke, 1973). Cities located in temperate climates typically have higher maximum ΔT_{u-r} compared to tropical, subtropical, and highland cities, even with population size held constant. Of interest is that tropical cities experiencing more precipitation appear to have lower observed maximum ΔT_{u-r} compared to tropical cities with pronounced dry seasons, which suggests a strong impact of surface moisture on attenuating UHI intensity (Roth, 2007).

Water bodies and topographic features located next to a city can also affect UHI distribution and intensity. Cities located by the coast are subject to a regular daily cycle of land and sea breezes. Typically, moister and cooler air masses over the ocean are advected into warmer urban areas in the early evening, resulting in a distinct temperature gradient seen in cities such as Vancouver, British Columbia (Runnalls and Oke, 2000). Similarly, cities located in valleys may be subject to daily cold air drainage flow patterns. These flows—also known as **katabatic flows**—are

Katabatic flows 1) Most widely used in mountain meteorology to denote a downslope flow driven by cooling at the slope surface during periods of light larger-scale winds; the nocturnal component of the along-slope wind systems. 2) Cold air flowing down a slope or incline on any of a variety of scales (AMS, 2017).

usually strongest at night and can directly affect urban temperatures in complex topography.

The temperature of the Earth's surface is controlled by solar energy or **solar radiation** from the sun. Solar energy accounts for approximately 99.97 percent of energy entering the Earth's atmosphere (other energy sources include geothermal, tidal, and anthropogenic waste heat). The interactions in the long wavelengths of radiation to and from the Earth are critical to the planet's sustainability, as the atmosphere acts as a greenhouse that keeps the planet warm. Scientists have calculated that without this greenhouse effect, Earth's temperatures would be some 30° Celsius (C) cooler than present.

Solar Radiation The total electromagnetic radiation emitted by the sun.

This energy enters the Earth's atmosphere in the form of shortwave radiation where the energy is either absorbed or reflected back into the atmosphere. The energy reflected from the earth back out to space is called longwave or infrared radiation. **Albedo** refers to the percent of incoming solar radiation that a body or surface reflects.

Albedo The fraction of earth's incoming solar radiation that is reflected back to space.

Albedo values range from zero (no reflection) to 100 (complete reflection) and vary depending upon surface materials. Light surfaces such as snow and ice are highly reflective and have albedo values in the range of 80–95 percent. Alternatively, dark-colored materials such as asphalt have low levels of solar reflectance and have albedo values in the range of 5–10 percent. Understanding the influence of albedo is important in **sustainability science** for two reasons.

Sustainability science A field of research dealing with the interactions between natural and social systems, and with how those interactions affects the challenge of sustainability.

The first is that surfaces with low albedo values absorb large quantities of solar energy; while high albedo surfaces reflect sunlight back into the atmosphere. Second, the albedo values of various building materials and LULC are critical when introducing or replacing features in urban environments. For instance, converting an asphalt parking lot (a low albedo surface material) to a park or open green space (a moderate albedo surface material) will increase the solar reflectance of that parcel of land and thus reduce local heat-storage capacity.

Impacts of the UHI on Cities

While the UHI effect is a near-universal phenomenon in cities, its resulting impacts—and whether such impacts are "good" or "bad" for urban residents' quality of life—depend on factors such as geographic location and season. Increased urban warmth can be beneficial for residents in high-latitude cities (e.g., Moscow) with cold winters, but detrimental to city dwellers in hot subtropical cities (e.g., Las Vegas, Phoenix) with hot summers. This section discusses three UHI-derived impacts on urban residents, 1) thermal discomfort and vulnerability to increased heat stress; 2) changes to urban energy and water use; and, 3) impacts on urban economic output (Kunkel et al., 1999).

The UHI effect, coupled with naturally occurring summer **heat waves**, is linked to negative impacts on human health and well-being, such as morbidity (illness) and mortality (death) (Rosenzweig et al., 2005). UHI increases thermal discomfort particularly when summer temperatures are elevated and sustained over multiple consecutive days. Excessive exposure to heat , especially in unshaded locations, increases thermal discomfort (Middel et al., 2016) and already accounts

Heat wave A prolonged period of **excessively** warm weather.

for more deaths in the US than any other weather-related phenomenon (CDC, 2006; Kalkstein and Sheridan, 2007). Research shows that mortality rates and hospital admissions for cardiovascular, respiratory, and other preexisting illnesses increase in conditions of very hot weather as well (Semenza et al., 1999).

The dangerous impacts of excessive exposure to extreme heat are evident from two historic heat waves. The first was the Chicago, Illinois heat wave in July 1995 that claimed over 700 lives (Semenza et al., 1996; Klinenberg, 2002). The second was the 2003 heat wave that gripped Western Europe resulting in between 22,000 and 52,000 deaths, many of them occurring in large cities (Larson, 2006).

A warmer urban climate can also translate into increased demands on energy and water. This is very likely to occur in cities subject to seasonal high temperatures or located in equatorial climates with relatively hot year-round temperatures. For example, increases in the UHI intensity in metropolitan Phoenix, Arizona have coincided with increased total and peak-energy demand for residential and commercial cooling from the period between 1950 and 2000 (Golden, 2004).

The demand for energy increases because air conditioning is used more often and for longer periods. Although air-conditioning systems provide interior relief from summer temperatures, they also release large quantities of anthropogenic waste heat into the surrounding outdoor environment. They also produce greater amounts of air pollutants, thus worsening urban air quality as well as contributing to urban warming.

Warmer temperatures in metropolitan Phoenix have been shown to increase demands on local water systems as well. Higher energy demands translate to increased demand on water resources because of the large quantities of water consumed when producing electricity. In addition, demand on local water resources is greatest during summer months when residents irrigate outdoor vegetation and fill their pools to provide relief from intense summer temperatures (Guhathakurta and Gober, 2007; Wentz and Gober, 2007).

UHI Mitigation and Adaptation Strategies

Mitigation and adaptation strategies help reduce the UHI effect. Table 3 provides a list of various strategies that cities and individuals have employed. These include the creation of urban forests that offer various environmental benefits, such as sequestering or storing carbon, retaining storm water during rain events, and improving air quality. Urban forests also provide benefits to residents. These benefits are often described as **health co-benefits**.

Health co-benefits Multiple benefits of a program whereby one of the domains benefits the public health sector, such as providing shade and increasing thermal comfort.

These co-benefits include providing shade and increasing thermal comfort, while simultaneously creating green spaces for recreation and relaxation. Improved air quality from urban forests, increased green spaces, and reduced fossil fuel consumption also help restore ecosystems and provide a healthier living environment, which has been found to stimulate physical activity among residents. A more active lifestyle translates into lower rates of obesity and reduced cardiovascular disease, as well as improved mental health (Besser and Dannenberg, 2005; West et al., 2006; Nurse et al., 2010). In addition to direct co-benefits, there are several wide-ranging indirect benefits.

Table 3 Illustration of mitigation and adaptation strategies to reduce the urban heat island effect with health co-benefits. Modified from Harlan and Ruddell, 2011.

Mitigation and Adaptation strategy	Environmental outcome	Examples	Human health Co-benefit
Urban Greening	Carbon sequestration, enhanced stormwater management and water quality, air quality improvement, reduced energy use	**Urban Forest**: Chicago, IL has implemented an urban forest to improve air quality while reducing the UHI effect	Increased shade and thermal comfort, less heat-related illnesses, improved quality of life
		Community Garden: Boston, MA is using community gardens to capture carbon, increase green space throughout the city, and provide fresh produce locally	Improved nutrition, increased community engagement
		Green Roofs: Toronto (Canada) utilizes green roofs to help reduce local outdoor temperatures while providing insulation and cooling capacity inside buildings	Less heat-related illness
Urban Fabric Modification	Reduced energy consumption, air pollution and GHG emissions by lowering energy use	**Cool Roofs**: The City of Phoenix implemented a Cool Roofs program to coat 70,000 square feet of the city's existing rooftops with reflective materials	Less heat-related illness, improved human comfort
		Cool Pavement: Sacramento's pervious concrete parking lot at Bannister Park is one of the first uses of cool pavement in the state of California	Less heat-related illness, lower tire noise, better nighttime visibility, improved human comfort
Urban Structure Modification	Improved air quality, ventilation and solar access	**Building orientation**: Singapore's Marina Bay Financial Centre aligns its arterial roads along the predominant wind direction to improve wind flow at street level, and staggers building heights to increase wind downwash that removes air pollutants	Reduced exposure to air pollutants, increased shade, and improved thermal comfort

Mitigation and Adaptation Challenges

Despite the extensive environmental and human benefits associated with the strategies outlined in the table, several challenges exist to successfully implementing these tactics. For example, implementing any mitigation or adaptation strategy requires an initial financial investment, which many individual homeowners and city officials may be reluctant to make, particularly in a depressed or flat economy.

A second challenge is the issue of sovereignty. For instance, individual homeowners are often limited by the covenants, codes, and restrictions (CCRs) governing their community or neighborhood that are enforced by Homeowners Associations. These may prevent homeowners from installing solar panels on their roof or solar hot-water heaters, as they may violate the CCRs.

Another challenge is the suitability of a given UHI mitigation or adaptation strategy in a local geographic context. A strategy that is perfect for residents living in Portland, Oregon may be inefficient or counter-productive when implemented in Phoenix, Arizona. Thus, UHI mitigation and adaptation strategies must be considered by individual cities based on the resources and natural advantages of a given city. Perhaps the biggest challenge to implementing a given UHI mitigation or adaption strategy is conducting a cost-benefit analysis.

For instance, the many benefits of expanding an urban forest need to be evaluated against the costs, such as increased pollen (allergies), tree maintenance services, and a greater demand on water resources, which is highly problematic in arid environments (Gober, 2006; Shashua-Bar et al., 2009). The use of alternative energy systems is an increasingly popular method for meeting residential and commercial energy needs with reduced GHG emissions. However, many of the current systems such as solar thermal energy (STE) systems are highly water-intensive, which must be considered when siting STE plants and allocating water resources (Hu et al., 2010).

Advancing sustainable solutions for one problem area often supports other benefits through a wider articulation of goals. For example, reducing UHI by additional vegetative cover also provides important urban wildlife corridors, rainfall runoff management, and urban farming opportunities, thus making for healthier cities. The concept of redesigning cities to reduce or adapt to UHI needs to be broadened to better understand both the benefits and challenges. Sustainability science and sustainable development are mechanisms for doing so.

Conclusions

Sustainability science and sustainable development offer valuable tools and insights for addressing environmental and social concerns associated with UHI. Although there are numerous definitions for both sustainability and sustainable development, it is generally agreed that there are three components of sustainability. These are 1) environmental conservation, 2) equity (inter- and **intragenerational**), and 3) economic development. Anthropocentric paradigms often exploit one sector

Intergenerational Occurring or existing between members of one generation; occurring during the span of one generation.

for the benefit of another, such as harvesting a resource beyond its natural rate of production for economic gain. Sustainability paradigms aim to simultaneously maximize all three.

Why is the larger framework of sustainability and the idea of maximizing the three values of sustainability—environment, equity, and economics—important for UHI? How we design cities, the materials used for construction and infrastructure, the types of transportation, and the extent of vegetative open space among many others affect the urban environment and UHI intensity. Sustainability science has shown that we can reduce the impacts of UHI by constructing and/or redesigning neighborhoods and individual buildings as well as establishing mitigation and adaptation policies. The degree of change in UHI intensity and spatial coverage will depend on the ways in which a given city is built (or rebuilt) and the materials that are used to construct various features within the urban environment.

The variability of temperatures within a given city present challenges for both inter- and intragenerational equity. For instance, increasing minimum temperatures in the urban environment compromises the ability of future generations to enjoy a comparable climate. It creates intergenerational inequity. In the case of UHI, some parts of a city are significantly warmer compared to other areas of the city. The distribution of temperatures within a given city is often concerning because the burden of temperatures is not evenly distributed among social groups. Research shows that this intragenerational inequity most often impacts minority and low-income populations. These populations are often exposed to the worst environmental conditions yet possess the fewest resources to cope with UHI impacts.

The economic sector is also significantly impacted by UHI. Urbanization is a significant contributor to UHI formation and the expansion of urban areas into sprawling megacities is largely driven by economics. Rather than introduce a growth boundary, redevelop an existing lot, or increase density (through vertical development), residential development often occurs on the urban fringe where land is relatively cheap. Although economics contributes to UHI development, this sector also offers opportunities to increase the efficiency of cities. For instance, the introduction or enhancement of public transportation systems provides local employment opportunities, commuting options for residents and visitors, and reduces dependency on automobiles and fossil fuel consumption. It also has the co-benefit of improving air quality. There are also wide-ranging opportunities for the private sector to provide services and innovative solutions for sustainable development, such as solar technologies.

Select Cases from around the World

The following section examines select cities throughout the world to better understand UHI patterns and characteristics in diverse settings. The cities selected in this analysis are among the 100 most populous urban areas in the world and are stratified to represent different regions and climate classes.

Table 4 Overview of Selected Cities

| City | Country | Population (Metro) | Area (Sq. mi) | Population Density (per sq. mi) | Koppen Climate Classification | | Temperature Ave Monthly | | Precip (inch) |
					Class	Description	Jan	July	Ann
Beijing	China	21.1 mil	6,336	3,396	Dwa	Humid Continental	−3.7°C	26.2°C	22
Lagos	Nigeria	21 Mil	1,045	15,369	Aw	Tropical Savanna	27.3°C	25.2°C	59
Mexico City	Mexico	20.4 mil	573	15,663	Cwb	Subtropical Highland	14.6°C	18.2°C	33
London	United Kingdom	14 mil	3,236	2,680	Cfb	Oceanic	6.8°C	19.6°C	23
Singapore	Singapore	5.5 mil	278	19,910	Af	Tropical Rainforest	26.0°C	27.4°C	92
Phoenix	United States	4.5 mil	1,147	3990	Bwh	Hot Desert	13.5°C	34.7°C	8

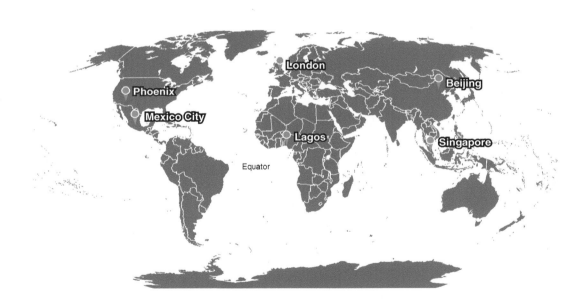

Case 1: Beijing, China

Established in 1046 AD, Beijing, China is located at 39°55'N, 116°23'E and is 144 feet above sea level. It has over 21 million residents and covers over 6,336 square miles comprised of high-density development. Beijing's Köppen climate classification is Humid Continental (Dwa) where the continental climate results in cold and dry winters with hot and wet summers. For example, the average monthly temperature in January is −3.7 °C (25.3 °F) in contrast to July where

© SJ Travel Photo and Video/Shutterstock.com

the monthly average is 26.2 °C (79.2 °F). The city has a seasonally concentrated precipitation regime that produces an annual average rainfall of 22 inches, with close to almost three-fourths of that total falling from June to August.

Research on temperature variability within the Beijing Municipality has measured UHI maximum intensity at 3.37 °C (or 6.1 °F) (Yang et al. 2013). This is attributed in large part to Beijing's rapid urbanization since 1980. While Beijing's UHI provides warming effects during the cold winters, the warm and humid summers increase energy demands as residents rely on air-conditioning for thermal comfort. Beijing's rampant air pollution also contributes to the UHI effect.

Case 2: Lagos, Nigeria

Lagos, Nigeria was founded in the 15th century and is located at 6°27'N, 23°2'E with an elevation of 135 feet. An important trade and administrative center, Lagos

© Bill Kret/Shutterstock.com

is among the world's fastest growing cities with a population of 21 million residents. The Köppen climate classification for Lagos is tropical savanna (Aw). The city's location near the equator in the tropical rainforest belt results in a temperature profile that remains constant throughout the year with only modest differences between the warmest (March; 28.5 °C or 83.3 °F) and coolest (August; 25.0 °C or 77 °F) months. The wet season, lasting from April to October, produces an annual average rainfall of 59 inches.

Current research shows a pronounced UHI effect within Lagos. The rapid population growth and establishment of formal and informal settlements throughout the city has resulted in maximum UHI intensity more than 7 °C (or 12.6 °F) (Ojeh et al. 2016). Lago's chronic exposure to modestly warm temperatures results in elevated energy demands year round.

Case 3: Mexico City, Mexico

The capital city of Mexico, Mexico City, was founded in 1325 and is located at 19°26'N, 99°8'E. Mexico City is currently listed as the 4th most populous city in the

© Andrea Izzotti/Shutterstock.com

world and is comprised of high-density urban development. A high-elevation city, the City resides at 7,380 feet above sea level and is classified as subtropical highland climate (Cwb) using the Köppen classification system. Average annual temperatures show modest variation throughout the year. Average monthly temperatures in January are 14.6 °C (58.3 °F) and July are 18.2 (64.8 °F). Temperatures rarely dip below 3 °C (37.4 °F) or exceed 30 °C (86 °F). The average annual rainfall is 33 inches, which falls primarily from June through October.

Research shows that temperatures within Mexico City vary considerably with maximum UHI intensity reaching 7.8 °C (or 14.0 °F) during a calm clear night in February (dry season) (Jauregui 1997). Thus, Mexico City demonstrates a significant maximum nocturnal UHI intensity. However, the modest average temperatures of the high-altitude city modulate temperature variability.

Case 4: London, United Kingdom

Founded in 43 AD and located at 51°30'N, 0°7'E, London is the capital of England and a global center of economic and political power. It sits just above sea level

©Elena Rostunova/Shutterstock.com

at 115 feet and has a population of over 14 million residents, which reside in low, medium, and high-density development. The Köppen climate for London is oceanic (Cfb). Seasonal temperatures range from average monthly temperatures of 6.8 °C (44.2 °F) in January to 19.6 (67.3 °F) in July. It receives an average rainfall of 23 inches annually.

London has been found to experience temperature extremes ranging from 38.1 °C (100.6 °F) at Kew during August 2003 down to −16.1 °C (3.0 °F) in January 1962. Summers are generally warm and sometimes hot. London's average July high is 24 °C (75.2 °F). On average, the city sees 31 days above 25 °C (77.0 °F) and 4.2 days above 30.0 °C (86.0 °F) each year. During the 2003 European heat wave, there were 14 consecutive days above 30 °C (86.0 °F) and 2 consecutive days where temperatures reached 38 °C (100.4 °F), leading to hundreds of heat-related deaths.

Perhaps the longest studied UHI, London has considerable temperature variability within the city which was recently measured at 4.5 °C (8.1 °F) (Bohnenstengel et al. 2011). UHI tendencies include warm days during the summer season that result in high energy demands and potentially harmful heat wave episodes (as observed in 2003).

Case 5: Singapore, Republic of Singapore

Established in 1299, the island city-state of Singapore is located at the equator (1 °N, 104 °E) with an elevation of 45 feet and a small footprint of 278 square miles. Since gaining independence in 1965, Singapore has observed tremendous physical and demographic transformations where concurrent rates of physical expansion through land reclamation, and population growth and urbanization

© mybeginner/Shutterstock.com

have resulted in an urban population density being among the highest in the world (about 19,910 people per square mile in 2015). With a Köppen climate classification of tropical rainforest (Af), the city's equatorial location provides little fluctuation in temperatures throughout the year where the average monthly temperatures of 26.0 °C (78.8 °F) in January and 27.4 °C (81.3 °F) in July. Singapore experiences high relative humidity with averages around 79% in the morning and 73% in the afternoon. It receives heavy annual rainfall with an average of 92 inches.

Accompanying Singapore's rapid urban development is a growing UHI, which scholars have measured maximum intensity at 3 °C (5.4 °F) in 1965 to more than 7 °C (12.6 °F) in 2004 (Roth and Chow, 2012). While the dense urban mosaic of concrete and asphalt can explain most of this increase, another likely cause is the extensive waste heat generated by its residents. The use of air conditioning to reduce thermal discomfort is prevalent year-round, especially given Singapore's equatorial location. The characteristically high daytime and nighttime temperatures combined with high ambient humidity present a continuous demand for energy.

Case 6: Phoenix, United States

Established in 1867, the Phoenix, Arizona metropolitan area (33°27'N, 112°04'E) is an ideal city to investigate UHI effects due to its hot desert climate (Köppen classification Bwh) coupled with the rapid increase in population and LULCC (Chow et al., 2012). Phoenix's elevation is 1068 feet. It has an average annual rainfall of 8 inches. It has rapidly transformed from a native desert landscape to the fifth largest city in the United States with a population of 4.5 million and growing. The resulting alterations in the local urban climate system present new challenges and vulnerabilities to residents.

The city experiences long and extremely hot summers with short and mild winters with average temperatures of 13.5 °C (56.3 °F) in January and 34.7 °C (94.5 °F) in July. Phoenix's all-time high recorded temperature is 50 °C (122 °F),

© Tim Roberts Photography/Shutterstock.com

observed on June 26, 1990. High average temperatures make it one of the hottest of any major city in the US during the summer. On average, there are 107 days where the daily maximum temperature is greater than or equal to 38 °C (100 °F). It also experiences an annual average of 18 misery days.

The UHI effect in Phoenix is quite pronounced with maximum UHI intensity measured at 7.3 °C (13.1 °F) (Hedquist and Brazel, 2006). The rapid population growth and sprawling urban development in the naturally hot Sonoran Desert has resulted in the following UHI characteristics for Phoenix: exposure to very high temperatures during the long and intense summer season; significant demands on air conditioning and energy use to provide cooling for residents; intensive water demands for thermal cooling, vegetation, and energy generation; and human vulnerability to heat stress.

Misery days A temperature threshold of thermal comfort. A misery day in Phoenix, Arizona is when the daily maximum temperature is greater than or equal to 43.3 °C (110 °F).

Summary of Case Studies

Understanding the predominant climate regime of a city is extremely important when conducting UHI research. For instance, cities classified in Köppen climate Group A (tropical/megathermal) experience high precipitation and relatively low daytime (diurnal) temperature variation resulting in chronic heat exposure (e.g., Lagos, Singapore). Alternatively, cities in Köppen climate Group B (dry) experience more drastic diurnal temperature variation and greater seasonal temperature variation which has potential for seasonally acute exposure to heat stress (e.g., Phoenix). This has wide-ranging impacts on energy demand, water demand, air pollution, and human vulnerability to heat stress, among other considerations. Köppen climate Group C (temperate/mesothermal) cities report characteristics of moderate temperatures but experience seasonal weather anomalies that can result in human vulnerability to heat stress, such as the 2003 European heat wave.

Can We Teach an Education for Sustainable Development? Measuring the Fourth *"E"*

Chad P. Frederick and K. David Pijawka

Introduction

This chapter explores why learning outcome measurements in education for sustainable development are necessary and how they can be approached. It illustrates the distinction between an education for sustainable development (ESD) and the traditional model of education. While ESD and traditional education are similar in some respects, there are important differences between them. In both approaches, students accumulate facts about the world. These facts allow students to have the information necessary to be competent citizens in society. Students also develop mental skills, such as mathematics and language. This allows them to be competent workers in the economy. The differences are found in how we situate the student in relation to the facts and the skills for which we try to develop competencies.

In order to study complex systems, traditional education has a tendency to start the learning process by simplifying and characterizing systems as collections of discrete elements. This method has been successful in the natural and physical sciences, such as chemistry and physics, where reducing systems into elements allows students to observe critical interactions. However, it has been applied to other academic fields, including the ecological and social sciences, with mixed results. In the traditional approach, people generally are treated as distinct and separable from the systems in which they live and independent of the goods and services they consume. Although there may be educational goals achieved by this approach, the separation fosters a disconnect between our consumption and the social and ecological environments where these goods and services are produced and consumed—a principal sustainability idea. Furthermore, it allows us to treat as separate whatever impacts the remains of our material consumption may have after we discard them. It even allows us to think of the products as separate from the processes used to produce them (Jensen, 2008, pp. 293–305).

Systems thinking The ability to recognize the existence of complex networks, and to think in terms of multiple relationships within networks, as opposed to being limited in thought to isolated actions and reactions.

Affective competencies Skills and abilities beyond cognitive or physical training which require a level of emotional and psychological self-mastery.

In contrast, ESD promotes **systems thinking** for our students. In systems thinking, products and services are not only seen as interconnected to their production and consumption, but also to their consequences throughout their lifecycle. This enables students to critically inspect their own relationships to the socio-ecological systems in which they live and those which they affect (Colucci-Gray, Camino, Barbiero, & Gray, 2006). For example, automobiles are seen not only as products providing utility for their owners, but are also a source of pollution and a force for shaping urban forms which, in turn, has accessibility and equity implications for those without automobiles (Boone, 2014).

Traditional education mostly focuses on acquiring technical skills, while ESD also strives for **affective competencies**. Thus, ESD requires that we consider not only knowledge but also attitudes, ethics, values, and behaviors. This is because it has become evident that people use their attitudes, ethics, and values when analyzing facts and making decisions (Kollmuss & Agyeman, 2002). Reflecting on our ethics and values allows us to see if they are accurately aligned with our habits and actions and if they are helping us to make informed decisions. This distinction is important because it calls into question the very purpose of education and can lead toward the empowerment of 'actively engaged citizens' (Jickling & Wals, 2008).

There is a growing literature on learning outcomes related to sustainability education. In this chapter, we address the dimensions of sustainability education outcomes and how these can be achieved. Utilizing one undergraduate course in a sustainability program, *Sustainable Cities,* we address the question of how to measure these outcomes using pre– and post–course surveys as the methodology. We begin the chapter by articulating the meaning of the learning outcomes expected in education for sustainability and ways in which to achieve them and then we demonstrate a case where the outcomes are measured for a particular course.

Sustainable Cities is one of the largest courses in the School of Sustainability at Arizona State University, teaching approximately 1000 students annually. The course is deliberately interdisciplinary using co-instructors from different disciplines, lecturers from various disciplines on campus and outside campus, and incorporating collaborative student team exercises/assignments with students from different disciplines and levels of educational achievement. Various methods of instruction are introduced including assignments that require individual responses and focus on ethical issues and complex systems understanding as well as discussions of how to change current consumer patterns toward sustainable solutions. Assignments also include community-based research at the undergraduate level through a student Honors section. Critical Thinking is introduced throughout each topic covered in the course via questions on the readings and specially developed assignments. The course is built around sustainability education outcomes and we offer insight into how we have applied them and measured the level of success throughout this chapter.

Learning outcomes The set of knowledge and abilities that students aspire to master and that educators aspire to deliver.

Learning Outcomes

Before exploring goals of the *Sustainable Cities* course, it may be helpful to review what is meant by **learning outcomes**. Learning outcomes can be found in three domains: cognitive, affective, and psychomotor. While often overlapping, there are some distinctions among these domains. Cognitive skills are fundamentally about thinking, while affective skills employ feelings. Psychomotor skills involve physical movement. Many learning outcomes involve all three (Kraiger, Ford, & Salas, 1993). How we perceive our capacity to perform cognitive skills often affects our overall ability to execute them (Dweck, Mangels, & Good, 2004). The good news is that all three domains can be trained for improvement.

Learning Outcomes in Education for Sustainable Development

Most learning outcomes in traditional education involve the performance of cognitive competencies such as recalling the content learned in class or demonstrating abilities such as applying mathematical operations and constructing coherent, convincing paragraphs. Unlike traditional education, ESD calls for students to use affective skills in addition to cognitive skills and applications of knowledge as the complex nature of many sustainability problems require affective competencies to solve them (Wiek, Withycombe, & Redman, 2011). For example, a sustainability problem might call for specialists from three distinctly different fields to work together and apply their expertise in finding a resolution. Such **interdisciplinary** teamwork means that the specialists must respect each other's input, which is largely an affective capacity.

> **Interdisciplinarity** The practice of distinct fields of knowledge working together to produce new knowledge that cannot be produced by the fields on their own.

As ESD is still a young field, it is challenging to provide a comprehensive list of appropriate learning outcomes (Sterling & Thomas, 2006). Indeed, sustainability educators recognize that ESD learning outcomes will evolve over time. Unfortunately, there is little research or theory in this area. Nonetheless, ESD scholars have identified a wide range of beneficial outcomes for which to strive and have developed useful frameworks for pursuing these outcomes. Furthermore, much like the need to master basic mathematical skills before attempting more advanced operations, it is likely that some affective competencies provide a sound basis for more advanced skills (Shephard, 2008).

Although ESD competencies have been described as unwieldy, Wiek, Withycombe, & Redman (2011) have reviewed these frameworks and identified five general competency categories: systems thinking, anticipatory, normative, strategic, and interpersonal competencies. Further, de Haan's concept of *Gestaltungskompetenz* (2010) views ESD problem-solving as a general competency which involves the interaction of twelve cognitive and affective sub-skills. These sub-skills include **self-reflection** and interdisciplinarity, as well as increased empathy and motivation. Svanström, Lozano-García, and Rowe (2008) also outlined a set of learning outcomes drawn from various institutions supporting sustainability in higher education. For them, knowledge is supplemented by characteristics, such as interdisciplinarity, reflectivity, systems thinking and ethical reasoning, for change agency in ESD. Indeed, sustainability scholars Sterling and Thomas (2006) remarked regarding ESD learning outcomes, "[while] authors often use different words, there is often a strong similarity in the meaning."

> **Self-reflection** The process of critically reflecting on your own personal ideals, goals, and actions, and objectively evaluating the impact those actions have on the world around us.

The *Sustainable Cities* course has adapted elements of these frameworks to satisfy its unique goals. In this course, we adopted another Sterling and Thomas idea that, in addition to affective learning outcomes, it is also critical students develop a personally meaningful *definition of sustainability*. We encourage scholars to draw on these resources as needed to support their own specific pedagogic goals.

We agree with Wiek et al. (2011) that attempting to establish all five domains of learning outcomes would be overwhelming for both the student and instructors, especially in one course. Nevertheless, these outcomes need to be driving factors for a learning program in sustainability (Pijawka et al., 2013). Therefore, it is critical to strive for those learning outcomes that are achievable. After all, each individual outcome presents a complex challenge. Boix-Mansilla & Dawes-Duraisingh (2007) dismantle interdisciplinarity into three components: 1) appreciating and understanding the methods of different disciplines, (2) understanding and valuing the contribution of disciplines as they relate to the students' discipline, and (3) understanding the limitations of their own disciplines as well as others. Untangling the complexities of individual learning outcomes assists in constructing more effective instruments with which to estimate their delivery to students.

Essential Outcomes and Assessments in *Sustainable Cities*

Achieving greater urban sustainability is considerably more than a technical exercise. It requires people who have the cognitive and affective skills necessary to apply sustainable solutions (Murray & Murray, 2007). Therefore, *Sustainable Cities* strives to enable the affective outcome of **environmental consciousness**, which encourages positive environmental behavior. By adapting various frameworks of learning outcomes in ESD, we have identified five central outcomes that support the development of positive environmental behavior and can be achieved over the course of a semester.

Environmental consciousness Awareness of and concern for the protection of the natural world and the environment in which we live.

Concept vocabulary. One of the foundations of creating sustainable policies and environments is the ability to effectively discuss sustainability (Orr, 1992). Since sustainability has developed independently in several different academic fields over the past 50 years, attaining a common vocabulary has been difficult. Without a shared understanding of the concepts that support sustainability however, people have difficulty working together successfully. Therefore, the course first develops a core vocabulary for students to use in articulating their goals, and importantly, to bring with them into other academic fields.

Interdisciplinary understanding. Working together on sustainability is essential, as problems are unlikely to be solved by a single academic or occupational specialization. Yet, few professionals have experience working across disciplines and are only rarely taught how to work collaboratively across fields. In *Sustainable Cities*, we emphasize the need for cooperation by illustrating how a multidisciplinary team can produce results amounting to more than the sum of its parts. This process involves illustrating the multi-dimensional nature of sustainability problems by linking it to the diverse perspectives that a multidisciplinary team brings into focus.

Ethical conceptualization. Working together by itself is not enough to solve sustainability problems. To tackle the complex problems challenging society, we need people with the ability to reason ethically from a sustainability perspective (Tilbury, 2004). Without ethical reasoning, we may be tempted to solve only the immediacy of the problem, and not the underlying cause, or view it as intrinsically an environmental problem, when in fact it may be a social problem as well. Instead of developing a solution to the problem, this tends to move the problem geographically. This means that the crisis merely becomes someone else's problem. Of course, many sustainability problems are difficult to contain, or are global in nature regardless of where they are geographically concentrated. Ethical reasoning recognizes our interconnectedness, and thus permits the consideration of people's welfare everywhere as well as that of future generations.

Definition of sustainability. While it is understandable that some desire a strict definition of sustainability, we recognize that such strictness limits our capacity to adapt to new situations and new considerations. Given the varied needs faced by those engaging with sustainability issues, the definition of sustainability must remain flexible. That said, resolving sustainability problems does require at least a basic and mutually understood basis bounded by vocabulary, interdisciplinary efforts, and ethical reasoning. When a student reflects on his/her values for ethical reasoning and interdisciplinarity, it helps them develop an enhanced personal concept of sustainability.

While having a personal definition of sustainability seems to conflict with achieving broader goals (Djordjevic & Cotton, 2011), it actually allows for sharing and adapting several perspectives, increasing effective communication and diversifying the ability to recognize problems. As McKeown and her colleagues (2002) conclude, many of the great concepts in human society, such as democracy and justice, are difficult to define or are culturally relative. This neither reduces their importance, nor does it prevent their further exploration. In essence, what is most important for sustainability education is that an informed, personally relevant definition supports a meaningful basis for environmental consciousness.

Environmental consciousness. One central goal of *Sustainable Cities* is to provide students with the opportunity to engage in positive environmental behavior, or to become change agents for improving their world. While environmental consciousness alone is insufficient to change behavior, it supports the process. Some scholars have cautioned against promoting environmental consciousness, suggesting it undemocratically removes choice. However, allowing for the possibility for change and demanding it are separate notions. Not providing students with the skills and information to behave sustainably limits their choice to one—acting unsustainably, and, it could be argued, is far less democratic. With ESD, students actually have more choices that range from business-as-usual to varying degrees of environmental consciousness and change agency. We are confident that taken together interdisciplinarity, ethical conceptualization, self-reflection and environmental consciousness open the door for change.

Measuring Cognitive and Affective Learning Outcomes

Another central difference between traditional education and ESD is that, because they deliver primarily cognitive learning outcomes, traditional courses typically measure success in passed examinations, successfully completed homework assignments, and final grades. Unfortunately, there is very little research on how to accurately measure affective learning outcomes. Available research points toward cumbersome instruments of unclear validity. In this regard, ESD is breaking new ground. Drawing on the lessons of diversity, the faculty of *Sustainable Cities* has taken a different approach focusing on the research principle of *triangulation*. Since direct measurements with single instruments are largely unavailable, a diverse set of limited tools can be used to develop a sense of the impact the course has on student learning outcomes (Buissink-Smith, Mann, & Shephard, 2011). This allows faculty to more nimbly adjust their toolbox of measurement instruments.

While there are numerous ways to analyze student learning outcomes, we present a few that can be easily incorporated into course construction and evaluation or that can be used to develop new assessment techniques. While measuring qualitative and affective characteristics can prove difficult, this does not mean it should not be attempted. Rather, the results should be interpreted very carefully. Improved measurement techniques are unlikely to occur without attempting to develop them, however limited the initial attempts prove to be (Rode & Michelsen, 2008).

Surveys

One of the goals of ESD is to increase change agency. The 'before and after' survey design is an ideal way to measure change, especially attitudinal change (Kraiger et al., 1993). Developing accurate instruments can be complicated and requires some background, however. If an instructor does not have training in constructing survey instruments, it is advisable to have a colleague review them before having students complete them or have someone assist in their development. Simple surveys that are relatively short, straightforward, and ask carefully constructed questions are recommended. Students should complete the first survey before the first lecture, to avoid biasing the after survey results.

Asking a series of questions at the beginning of the course can help capture students' opinions, skills, and feelings at that time. Repeating those questions at the end of the course can give an instructor an indication of how much the course impacted the students. Of course, many other influences can transpire over the course of a semester, but very few can be expected to happen to all students. Those that do are rare, history-making events, such as the fall of the Berlin Wall or the landing of astronauts on the Moon, which cannot be anticipated.

Journals

Having students turn in periodic electronic journal entries is a good way to amass considerable data that can be easily analyzed. These can also be completed at the end of each class to assess what students found to be the most important lesson of the lecture (Abbott, 2012). Electronic submission enables instructors to scan for keywords and measure how the use of the words change and what words are associated with them. There are advanced software packages available for qualitative research such as Atlas Ti, but oftentimes simple programs such as Microsoft Word can be used effectively if the journal entries are designed appropriately.

Measuring Learning Outcomes in Sustainable Cities

Concept Vocabulary

Concept mastery is foundational for student understanding. Measuring student mastery in examinations is not always accurate, as the pressure of the exam can alter their performance. Furthermore, most exams are not conversational. Whereas one of the main purposes of developing a mastery of key terms is to be able to communicate with one another and develop solutions. Nevertheless, vocabulary is largely a cognitive skill and therefore fairly easy to measure.

One of the simplest methods is to ask students how confident they are in explaining a concept. This can then be compared to how much lecture time is devoted to each concept as a means for determining how much time is needed to teach each concept, and whether or not better methods of presenting the terms are needed. Factors contributing to understanding can also be determined. For open-enrollment classes like *Sustainable Cities,* this can be quite informative.

Case Study

Raw data was collected during one semester of Sustainable Cities to evaluate student understanding of key concepts and determine what factors play a role in their mastery. Six terms were chosen: ecological footprint, food desert, food-miles, ecosystem assets, natural capital, and resilience (see Figure 1).

Using regression models, it was found that a student's academic year played the largest role in student concept acquisition. Seniors had a 19% greater likelihood of mastering an additional concept over juniors. Sophomores had a 9% chance of mastering an additional term than freshmen. Academic major also played a significant role. For instance, Art and Architecture students had a nearly 16% greater likelihood of acquiring an additional term. Other factors such as a student's initial self-reported environmental conscientiousness also had an impact. The main takeaway is that knowing student class levels and majors is important for helping to structure the class in ways that help effectively deliver the concept vocabulary learning outcome.

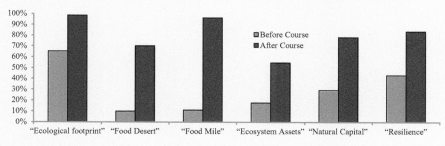

Figure 1 Change in student comfort in defining concepts – before and after course

Interdisciplinary Understanding

As a competency, interdisciplinarity is both a cognitive and affective skill. It requires having an ability to work collaboratively with other disciplines and the desire to do so. As a cognitive skill, it entails both understanding what interdisciplinarity is and how to approach it. As an affective ability, it requires respecting the contributions and mind-sets of others and being able to envision the multidimensionality of sustainability problems. Measuring student understanding of interdisciplinarity is different from, and much easier than, assessing their skill at using it.

Research in this area is limited and several factors make this type of measurement problematic. First, some disciplines are considered 'farther apart' than others. In other words, art and architecture students might find it much easier to work together than art and math students. In addition, different disciplines hold various methodologies in differing levels of esteem. Despite these constraints, it is not impossible to measure students' cognitive and affective skills.

For assessments of cognitive skills, inferences can be made as to the degree of cooperation that has been achieved. Specifically, the level of cooperation can be determined by a) analyzing how much students achieve when working to resolve an issue, b) determining each student's unique contribution, and c) establishing the team's disciplinary composition. This exercise can be repeated over the semester to gauge changes in these skills.

For assessments of affective skills, student attitudes regarding various disciplinary contributions toward understanding sustainability can be compared before and after the course. For example, you might use a survey instrument to simply ask students to rank or rate the contribution of different lectures toward their understanding of the interdisciplinary nature of sustainability. A well-delivered course should show

Case Study

Students were asked if they considered altering their academic plan. This question was followed by another open-ended question asking if students considered adding a minor or certificate to their course of study. We found that up to 25% of students considered adding a minor, and an additional 15% would have considered it but were too far along in their studies to alter their plan (see Figure 2). While this is not a measure of interdisciplinarity, it demonstrates whether or not the course creates a new acknowledgement of the utility of other disciplines and on whom it has that effect. When combined with and compared to other data, it can help instructors gauge the impact of the course.

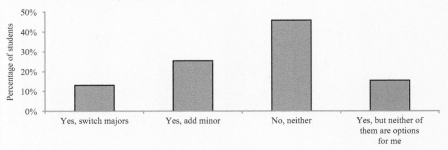

Figure 2 Impact of course on academic plan

an increase in student regard for the contribution of different disciplines. If students largely dismiss the contribution of disciplines that are a key component in the resolution of a sustainability problem, instructors should consider a different approach that promotes a better understanding. One approach could involve asking students to select an ideal team of three or four specialists for tackling different sustainability problems and describing how each member would contribute to the solution.

Ethical Conceptualization

A student's ethical contribution toward resolving a problem in sustainability is also difficult to ascertain. Assessment instruments can be extremely simple and yet still provide a wealth of information. Although there is more research on this topic than exists for measuring interdisciplinarity, the available research is either difficult to implement in a classroom setting or of mixed value. Like assessing interdisciplinary understanding, however, this should not deter us from trying to develop an understanding of the impact our efforts have but, rather, that we should be cautious of the results.

Assessments can be useful when built into an assignment. They should be embedded in such a way as to not draw attention to the instrument itself as undo attention will corrupt the results. Assignments can be designed to measure the level of ethical reasoning that students apply toward resolving a set of sustainability problems by positioning ethical dilemmas as merely part of the overall solution. For example, students can be asked to describe in which ways the problems are ethical, environmental, or economic in nature. Furthermore, students can be asked to what

Case Study

As part of an assignment for using the online Ecological Footprint tool, students were asked whether or not people have a 'right to consume' products and services as they wished. Since, it would not be highly informing to ask outright for a 'yes' or 'no' answer and have the students simply check a box, they were required to reply with a well-thought out paragraph justifying their answer. This allows students to unpack their ethical positions and gives instructors a more valid response. The teaching assistants (under supervision) formed a rubric, read a random sample of responses together to provide a level of continuity, and then coded the remainder according to the rubric. The results were surprising. We found that students overwhelmingly responded with "no" with only a small percentage responding unequivocally "yes." 27% responded that while people have a right to consume as they please, such attitudes are irrational, harmful, or detrimental to the public well being (see Figure 3).

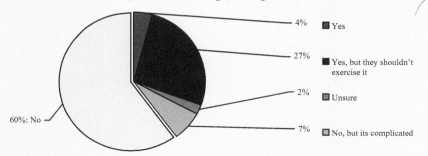

Figure 3 "Do people have a 'right to consume'?"

degree are each of the "three e's" (i.e., equity, ecology, and economy) involved in their resolution and why. The important part is to follow up afterward with students as to why the problems might be ethical when they seem to be more straightforward.

Definition of Sustainability

Determining the changes in students' definition of sustainability is much easier, although it can be a bit more work, and almost certainly involves qualitative coding. However, results are definitive. In both the before-course and after-course surveys, students were asked to give a brief definition of sustainability using a sentence or two. (Alternatively, they might be given a word count of 30 or 40 words). If the course is generating a personal, ethical definition, then a shift in the types of words used is to be expected. For example, in a recent semester of *Sustainable Cities,* words reflecting environmental concerns decreased dramatically from before to after the course and were replaced by a wider range of words concerned with equity and economics (see Figure 4). These results demonstrate a broadening of the meaning of sustainability to encompass more than core environmental

Student Definitions of Sustainability: Frequency of Change in Key Words1

	Before count	Adj. count[2]	After count	Percent change
future	73	64.97	205	215.53%
consci-*	8	7.12	19	166.85%
generation	54	48.06	125	160.09%
lifestyle	8	7.12	15	110.67%
change	12	10.68	18	68.54%
socie-*	16	14.24	23	61.52%
social*	30	26.7	36	34.83%
action	17	15.13	19	25.58%
preserve	26	23.14	29	25.32%
planet	20	17.8	22	23.60%
world	42	37.38	45	20.39%
plan*	32	28.48	32	12.39%
conserve*	16	14.24	16	12.36%
survive	15	13.35	15	12.36%
impact	20	17.8	19	6.74%
econo-*	23	20.47	21	2.59%
health	17	15.13	14	−7.47%
earth	51	45.39	41	−9.67%
sustain*	311	276.79	239	−13.65%
human	41	36.49	30	−17.79%
enviro-*	123	109.47	84	−23.27%
resource	158	140.62	107	−23.91%
system	21	18.69	14	−25.09%
natur-*	51	45.39	34	−25.09%
maint-*	38	33.82	22	−34.95%
green	20	17.8	11	−38.20%
effici-*	33	29.37	17	−42.12%
produc-*	18	16.02	8	−50.06%

	Before count	Adj. count²	After count	Percent change
material	15	13.35	5	−62.55%
build	15	13.35	5	−62.55%
waste	15	13.35	4	−70.04%
tech-	16	14.24	4	−71.91%
energy	36	32.04	9	−71.91%
renew*	22	19.58	2	−89.79%

1 words with a count over 15 in either the before- or after course surveys

2 Adjusted count to reflect decrease in total word count

* Root words used to capture student intent. Specific contexts were evaluated.

Italicized words reflect an overall increase in usage.

Figure 4 Changes in the types of words used from before to after course in articulating students' personal definitions of sustainability.

issues. Word count dropped as well, from over 26 words per response in the before course survey to less than 23 in the after course response; a drop of 12%. Looking closely at the responses, it was determined that students were able to resolve the matter with more precision.

Environmental Consciousness

By incorporating research from the cognitive sciences and environmental psychology, we know that behavioral change requires more than a conscious desire. It also requires the development of new habits. However, asking students to develop new habits is beyond the scope of the *Sustainable Cities* course. Instead, students are asked to report how environmentally conscious they consider themselves to be at the beginning of the course and compare it to their answers at the end of the course (see Figure 5). Survey prompts are categorized in bins that feature short descriptions of each level (e.g., how much consideration they give the environment when making decisions, various levels of consumption, buying new vs. used, and engaging in socio-ecological organizations). Results are compared with parallel data to address two major issues. First, many students initially rank themselves as highly conscious only to learn during the semester that in reality they were

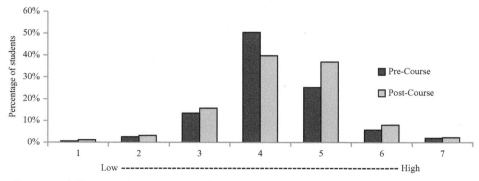

Figure 5 Shift toward environmental consciousness when asked, "How environmentally conscious do you consider yourself?"

not as environmentally conscious as they initially thought. The second concern is the opposite: some did not know how environmentally conscious they were before taking the course. While there was a statistically significant shift overall toward greater consciousness using this one metric and despite these concerns, additional questions proved to be valuable for uncovering these two concerns. A simple question in the after-course survey asking how accurately they think they answered the before-course question about consciousness would be helpful as well.

Conclusions

In order to be effective, it is important that ESD course instructors ask themselves, "Does this course teach education for sustainable development and how well does it do it?" This can be accomplished using existing, readily available, and easily implementable diagnostic methods. Student surveys provide critical data that keeps course content suitable to the important societal events students experience. Surveys also inform instructors as to slight changes that emerge in the composition of student demographics. As we have just exited the United Nations' *Decade of Education for Sustainable Development* (2005–2014) that impacted student awareness of sustainability issues in K-12 education nationwide, reflexive surveys will help determine if content should be adjusted as more students enter college with a basic knowledge of sustainability concepts.

In the absence of simple and verified direct measures of affective learning outcomes, we suggest developing a toolbox of measurements and using them in concert to understand the course's efficacy. This "triangulation of methods" can easily adapt to course content changes, student demographic changes, evolving learning outcomes, and other adjustments.

In short, sustainability instructors should not dismiss the impact of collecting triangulated data that informs their pedagogy. In this chapter, we have provided a modest sampling of the techniques used in Arizona State University's *Sustainable Cities* undergraduate course to measure the impact of the course and the achievement of students' affective learning outcomes. We hope that this brief excursion into our methods will aid in the development of individual, customizable toolboxes of instruments for the measurement of affective and cognitive learning outcomes.

Chapter 16

China's Pursuit: Smart Sustainable Urban Environments

Douglas Webster, Feifei Zhang, Jianming Cai

Introduction

This chapter outlines the issues facing China's cities as they pursue smart, sustainable futures in an economy moving from manufacturing to services. Ongoing sustainable initiatives of both national and local (primarily municipal) governments are described and assessed. A major Chinese city at the forefront of locally driven sustainability efforts, Hangzhou, is profiled to provide more on-the-ground detail. This permits comparisons with other sustainable international cities as profiled in Chapter 4. Because the national government of China is an important innovator and supporter of sustainable urbanization, an important aspect of this chapter is documenting the role of China's national government in propelling sustainability momentum at the local level, especially during the current 13th national development period from 2016 to 2020.

Context

China's spectacular economic growth and rise to become the second largest economy in the world (after the US) has been one of the great global success stories of the late twentieth and early twenty-first centuries (China Internet Information Center, 2007). China's urban areas, including their peri-urban industrial peripheries, became the "factory of the world" propelled by the market reforms that started around 1980. This resulted in massive migration of over 250 million people to the nation's cities and instigated a **rural-urban transition** that is ongoing. According to the World Bank, this growth has lifted over 500 million people out of poverty since 1978—an unprecedented feat (Chinese Ministry of Environmental Protection, 2016). Rural-to-urban migration raises the income of a worker by at least four times immediately. The economic growth and poverty alleviation is almost entirely the product of urbanization. In fact, China's rural agricultural sector remains very inefficient. Now, China is becoming a service economy. In 2013, the output of the service sector became larger than manufacturing for the first time since the Communist Party of China took power in 1949.

Rural-Urban Transition The process by which a country moves from being primarily rural (an urbanization level of 20% or less) to being primarily urbanized (urbanization levels above 75-85%).

However, China's urban-propelled economic rise did not come without cost. China's manufacturing has been highly dependent on coal for generation of electricity and steeped in energy-inefficient production. The country has also had lax emission controls and rapid motorization (China is the leading producer and consumer of automobiles in the world). The result is severe water and air pollution, soil contamination, and the exacerbation of water shortages, especially in the north. An emphasis on quantity (e.g., massive production of housing and urban infrastructure) rather than quality of housing and community design has resulted in less livable communities as well (Koleski, 2017).

The costs of China's rapid urban-centered economic growth to Chinese residents have been high. Over 500,000 Chinese urban residents die prematurely each year from air pollution (People's Republic of China [PRC], 2016). In 2007, the World Bank reported that 12 of the 20 most air-polluted cities in the world were in China. However, this appears no longer to be the case. 2012 data released by the World Health Organization (WHO) indicates that India's cities are, in general, more polluted. Yet, this is not cause for celebration. Only 1 percent of China's urban population live in urban areas that meet the EU's minimum air quality standard (People's Republic of China [PRC], 2003). Globally, China produces more Greenhouse Gases (GHGs) than any other nation.

River pollution from upstream industrial emissions and spills, human waste, and agricultural pollutants (i.e., animal waste, pesticides, and fertilizers) have made most of China's great rivers severely polluted. This has led to the complete shutdowns of urban water supplies on occasion, such as in Harbin in 2005 and Lanzhou in 2014. The Yellow River, birthplace of Chinese civilization, is so polluted for most of its length that it is not safe for swimming.

As cities move from primarily agricultural to industrial, more resources are extracted and used by more people. So, there is nothing new about the severe urban environmental problems and shortcomings in terms of urban quality of life that China has experienced while becoming a great industrial power. A similar trajectory has been seen during other industrial and subsequent motorization revolutions. For example, in the 1950s in London, killer fogs were frequent – people could not see their hands in front of them, and thousands of deaths occurred. In the 1970s, Los Angeles experienced deadly smog (ozone based) primarily associated with automobile emissions under inversion meteorological conditions. Rivers caught fire in the US, e.g., the Cuyahoga River outside Cleveland in 1969. Japan experienced the mercury poisoning (Minamata Disease) incident in Niigata Bay in 1965.

Environmental Trajectory A useful conceptual framework that indicates that cities and countries experience the worst environmental conditions when they have recently industrialized and are in the lower-middle income stage. Afterwards, urban environmental conditions usually improve as cities transform into service economies, then post-industrial amenity-based economies.

In both the US and Japanese cases, such incidents were catalysts that led to a massive regulatory push to clean up the environment. This included the creation of agencies with strong mandates, such as the Environmental Protection Agency (EPA), established in the US in 1970. This relationship between economic development and environmental problems and conditions is known as the **environmental trajectory** (see Figure 1).

What is significant about the Chinese case is that its industrial revolution was compressed into about 40 years, a process that took approximately 150 years in the Western experience. Secondly, the numbers are much bigger, both in terms of levels of production and population. Therefore, the impacts are proportionately higher. China's population is about four times as large as the US population

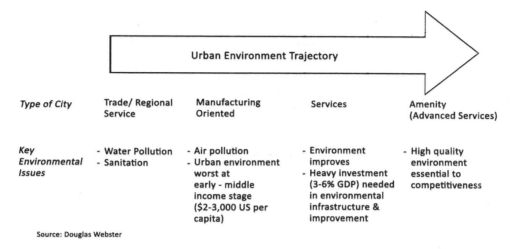

Source: Douglas Webster

Figure 1 Urban Environmental Trajectory.

and its economic growth rate about three times as large. In 2017, China's population was about 1.39 billion, of which 59 percent, or 820 million people, live in urban areas. This is approximately 2.6 times higher than the entire population of the US.

The good news is that as countries move up the environmental trajectory, they build the capacity to address environmental problems, and residents and investors demand higher quality environments. In a nutshell, almost invariably urban environmental quality deteriorates rapidly to a very low point during its middle-income stage as a country industrializes, then people and governments react, using financial resources acquired from economic success to dramatically improve the quality of the urban environment. China's transition to a service economy is enabling it to address its environmental problems.

The current situation is both sweet and sour. As noted, China's cities are neither currently healthy, nor sustainable. However, much is being initiated by both the private sector and government that is very encouraging. New Chinese companies, such as BYD, a Chinese manufacturer of automobiles and rechargeable batteries, are pioneering new environmental technologies. China has more installed solar capacity than any country, growing at an annual rate of 89.5% between 2010 and 2015. In fact, China is the world's largest producer of solar panels, exporting inexpensive solar equipment to the world. The country also has installed significant wind capacity, 169 gigawatts (2017), compared with 82 gigawatts in the US (People's Republic of China [PRC], 2013).

Both national and local governments are taking dramatic actions to improve the situation, e.g., phasing out the use of coal in Beijing (People's Republic of China [PRC], 2016). creating financial incentives for the purchase of electric cars; undertaking investments to increase the percentage of future electricity generation from nuclear, natural gas, hydro and alternative sources dramatically; applying hi-tech in urban waste treatment; and relocating polluted manufacturing firms into industrial zones where they are required to adopt cleaner technologies and become more energy efficient or be shut down. Throughout China, Smart Cities and sustainability are the new mantra.

Sustainable cities are gaining momentum due in large part to the shift in the country's economy. China is in the process of moving from a heavy reliance on often energy and resource inefficient manufacturing (generally of low cost consumer goods) to a higher-value service economy. This includes personal consumption of services such as medicine, leisure and tourism, business and professional services as well as knowledge activities such as higher education and research. Service economies almost invariably consume much less energy and produce considerably less pollution per unit of economic output, thus making it much easier to achieve high and sustainable urban environmental quality.

Smart Cities

Smart Cities An urban development strategy that leverages human and social capital and integrates information and communication technology (ICT), and computing, mechanical, and digital infrastructure to promote sustainable economic development and high quality of life.

Internet of Things A network of internet-connected devices capable of collecting and exchanging data using embedded sensors that enables consumers, businesses, and governments to analyze and store data, enhance security, and control systems such as lighting, heating and cooling, and appliances. Large-scale systems include smart grids, virtual power plants, smart homes, intelligent transportation and smart cities.

A **Smart City** is characterized by the extensive use of information technologies to achieve efficient urban outcomes. They extensively deploy sensors to monitor the status of the city in real time and adjust sub-systems accordingly. Their technologies frequently incorporate "**Internet of Things**" (IoT) sub-systems and often rely on "big data." Smart cities also frequently use technology to achieve efficient traffic flows, public lighting, irrigation, police surveillance, and e-governance, among others.

Until recently, China's urban populations accepted high levels of pollution given the benefits of rapid household income growth and poverty alleviation associated with rapid industrialization. These benefits were perceived to outweigh the costs of pollution and less attractive cities. The majority of urban Chinese understood pollution was a sacrifice to enable China to develop its economy. Now that most Chinese cities have achieved middle-income status, and leading coastal cities such as Shanghai are essentially first-world cities, public attitudes are changing. The nature of this change reflects a cultural preference in China (and East Asia) to associate environmental quality with public health rather than with bio-diversity and wilderness as in the West.

Since the extreme air pollution crisis in Beijing in January 2013, the population is demanding higher quality urban environments, especially in terms of air quality and the livability of their immediate residential neighborhoods. This change in public opinion coincides with increased national government emphasis on urban environmental quality that began around 2011. Three major drivers underlying this change can be identified.

First, China has developed a very large urban middle class over the last three decades. According to the McKinsey Institute, China's population reached over 460 million people in 2013 and is forecast to rise to 650 million by 2022. Much of this growth is a result of the country achieving middle-income status. As has happened virtually everywhere that urban middle-income status has been achieved, this more affluent, better-educated cohort exhibits changed values. Although household income, material goods, and opportunity (especially for their children) are still valued highly, there appears to be a shift in values and preferences among China's urban middle class toward less tangible goods and services, e.g., clean air, less time wasted traveling in cities, attractive and interesting neighborhoods, and public spaces, especially parks. However, no substantive research has yet been done on this topic.

The second factor is the exponential growth in the use of social media in China (e.g., "We Chat," the equivalent of Twitter; tens of millions of Micro Blogs; etc.) especially among the rapidly expanding urban middle classes. When air quality in Chinese cities plummets, tens of millions of people take to the social media complaining and demanding action, primarily from government. The government takes such "direct democracy" seriously.

Third, there has been significant political and administrative will and action. Increasingly, the leadership of the Communist Party of China is placing a priority on cleaning up the environment, particularly in urban areas. This change was noted as early as the country's 11th five-year national development plan period (2006-2010) and was very pronounced in its next reiteration (2011-2015). The current period (2016-2020) places even more emphasis on improving China's environment. The current President of China, Xi Jinping, has forcefully indicated that the country needs to restructure its economy, and reorder its values, consistent with the Government's priority on environmental quality and sustainability.

It will not be easy for the National Government to achieve sustainable urbanization, as there are deep structural constraints to changing environmental behavior. Even the President, the Politburo, and the State Council (equivalent to a Western national cabinet) will struggle hard to overcome vested interests to implement environmental reform. One of the most serious is that civil servants are promoted (or not) based on economic growth (GDP) in their jurisdictions. Despite some change in this regard, economic growth is still the standard by which local officials are primarily judged. Secondly, although the National Government formulates progressive guidelines and laws to improve the environment, based on international best practices (China is very open in international expertise, e.g., from the World Bank), the fiscally powerful local governments often choose to largely ignore them. In China it is said, "the mountains are high and the emperor is far away." Lastly, but very important, State-Owned Enterprises (SOEs) have grown very powerful as China's economy has boomed, further strengthened by the SOE reforms of the 1990s. These institutions (e.g., Petro China) are among the most powerful institutions in the country and may ignore environmental dictates of the national government.

What are the Key Environmental/Sustainability Issues Facing China's Cities?

The most critical issue in China is air pollution. Over 500,000 people in the major metropolitan regions are estimated to die prematurely each year due to severe air pollution according to scientific Chinese sources. The burning of coal, primarily to generate electricity, is the main source of air pollution and associated GHGs followed by vehicle emissions and industrial emissions. The relative importance of these different sources varies city to city. Air pollution in Beijing rose to a record high on January 12, 2013, sparking criticism of the Government's handling of air pollution. Although there has been a trend toward moving coal-fired electrical generation plants away from large cities, this has no effect on GHG gas emissions and does not improve urban air quality as much as might be expected. For example, relocation of coal-fired thermal plants to less settled regions north of Beijing, still results in pollution impacts on Beijing when the wind is blowing from the north.

Water pollution, particularly of rivers, is another major issue. Coastal water pollution occurs as these rivers empty into the ocean. Urban water pollution problems are exacerbated by the fact that a significant proportion of river pollutants are from point sources upstream (e.g., chemical plants), which are often located in different municipalities or even provinces. Upstream-sourced human and agricultural waste worsen these water pollution problems. This makes remediation difficult because polluters are outside the jurisdictional control of the impacted city or municipality. Although there are national standards, enforcement of these standards varies by jurisdiction as well. Furthermore, municipalities with large, wealthier cities tend to have higher standards than smaller municipalities or their rural equivalent, Prefectures, which may lack technical capacity.

Making matters worse is the fact that China overall is short of fresh water, especially in the north, where the Capital city of Beijing is located. More expensive treatment costs of polluted surface water make it more expensive to supply Chinese homes and workplaces with safe potable water. Groundwater under and near most cities is being harvested beyond sustainable rates and groundwater supplies are often polluted, especially in metropolitan areas. For example, 44.1% of the groundwater under the North China Plain is polluted to various degrees.

Beyond environmental issues, energy inefficiency is a major threat to China's urban sustainability. The problem comes from both the supply side (as noted, China is the most coal dependent nation on earth) and the demand (consumption) side. Governments in China recognize the coal problem and are acting to reduce coal dependency. For example, Beijing is phasing out the burning of coal. On the demand side, national and local governments are increasing their commitment to improving the efficiency of industrial processes and energy efficiency of residential and, to a lesser extent, commercial buildings. Industries and firms are being required to meet higher energy efficiency standards, especially in the case of new industrial parks known as Economic and Technological Development Zones. Residential buildings are subject, both through regulation and voluntary compliance, to higher standards of the China's 3-Star Rating System GBEL (The Green Building Evaluation Label) and the US Green Building Council's (USGBC) LEED standards.

The low energy efficiency of industry and residential stock has historical roots. In the case of industry, China's economy competed on large-scale, low-cost production, which often led to overlooking inefficient energy consumption and other **environmental externalities**. In the case of the urban housing stock, construction was typically low quality during the 1960s and 1970s. Starting in the 1980s, rapid urbanization led governments and property developers to emphasize the quantity of housing rather than quality including energy efficiency. Much of China's residential building stock is not well insulated, which is especially inefficient in cities with extreme climates (e.g., Beijing, Harbin) with very cold winters.

Although the aforementioned environmental issues are serious, there is considerable room for optimism from a futures perspective. As noted, China is already the largest generator of solar and wind energy in the world, and capacity continues to grow very rapidly. Twenty nuclear reactors are under construction plus 37 are operational, a strategy to reduce coal dependence. In addition, it is possible that China could exploit its shale resources, as North America is doing, to produce more less-polluting natural gas. A major constraint to shale gas production in China is that the extraction process is very water consumptive and the country is already facing water shortages.

Environmental Externality Environmental Externality refers to benefits or costs from an investment or activity that are not reflected in the pricing of a product. For example, a coal-fired thermal plant may damage the health of nearby residents causing them costs, but this externality is not reflected in the price of the electricity sold. Or, a new park in a city may increase the value of residential buildings surrounding it, but the beneficiaries (the home owners) do not have to pay for the benefit.

Other problems directly and indirectly related to these problems include soil contamination, rampant motorization, and inefficient land use. Soil contamination is a problem in both rural and urban areas nationwide. The problem has gained a high profile recently with the release of a major report.

In urban areas, soil pollution is especially a problem in former industrial areas where heavy metals may be present. Rural soil contamination tends to be related to fertilizer and pesticide use. Because urban land markets did not exist prior to the 1980s, factories often were in the center of cities. Although virtually all such factories have been moved to peri-urban areas, the high-value land left behind often has high levels of pollution. For example, the Shanghai 2010 World Expo site along the Huangpu River (the city's historic and contemporary axis) was a major industrial site for close to a century. It had to be cleaned up before the fantastic array of Expo pavilions could be constructed. The Taopu Smart City, to be constructed in northwest Shanghai, represents a similar case. Formerly known as the Taopu industrial area, the area was heavily industrialized with large petro-chemical complexes. There is a major soil rehabilitation effort underway before development can begin (see Image 1).

Land clearing underway in the Taopu are of Northwest Shanghai. The formerly heavy industry area is being redeveloped into a "Smart City."

Courtesy of Douglas Webster

Integrating urban transport and land use efficiently is key to urban sustainability. For example, transport generates over half of the GHG emissions, but this can be reduced below 10% with skillful planning. In this realm of urban sustainability, sometimes referred to as **Smart Growth**, China is a world leader. The Ministry of Housing and Urban-Rural Development (MOHURD) recommends a **population density** for China cities of 10,000 persons per square kilometer. This is much higher than North American or even Western European urban densities. For example, the average density of US cities is 905 persons per square kilometer. The densest city, Los Angeles,[12] has a density of 2,702 persons per square kilometer.

Densities on the suburban periphery of major Chinese cities are high. In Beijing, the suburban periphery is about 7,000 persons per square kilometer. Core densities are even higher, around 8,000-12,000 persons per square kilometer. High densities are extremely conducive to sustainable urbanization, enabling lower unit costs of infrastructure, higher building energy efficiency, and reduced trip distances. They can also positively affect urban transport mode share toward mass transit and walking.

Learning from the dramatically positive outcomes in Western Europe, Japan, and South Korea, China is nearing completion of the world's most extensive High Speed Rail (HSR) system. This will enable enormous travel-based energy savings between cities and within **Megapolitan Regions**. Disappointingly, the US has been very slow to adopt this transport mode, which is key to megapolitan and inter-urban sustainability. An important exception is the start of construction of a California HSR system which is expected to have 800 miles of track by 2029. The goal is to connect major cities between San Francisco and Anaheim in the first phase then extending it to Sacramento and San Diego in a secondary phase.

Smart Growth Land use patterns (e.g., contiguity, higher densities, compactness) and land use-transportation integration (e.g., transit-oriented development) that minimize consumption of urban land and contribute to less energy consumption, pollution, and shorter travel times.

Population Density The number of people per unit of land. In most of the world, the standard measurement is number of residents per square kilometer. In the United States, this is usually measured as number of residents per square mile.

Megapolitan Regions Consists of at least two metropolitan areas that are strongly inter-connected through transport, communications and economic systems, and has a population of at least 10 million people, such as the Pearl River Delta Region of China which contains three large metropolitan areas (Hong Kong, Guangzhou, Shenzhen), all of which are within one hour's travel time of each other using HSR; the Region has a total population of 66 million people.

China is also currently engaged in a major nationwide mass transit subway initiative. The 2,400 kilometers of subway routes in 2013 will increase to an astounding 6,100 kilometers by 2020. Shanghai, with its **Saturation Mass Transit** system, has already surpassed cities such as Tokyo, Seoul, New York, and London in having the most extensive subway system in the world. Beijing and Guangzhou are not far behind, and second-tier cities such as Wuhan are building extensive mass rail transit systems. Since rail-based transportation is extremely energy efficient (even more efficient than bicycling), China's cities could become the global pace-setters in terms of sustainable transportation. Of course, all is not perfect regarding the transport-land use interface in urban China.

While the mass transit mode share is surprisingly high in the leading cities—40.9% in Beijing, 49% in Guangzhou, and 33% in Shanghai (2011 data), the shift to electric and alternative fuel vehicles is slow. However, the government has introduced policies to support use of electric cars, and BYD's production of electric cars and hybrid batteries. Another major problem is that **Transit-Oriented Development** (TOD) is not as developed as it could be. There are three specific issues that need improvement (i) population densities could be higher, (ii) land and building use could be mixed more effectively, and (iii) micro-connectivity could be more intense around mass transit stations. This should include moving sidewalks and escalators, small vehicle feeders, and removal of pedestrian obstacles.

A significant driver of urban land efficiency in China is the fact that a strict agricultural land protection quota has been put into effect. This policy dictates that China must retain at least 120 million hectares of agricultural land. The importance of this threshold is indicated by the fact that it is referred to as the "red line." Despite the quota, this line is being threatened by a variety of factors, including urbanization involving extensive rural-urban land conversion of some arable lands. Thus, future urbanization in China will need to be land efficient and steered away from arable land, if the red line is not to be crossed. This policy to protect agricultural land from urbanization is very important to China because arable land constitutes 12 percent of the country's area at most. This is considerably lower than the US, where arable land accounts for 18 percent of the surface area. This is a challenge since China's population is approximately four times that of the US.

Saturation Mass Transit A system in which over 80% of residents of a city live within 600 meters of a rapid transit station, e.g., a subway station.

Transit Oriented Development (TOD) Characterized by mixed land/building use, extreme connectivity (e.g., pedestrian overpasses, escalators, local feeder vehicles, and higher employment (workplaces), and higher residential densities than the area norm, surrounding a rapid transit station. Normally, TOD developments are planned, and are frequently found where mass transit lines intersect or where stations serve an existing vital urban sub-center.

Sectoral Hierarchical Institutions Set the Framework for Environmental Policy

In China, governance and policy implementation is hierarchical with principles and guidelines established at the national level and implementation occurring at the Provincial, Municipal, District (Urban), and County (Peri-urban and Rural) levels, respectively. At each of these levels, equivalent bureaus to national ministries usually exist. For instance, Provincial and Municipal Development Reform Agencies are the equivalent of the National Development and Reform Committee [NDRC] and Environmental Protection Bureaus. The most important levels are the National levels, because of their authority and technical capabilities, and the

Municipal, because Municipalities are large, covering most metropolitan regions and smaller urban agglomerations as well as being the site of headquarters of major economic engines/players and massive financial resources, both public and private.

This vertical governance hierarchy is not as simple as it may appear because the key ministries and their local counterparts operate independently from each other to a certain extent. In other words, they are "siloed." The key Chinese agencies in terms of urban environmental quality and sustainability, represented at most spatial levels, are MOHURD, formerly the Ministry of Construction; the Ministry of Environment Protection (MEP), established in 2008; and the National Development and Reform Commission (NDRC). Overall national development policy, including environmental policy, is set forth in five-year development plans. China is currently in its 13th national development plan period, known as the 13th Five-Year Plan (FYP) or 13th FYP.

During the 11th and 12th FYPs, the focus shifted to local sustainability initiatives in existing metropolitan areas, e.g., Hangzhou, or for initiatives, particularly Smart Cities, e.g., Minhang District in Shanghai and Shunde District in Foshan. New Smart City initiatives continue to spring up, e.g., the proposal to create a Taopu Smart City in Putou Urban District of Shanghai. This national plan has become much more strategic, rather than **command and control** (as was formerly the case), reflecting China's change to a nationally guided market economy in the 1980s.

As a result, the Provincial and Municipal Bureaus set more detailed guidelines, policies, and targets for their areas of responsibility. MOHURD is responsible for the built environment, establishing standards for buildings, highways, and civil infrastructure such as sewer and mass transit systems. Through subsidiary companies at different levels it is also indirectly involved in construction of the urban environment. An important reform introduced in the 13th FYP was to give local Environmental Protection Bureaus direct access to the national MEP. It is directly involved in establishing urban environmental targets, policies, initiatives, and monitoring. The NDRC establishes overall strategic guidelines for the development of China.

Siloed The isolation of departments or sectors in the same organization (i.e., government, higher education, etc.) propagated by disparate goals and objectives, lack of collaboration, and ineffective communication that reduces efficiency.

Command and control Direct governmental regulations, policies, and/or guidelines dictating mandatory requirements for industry or activities, such as environmental, on what is within the law and what is not.

Urban Environmental Policies of the Ministry of Housing and Urban-Rural Development (MOHURD)

As indicated, MOHURD is responsible for overseeing planning and regulating the built form of Chinese cities. While some of its guidelines are ambiguous, it plays an important role in establishing energy efficiency standards for buildings and influencing transportation mode choices in cities, which can have a very significant impact on the environment. Under MOHURD, Urban Planning Bureaus are active players at the Municipal level and other levels such as urban districts. They play an important role in pursuing land-use efficiency, which is highly important in addressing energy efficiency and pollution concerns. To date, there have been varying degrees of success across the country. The 13th FYP seeks to make sufficient progress in these areas.

The 12th and 13th FYPs are also much more focused on urban sustainability issues. Accordingly, the MOHURD has launched a series of theme-oriented objectives that include: (i) a Green City and Lighting Plan, (ii) an Urban Disaster Prevention and Reduction Plan, and (iii) an Urban Building Energy Conservation Plan. In addition, MOHURD has established guidelines to provide technical assistance to cities. These guidelines cover topics such as sustainable planning of resource-oriented cities (e.g., coal mining cities), methods for low-carbon eco-city planning, and historical preservation.

Urban Environmental Policies of the Ministry of Environmental Protection (MEP)

One of the most important functions of the MEP is to establish enforceable environmental standards and targets for environmental protection. The MEP strengthens China's Environmental Impact Assessment (EIA) processes established in the *Environmental Impact Assessment Law of The People's Republic of China* that went into effect September 1, 2003 (PRC, 2003). It targets major projects in key cities and explores development of Urban-Rural Environmental Master Plans, which focus on the dynamic peri-urban fringes and more far flung recreational and agricultural hinterlands of major cities. Figure 2 illustrates the 12th FYP MEP targets.

Key 12th FYP policies of the MEP related to urban environmental improvement include:

1. An Eco-Civilization Development Index shall be integrated into the evaluation systems of local governments and their staff. A major constraint to achieving sustainable cities in China is that cities and their staff are now evaluated primarily on economic growth criteria. Environmental protection should be accorded more power in local governance, including veto power over projects that will do serious environmental damage. Furthermore, officials who fail to meet defined environmental targets will be denied promotion.

2. Enhance environmental standards and improve the effectiveness of the regulatory system, including enforcing regulations through an enhanced environmental legal system.

3. Improve the technical quality of environmental economic policies, addressing key issue areas such as subsidies for electricity pricing and construction of waste water facilities. Develop "zero-discharge" policies. Relate environmental performance of firms to the banking system and credit access. A green rating system should be developed for use by banks and other financial institutions. Given that Chinese governments are multi-billion dollar consumers of products, promote government green procurement. Early steps in this direction are already underway, e.g., mandatory purchase of electric vehicles by some urban governments for certain functions and government financial incentives to purchasers of hybrid and electric vehicles. Still, only five million of the over 260

million vehicles expected to be on China's roads by 2020 will be hybrid, electric, or fuel cell.

4. Develop a major environmental protection industry. China already possesses a well-developed environmental protection industry in areas such as water treatment, wind and solar power, and meter technology. The objective is to grow this industry and move into more technologically sophisticated areas such as Smart City information technologies.

5. Improve access to environmental information and encourage public participation. Information on urban environmental quality, key pollution sources, drinking water quality, the environmental performance of enterprises, and nuclear power plant security should be freely available and well publicized.

6. Enterprises discharging toxic and/or harmful pollutants should be subject to compulsory disclosure. Clear complaint channels need to be established at all levels of government. Environmental prosecutions should be undertaken to resolve the most serious cases of environmental damage. This is a looming issue in that 20 nuclear reactors are now being built in China – compared with only 4 in the US and 1 in France.

The National Development Framework for Urban Environment Quality. The goal of the FYP is to ensure that population and economic activity are distributed in a more rational manner, and protecting agricultural land and fragile ecosystems from rampant urbanization. In practice, this means different pollution emission caps and environmental standards for different parts of the country. For example, a high **amenity** city such as Hangzhou (discussed below) would have higher environmental standards than the minimum national standards than a heavy industrial area such as the string of cities along the Yangtze River in

> **Amenity** Urban amenity refers to the attractiveness of an urban area. Frequent attributes of cities used to measure amenity include climate, scenery, environmental quality, culture, urban design, leisure facilities, and cuisine. The level of amenity of a place is of considerable concern to people and policy makers as it affects the level and mix of migration, and the types and level of investment in a city.

Anhui Province. This Wanjiang Region City Belt Industrial Transfer/Relocation Demonstration Zone, which covers nine cities, would likely have standards closer to the national minimums. These minimum standards, however, are higher than the environmental standards for their former locations on the coast. Given China's enormous manufacturing economy located in peri-urban areas, the 12th and 13th FYPs emphasize the importance of production efficiency, energy savings, pollution reduction, and environmental protection throughout China.

The 12th FYP (2011-2015) indicated that the guiding principles underlying sustainable urban development should be human-oriented and safe for residents, ensure land and energy savings, and promote the unique features of individual places, especially by preserving each place's cultural and natural legacy. Importantly, it recognized that some areas, particularly eco-functional

Courtesy of Douglas Webster

Recently relocated factories in Xuancheng in the Wanjiang Industrial Relocation Zone. In the moving process, factories upgrade their environmental performance.

Public Private Partnerships (PPP) A mechanism whereby government (usually local government) and a private firm(s) co-operate to construct and/or operate a needed piece of catalytic infrastructure, e.g., a toll expressway or LRT line, or large-scale development, e.g., a new urban sub-center in a city. Such co-operation may involve land (often from government), finance, and technology. For example, the redevelopment of the historic core of Foshan, China was a co-operative project of Shui On Land (a private corporation) and Foshan Municipality.

Urbanization Level The percentage of people in a nation (or other geographic jurisdiction such as a continent, state or province) who live in urban settlements, i.e., cities and towns. This term is not to be confused with the Urbanization Rate, which is the percentage increase in the urban population of a nation (or other geographic jurisdiction) year-on-year.

Growth Boundaries Designated boundaries around cities, restricting urban growth beyond the boundaries. They are usually enforced through zoning regulations and building codes. Similar, and sometimes utilized in conjunction with Growth Boundaries, but different in terms of implementation mode, are Service Boundaries. Service Boundaries restrict delivery of services, e.g., water and sewer supply, or garbage pickup beyond the Service Boundaries, but are not geared to land use per.

areas providing environmental services and/or valued for scenic value, were not included in the GDP criteria used to measure "progress" in China since the 1980s. In such areas, other indicators prevailed. These included preservation of scenery; value of environmental services; and integrity of nature and culture, including preservation of historical buildings. This is a major step forward.

In the past, high amenity and tourist areas such as Hainan Island (China's leading sub-tropical resort island) and Dali in Yunnan, were blighted by concurrent industrial development (often heavy and polluting) designed to raise local GDP. In the long run, however, this actually did the opposite, undermining the amenity economy (Tourism; Meetings, Incentive Travel, Conferences, Exhibitions [MICE]; amenity migration), which had much more potential economic value than heavy industry.

The Chinese Government continues to advocate relatively rapid urbanization as a positive contributor to both economic development and environmental improvement. This growth is mainly the result of rural-urban migration to China's cities, rather than natural increase. Better planning and effective early infrastructure investment (e.g., subway systems, waste water systems) will increase the carrying capacity of cities as they grow. To facilitate this growth and build capacity, the *National New-type Urbanization Development Plan* (UDP; 2014-2020), released in March 2014, indicates that "comprehensive" infrastructure is to be put in place. This correlates with the 12th FYP call for improvement in the quality of urban planning.

The UDP further indicates that planning should not be two dimensional, but rather three dimensional. A vertical dimension is important to China's cities since most of its largest metropolitan areas will achieve extensive, even saturation-level, rapid transit networks by 2020. This provides extensive opportunities to accommodate growth and build capacity along these transportation corridors. For example, TOD encourages complex, highly connected mixed-use urban development below and above ground near transit stations. This strategy is important for China's urban planning, ensuring effective infrastructure investment. The UDP called for urban construction investment and financing system reform, incorporating innovative financing approaches such as **Public-Private Partnerships** (PPP) as a means of enabling TOD and similar strategies.

At present, most urban planning services are provided by Urban Design Institutes. The Institutes need to be more effective, locally oriented, and creative to meet the guidelines of the UDP and FYP. We believe that China would benefit from the growth of a private urban-planning consulting industry, which would likely result in more localized planning and higher quality designs from more competition within the profession. This is especially significant as China's **urbanization level**, currently at 59%, continues to rise by about one percentage point per year.

The 12th FYP importantly advocated that urban construction standards be improved in terms of energy efficiency and seismic (earthquake) resilience as well. In general, it determined that new urban construction should be more carefully regulated, monitored, and standards enforced. For instance, the UDP suggested that **growth boundaries** be designated and enforced around cities. Although China's cities are already dense, the 12th FYP calls for higher urban densities. Increased densities would lower unit costs of infrastructure, enable saturation mass transit (primarily subways and LRT), and reduce rural-urban land conversion on the peripheries of Chinese cities. Similar to efforts in Germany and some North American cities (e.g., Vancouver, BC, Canada and Portland, Oregon USA), Smart Growth policies are advocated in the UDP to increase land efficiency for both urban areas and peripheries.

Given the lack of arable land in China and the fact that the national agricultural land quota red line may soon be threatened, the 12th FYP pays particular attention to the rural-urban fringe, i.e., peri-urban areas. These peri-urban areas are transitioning away from manufacturing-based to more service-oriented economies. Thus, the UDP calls for redevelopment of these areas to reflect their new function.

Within urban areas, the UDP advocates for more public space, particularly green space. It promotes the redevelopment of urban villages (formerly collective rural villages enveloped by the growth of large cities). It also recognizes the close relationship between urban economic structures, including production processes and local environmental quality. Accordingly, it recommends that high energy-consumption economic sectors should be constrained in their growth.

Economic instruments are endorsed, including caps on overall emissions of major polluters to encourage firms to use energy more efficiently and reduce pollution. To facilitate the effectiveness of market mechanisms, the UDP emphasizes the need for Smart Cities to enable local e-governance, support economic development, and enable resource-saving city management. For example, it highlights the development and use of energy-saving technologies, e.g., sensor-based management of irrigation, traffic, and lighting.

On the commercial side, initiatives to reuse as much waste as possible are supported as well. This is important as China leads the world in industrial, construction, and agricultural solid waste. For example, the UDP sets an overall re-use rate of 72% for industrial solid waste. Given that most manufacturing and much of the R&D and scientific activity is in Industrial Parks, known as Economic and Technology Development Zones (ETDZs) and Science Parks, these areas should be carefully planned, built, and renovated (in the case of existing industrial zones) to support the cyclical economy, achieve intensive land use, generate energy through cogeneration, promote waste water recycling, and centralize treatment of pollutants.

On the consumer side, the UDP promotes green consumption and a green living style. According to the China Internet Information Center (the official government web portal), green consumption focuses on three key principles to achieve sustainable consumption in the country 1) green products "that are unpolluted or good for public health," 2) pollution avoidance through waste treatment "under special surveillance," and 3) changing "public understanding of consumption" and raising people's awareness. They also encourage people to take the needs of their descendants and future populations into consideration (China Internet Information Center, 2007).

Consumers are encouraged to buy energy- and water-saving products; energy-efficient and environmentally friendly automobiles, including alternatively powered vehicles, such as electric vehicles made by the Chinese BYD company; and energy- and land-saving housing. The UDP emphasizes that the use of disposable products should be reduced and over-packaging restricted. In addition, it suggests that treatment capacity for urban wastewater and solid waste should be improved so that the treatment rate will reach 85% and 80%, respectively.

Watershed According to the United States EPA, a watershed is the area of land where all the water that is under it or drains off it goes into the same place.

Airshed Part of the atmosphere that behaves in a coherent way with respect to the dispersion of emissions. As a geographical area within which the air frequently is confined or channeled, all parts of the area are thus subject to similar conditions of air pollution. Airsheds typically form an analytical or management unit.

The 12th FYP also recognizes the importance of regional scale management of **watersheds** and **airsheds**, as cities have little control over these larger atmospheric and hydrological ecosystems. Accordingly, institutional mechanisms are proposed related to the planning and management of key watersheds, airsheds, and transboundary water systems. It also proposes to reduce environmental risks by enhancing monitoring and warning systems related to natural hazards.

Finally, the UDP advocates that resource pricing mechanisms should be reformed to improve energy- and resource-use efficiency, and support a more equitable society. Water and other resources should be priced to reflect their real value (scarcity value). In support of more equitable outcomes, stepped pricing for basic human needs such as electricity and water is advanced, such that the first units of consumption needed to meet basic human needs are priced lower than additional units consumed. Waste treatment fees and up-front financial subsidies to build needed urban infrastructure, e.g., waste treatment facilities, should be increased. Environment taxes should be promoted, including those based on higher taxation of products that are damaging to the environment.

The recently released 13th FYP reinforces and deepens these measures. Among its priorities are upgrading the economic mix, the energy mix, and the advancement of environmental technologies. The Chinese government is also supporting the growth of domestic green industries (energy-saving and environmental protection industries) into globally competitive firms. It is mobilizing domestic and international investment toward these industries. Industries singled out for support include new energy vehicles, soil remediation, and advanced energy-saving technologies and equipment.

One of the main thrusts of the 13th FYP is to recognize that different places in China require different responses, based on their resource endowments, economic development situation, and strategic positioning. Thus, more customized environmental protection measures and targets are being introduced to replace "cookie-cutter" policies previously applied to the whole nation. This transition entails continuing efforts to expand the nationwide monitoring system and a series of specific plans focusing on improving the quality of air, soil, and water. These include the 2013 Action Plan on Air Pollution Prevention and Control, the 2015 Action Plan on Water Pollution Prevention and Control, and the 2016 Action Plan on Soil Pollution Prevention and Control.

For example, the 2013 Action Plan on Air Pollution Prevention and Control targets reducing $PM_{2.5}$ concentrations by 25%, 20% and 15% for the Jingjinji (Beijing-Tianjin-Hebei), Yangtze River Delta, and Pearl River Delta metropolitans, respectively. For Beijing, $PM_{2.5}$ annual concentrations shall be controlled below 60 $\mu g/m^3$ by 2017.

The 13th FYP also advocates and outlines a national pollutants cap-and-trade system, a national ecological safety warning system, a nationwide environmental monitoring system, a national water monitoring system, and an audit system of officials' performance in meeting environmental targets. These systems will allow the central government to hold local governments and firms accountable for environmental performance in their jurisdictions, and establish more localized targets in the future.

Locally Driven Sustainability Planning

Not all of China's urban environmental initiatives are driven by top-down planning and administrative processes. Cities, usually at the powerful municipal level, are increasingly developing strategies that reflect local geography, financial resources, culture, preferences, and levels of motorization. Typically, these locally driven strategies contain targets that are stricter and more ambitious than the targets and standards found in the national plans. They are often branded (e.g., Beautiful Hangzhou) with implementation occurring through policies, public behavior changes, and investment in catalytic environmental projects. These broad-based urban environmental and sustainability strategies are springing up in many Chinese cities across a wide variety of urbanization contexts. Contexts range from wealthy (first world in terms of economic development), high amenity cities (Shanghai, Qingdao, Xiamen, Chengdu, and Hangzhou) to cities that are known as some of the most polluted in the world (Taiyuan).

The Tianyuan Municipality with approximately 4.25 million residents is infamous throughout China for its polluted environment. Home to approximately 30% of China's coal reserves, the industrial-based economy, polluted environment, and lack of amenities have kept Taiyuan a third-tier city. However, in co-operation with the national government, the Municipality has declared its intentions to become Taiyuan "Eco Garden City." This new Taiyuan is envisioned as offering a high standard of living for residents with international brands and high-class amenities as well as a cultural destination, like Xi'an (home of the terra cotta warriors). It plans to accomplish this by renovating, marketing, and expanding historic areas such as Jinci Temple. Upgrading the image of Taiyuan includes rebranding the metropolis as an environmentally robust city. This will be achieved by the creation of a massive ecological preserve around Jinci Temple and protecting the water sources in the north that feed the Fen River.

Another example is Chongqing, a municipality in Western China of 32 million people with an urbanization level of 58% in 2014 known for its heavy industry and air pollution. In 2010, it announced that it would become a "National Environmental Protection Model City." Its strategic objectives seek to improve air quality, achieving over 300 days of blue sky annually; improve the water quality of the Yangtze, Jialing, and Wujiang Rivers, which flow through the municipality to meet the Chinese Class II water quality standards; achieve 100% public access to safe potable water; reduce urban ambient and road traffic noise to within 55 and 68 decibels, respectively; and engage in large-scale tree planting and ecological restoration.

At the other end of the urban development spectrum, Xiamen, a high amenity city, has introduced several initiatives to move it towards becoming a more sustainable city. Many of these initiatives focus on Xiamen's **carbon footprint**. For instance, $PM_{2.5}$ monitoring for the city proper was introduced in 2012, with a comprehensive air quality index made continuously available beginning in 2013. To improve performance against these and similar indicators, strong measures have been taken. Vehicles without Environmental Protection (EP) labels were banned from the city starting in 2012. The introduction of 409 energy efficient public vehicles was initiated in 2011, starting with 40 hybrid-powered buses.

Carbon footprint The amount of carbon dioxide and other carbon compounds emitted due to the consumption of fossil fuels during human activities, usually expressed in equivalent tons of carbon dioxide (CO_2).

In addition, construction of its LRT system was accelerated. These actions build on Xiamen's earlier green strategy, which won the 2003 Nations in Bloom Award in 2003 and the UN Habitat Scroll of Honor Award in 2004. This is in sharp contrast to China's model eco-cities initiative.

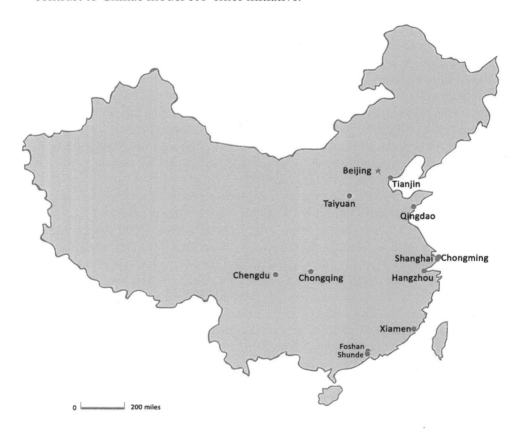

In the early twenty-first century, China's preferred route to local sustainability was through the creation of model eco-cities. These cities were intended to be almost environmentally utopian in nature and many were new developments rather than retrofits to existing cities. To a considerable extent, this reflected the previous Chinese development strategy of testing new policies in limited geographic areas. For example, the establishment of four "open special economic zones" in the early 1980s in South China to test market economy and opening principles. Unfortunately, these model eco-cities have a mixed record at best.

The Dongtan Eco-city on Chongming Island (see Map), an urban district of Shanghai, was one of these early attempts. Despite construction of a bridge and tunnel to the island in 2009, the project has essentially been abandoned. The city was to be comprised of carbon-neutral and zero-waste communities with attractive, high density, highly livable mid-rises. The forecast population for 2050 was 500,000. The causes of the failure of the Dongtan Eco City are not completely clear, however, confusion over funding and a corruption scandal are normally cited as the reasons for its failure.

The Sino-Singapore Tianjin Eco-city near Tianjin Metropolis is another model environmental city. Ground was broken in 2008 and the city is still being developed. While the government and development companies involved continue to reaffirm their commitments, it is too early to judge its success. Purposely built on a dumping ground for toxic waste along the coast, it is designed to house 350,000 inhabitants by 2020. If successful, the city will be low carbon. Among the alternative energy technologies being implemented are wind turbines (five installed), solar for lighting, and ground source heat pumps for energy that exploit differences in ground temperatures. Because it is a joint project between the Chinese and Singapore Governments, the Sino-Singapore Tianjin Eco City has a greater chance of success than other model eco-cities in China. Nearby, the Yujiapu Financial Center, under construction, is intended to be a low carbon model central business district, as part of APEC's green growth initiative. However, its future is problematic given its failure to become a leading Chinese financial center.

Beautiful Hangzhou

Hangzhou is home to beautiful West Lake, one of China's prime tourist attractions, and the southern terminus of China's historical Grand Canal. It is one of China's richest and most beautiful cities. The city's history is long. Starting as a prefecture in 589 BCE, it was named the capital of the state of Wu-Yue during the Ten Kingdoms (Shiguo) period (907–960) then the capital city during the Southern Song Dynasty. Today, it is a popular tourist destination and industrial center known for its entrepreneurship (it is the home of Alibaba, one of the world's leading IT firms). It is further along the urban development trajectory than many less-developed cities in the interior and western China. With a population of approximately 8.8 million, it is not one of China's largest cities. Nevertheless, it had the eighth highest GDP among municipalities in China in 2011. By Chinese standards, densities are relatively low with the city proper being 3,068 persons per square kilometer.

The Chonqing Xintlandi Retail/Lifestyle complex along the reforested valley of the Jialing River part of Chongquing's Environmental Protection Model City Initiative.

Courtesy of Douglas Webster

In 2013, President Xi Jinping gave a speech about "beautiful China" when he visited Hangzhou earlier in the year, and he encouraged Hangzhou to base its development on a "beautiful city" model. Hangzhou launched its Beautiful City strategy soon after. One of the major problems in China's urban sustainability efforts is that they tend to be similar. These "cookie cutter" plans are often the result of the lack of qualified local personnel to produce a plan that is crafted to meet the reality on the ground. Hangzhou wanted to avoid this outcome. Therefore, the strategy was crafted by

Courtesy of Zhang Feifei

River banks of Hangzhou have been restored, including wetland areas.

Courtesy of Zhang Feifei

Organic wastewater treatment in peri-urban Hangzhou.

the Policy Research Department of Hangzhou Municipal Government together with the Hangzhou Environmental Protection Bureau, in consultation with two top national think tanks in sustainable development. It contains progressive targets to achieve its goals.

The research team did a thorough study on the most pressing environmental issues affecting the city to establish targets to be achieved. The team also reviewed global best practices at metropolitan-scale sustainable development. To understand local concerns and opinions, a questionnaire was administered to 1,200 people, including locals, migrants, even tourists. Respondents were asked about their expectations regarding the six dimensions (see below) of a Beautiful Hangzhou to identify the gap between current reality and expectations. They were then asked to indicate what actions might be taken.

For example, 52.6% were most dissatisfied with air quality when asked about environment issues such as water, air, noise, solid waste and greening. They identified the most urgent things to be done as cleaning the air and improving drinking water quality. As with many municipal initiatives, the environmental standards in the Beautiful Hangzhou sustainability strategy are higher than equivalent national ones. They are higher than those of most other Chinese cities as well. Interestingly, indicators that are not included in the national plan are included in the Hangzhou strategy, e.g., swimmable and fishable water. This was to address a complaint heard in Hangzhou, and elsewhere in urban China, that "the official environmental indicators show improvement, but the environment feels worse."

Key goals of the plan are to have the spatial planning, strategic thrusts, and policies established by 2015. By 2020, eco-system restoration should be well underway, while real results in terms of the environmental quality of the metropolis should be evident. By 2030, the city hopes to achieve a natural-economic-social virtuous circle, its equivalent of sustainability.

The basic premise of the Beautiful Hangzhou sustainability strategy is that the city needs to have a higher quality sustainable environment to outperform other Chinese eco-cities. Thus, it is based on both an intrinsic rationale (a higher amenity, healthier Hangzhou for people) and an economic one. A prime objective was to create a sustainability strategy that truly reflected Hangzhou's context, issues, and its population's preferences. To accomplish this, care was taken to include the needs of migrants in the strategy and increase the plan's level of social inclusion as well.

The importance of Hangzhou's focus on sustainability goes beyond the local, however. Because it is one of China's most developed cities, other cities in China can learn from its experience. Furthermore, it illustrates the principle that once wealth is achieved, the development work is not over. The environmental quality, livability, and amenity of a city become a value and objective of equal developmental importance to GDP growth.

Plan Content

There are six dimensions to the Hangzhou strategy. The key dimensions and indicators/targets associated with each dimension are as follows:

1. *A complete ecosystem, protecting picturesque scenery of the mountains and rivers.* The forestry cover (percentage of Municipal land area) is to be stabilized at 65%. 90% of existing wetlands are to be protected.
2. *A healthy environment with blue sky and clean soil.* Objectives for drinking water source compliance rates with national standards (safe potable water) objectives are 100% for urban and 99% for rural areas within the Municipality. The ratio of city waters (lake, inner rivers, and the Grand Canal) achieving Grade IV water quality is to be 90% and the days when air quality is above Grade II should be over 300 per year.

3. *A green and low carbon industrial system.* Services will constitute 60% of the GDP. In addition, the output of ten major targeted desired economic activities will account for more than 55% of the GDP. Related to this shift in the structure of the urban economy, carbon emission intensity will decrease by more than 50% compared to 2005, while emission intensity of major pollutants such as CO_2 and SO_2 per 10 thousand GDP are to be under 1.5 kg and 1.0 kg, respectively.

4. *Pleasant habitat.* Buildings over 50 years old should be reviewed for historical protection before being torn down. The mode share of public transportation should be over 50%; the green community rate (number of neighborhoods) should be 50%; 550 "beautiful villages" should exist; and 75% of the building stock should meet the green building standard of the MOHURD.

5. *The education system will teach respect for nature.* At least 90% of Hangzhou's residents will be aware of key environmental concepts.

6. *Happy and harmonious quality of life.* The average (mean) labor force educational attainment rate is to reach 13 years. The Gini Index of urban income distribution is to be under 0.3. The average life expectancy is expected to be 82. Medical insurance should cover all municipal residents, both urban and rural residents. The green travel rate (mode share of walking, cycling, mass transit) should be over 80%.

Implementation

To smoothly implement the plan, Hangzhou established a "Building a Beautiful Hangzhou" institutional mechanism, which includes a committee that coordinates the main tasks. This committee takes responsibility for meshing the Beautiful Hangzhou plan with Hangzhou's economic and social development plan, annual budgeting, and the Master Land Use Plan. They are also responsible for integrating the plan with local economic restructuring, public input, and consultant's expertise. A three-year action plan and an annual implementation plan for each year were developed, and detailed implementation instructions are assigned to a wide variety of government departments. The performance of government departments and officials will include evaluation of implementation results relative to targets. Implementable priority policies include the following:

1. *Low Carbon City.* As noted, environmental quality in any city is as much a product of the economic structure as the regulatory environment. To restructure its economy, Hangzhou has identified ten economic clusters that will be strongly supported. These include culture and creativity, tourism and recreation, financial services, e-commerce, information and software, advanced equipment manufacturing, the Internet of Things, bio-pharmaceutical, energy-saving and environmental protection equipment, and new energy. Low carbon buildings will be incentivized, including solar roof-top installations and green roof-tops.

2. *Two banks of Three Rivers Ecological Planning.* The length of river banks in Hangzhou is 436 km. Corridors 200 to 500 meters from the river banks

will be protected from urban construction, a total area of about 602 square kilometers. Underway are river bank ecological rehabilitation projects and beautification projects at certain points along the river. A riverside scenic slow green pathway system (walking, biking, interpretative stations) will be developed to promote water front tourism and recreational industries.

3. *Public Bicycle System.* Hangzhou has the most sophisticated public bicycle system in the country. It built the system to solve the "last 1 kilometer" problem of the public bus system in 2008. By the end of 2011, there were 2,200 rental points distributing 65,000 bicycles. Based on research by the Chinese Academy of Social Science, the Hangzhou public bicycle system saves 7,500 tons of gasoline and reduces CO_2 emissions by 23,897 tons per year. Hangzhou was "one of the eight best public bicycle cities" in the world according to a 2011 BBC report. Interestingly, Hangzhou is currently the only city in China to sell carbon credits based on GHG reduction through bicycling. However, China has already introduced pilot regional carbon trading markets. It will establish a national market in 2016.

4. *Trash Sorting, Collecting, Clean Transporting and Recycling.* Citizens are encouraged to sort trash. Newly designed transfer stations and direct transport of solid waste are being developed, virtually eliminating leakage. Kitchen waste is reused through efficient composting.

5. *Wetland Organic Restoration.* Building densities in the Xixi Wetland have been lowered to reduce environmental pressure. Now Xixi wetland, with an area of eleven square kilometers, is the first urban wetland park in China. It has become an attractive destination for high-end tourists.

To fund implementation, the government encourages private enterprises and banks to invest in Beautiful Hangzhou development. An eco-compensation system is being established to compensate communities along the Qiantang and Tiaoxi Rivers for limiting their own development to protect downstream water quality and ecological integrity.

The Beautiful Hangzhou Plan is not without critics, however. Some wonder if it is naïve. Others question if the plan is wasteful of land with its emphasis on greening, massive forest cover, and beautiful peri-urban villages. Other questions include can urban governments in China, or anywhere in the world, really impact income distribution and the Gini coefficient; and would transformation to a much more energy-efficient post-industrial economy not occur through market forces without the existence of a formal sustainability strategy?

The Foreseeable Future: Chinese Sustainability

There is a need for bottom up and top down approaches to urban sustainability to coalesce initiatives in China. Wealthy municipalities such as Hangzhou, Shanghai, and Xiamen are developing customized sustainability strategies aligned with their problems and strategic competitive positioning. Many of China's less developed cities tend to copy the environmental sustainability plans of other cities, domestic and international, because they lack expertise or ideas. Currently, much of the sustainability expertise is in the national agencies in Beijing, where talent tends to concentrate and access to international expertise is more readily available. However, as talented university graduates in China increasingly settle in second and third tier cities, the talent pool in the urban environmental field is likely to improve in lower tier cities.

In addition, little research is done to determine the most effective mix of strategic thrusts and policies in local sustainability plans, nor is there enough attention paid to prioritization, given the magnitude of the environmental issues that Chinese cities face. This needs to be corrected over time with local (Municipal Governments) being more aggressive in customizing plans to local conditions, undertaking more research (evidence-based planning), and applying prioritization techniques informed by human resource, financial, and other realities. Central agencies, such as the Environmental Protection Bureau, will need to continue to play a strong supporting role.

Governments alone cannot create sustainable cities. They need substantial involvement of the citizenry and the private sector. Chinese cities lack well-developed civil society in contrast to its strong government and private sectors. So, it is yet to be seen if civil society will become populated by NGOs, creating a third voice. Or perhaps, China's urban civil society will be the discourse of social media, already very developed and dense in Chinese cities. Government ministries associated with urban environmental planning, especially the MOHURD, are formally advocating more public participation in plan making.

As noted in the Hangzhou case, extensive surveying was undertaken, and the last two cycles of the FYP have incorporated citizen feedback through the Internet. Officials were surprised by the high value citizens placed on green space during the 11th FYP preparation process. Currently, air pollution is at the top of citizens' urban environmental agenda, reflected both in official surveying and discourse on social media. The urban populace is no longer willing to tolerate the smog that envelops many Chinese cities.

Another high-profile issue is water quality, particularly of rivers, many of which are black and odorous as they wind through metropolitan regions. The urban population has concerns over food and drinking water safety as well. Another concern of the urban citizenry is garbage dumps (not well-managed landfills) that are increasingly overflowing with uncovered waste. This is particularly a concern in peri-urban areas, where the problem is most rampant.

At the more bureaucratic level, there is an increased concern with public health and increasing research on links between public health and the status of urban environments, e.g., GHGs and CO_2 impacts on health. Efforts to control emissions (air and water) will continue to be a priority in the 13th FYP period, based on more realistic regional spatial scales. Increased monitoring of emissions and disclosure of information to the public is likely to be advocated. We concur with this public health emphasis; it should be the first order of business.

Another area of urban environmental governance that will likely change is finance. China's economy is likely to grow slower over the next two decades, as it is harder to grow fast as a country moves up the economic trajectory. Growth will probably be in the range of 5-6% annually as opposed to the approximate 10% annual growth experienced over the last 35 years. This situation, positive in many respects, will require more evidence-based environmental project mixes.

There will also be an increased need to find the most cost-effective path to meet targets. Additionally, innovative finance avenues will be required. This must go beyond local bond issuance and dependence on "few questions asked" borrowing from domestic banks, which has created a national scale problem of local debt present in most Chinese cities. There needs to be more emphasis on PPPs as well.

This is necessary not only to tap the resources of China's capital rich private sector, but to benefit from private sector expertise as well. PPP approaches to city building, including in areas such as historical preservation, river front development, and knowledge zones is already underway but needs to be vastly enhanced. There are vast sums of capital in China associated with insurance and pension holdings that could be tapped. There are positive trends in this direction, e.g., the massive Ping'An insurance company is becoming involved in urban adaptive reuse historical preservation in cities such as Qingdao. However, there is still a tendency for Chinese local governments to cling to public sector projects.

The 12th FYP period advocacy of new environmental governance and financing mechanisms, such as PPP, has been reinforced in the 13th FYP. In the current reiteration, more emphasis is being placed on outcomes. Economic instruments are strongly advocated to address environmental problems. For instance, economic mechanisms are likely to go beyond conventional instruments such as polluters pay to embrace natural resource ownership and pricing, compensation for use of environmental resources or environmental damages caused, environmental and natural resource auditing, and lifetime responsibility for environmental damage.

Another area likely to receive attention is support for environmental protection industries. China's environmental challenges are a potential source of economic growth. Thus, there is much room for the growth of companies producing environmental products, given the scale of China's urbanization and the magnitude of the challenge. China's urban environmental initiatives will continue to be target oriented, and current targets will likely not only be adhered to but increased. Current targets are already ambitious; yet, there is room for growth and innovation.

For instance, Chinese cities are increasingly expected to aspire to be Smart Cities. Xiamen is a leader in this regard. Becoming a Smart City entails incorporating technology, sensors, and monitors as well as IoTs into sustainability plans. This is a positive trend in that sensor-based technologies can do much to save energy, water, and human time. As more cities embrace these technologies, costs associated with them will decrease enabling broader implementation in less developed urban areas.

In general, while past generations in China (from 1980 to the early twentieth century) valued economic growth and growth in household income above all, there is a profound shift occurring in the urban Chinese populace. "Growth at all costs" is no longer the dominant mantra. A major trend influencing inter-urban migration and urban sustainability planning in China is the growing importance of amenity, i.e., the availability and pull of attractive, stimulating, vital environments in which to live, work, and play. As populations become wealthier and basic environmental needs such as clean water and air are met, the focus of sustainability strategies tends to move in the direction of meeting amenity objectives. This orientation can be seen in cities such as Singapore, San Francisco, and Stockholm.

For example, Kunming in the Himalayan foothills of Southwest China is currently orienting its sustainability policies toward amenity. It is improving the water quality of Caohai Lake and making the lake accessible to tourists and residents (see Image 6). Hangzhou's emphasis on river corridors, slow bike paths, beautiful peri-urban villages, and appealing wetlands, reflects an awareness of amenity preferences among the city's increasingly well-off populace. Urban environmental governance in China will continue to change, but it is not clear exactly how.

Courtesy of Douglas Webster

North Caohal Lak in Kunming, China's City of "Eternal Spring" is being cleaned up. The lake shore is being beautified and made accessible as part of Kunming's amentity.

Conclusions

China's cities, along with India's, have the most challenging urban environmental conditions on earth. In both cases, the impact is multiplied by their very large, and increasing, urban populations. On the positive side, much has been achieved.

Most Chinese cities will soon have water supply and wastewater systems that cover their entire built up areas. This contrasts with some highly economically developed cities in East Asia (e.g., Bangkok) that still lack citywide sewer systems. China's cities now lead the world in total subway tracks. Shanghai has more subway line than any city on earth. Furthermore, China has constructed the most extensive high speed rail system in the world. These intra- and inter-urban rail systems make China potentially the most environmentally efficient people moving system in the world, if people can increasingly be induced away from private vehicles in cities and off highways and airplanes between cities. Furthermore, China has more installed solar and wind capacity than anywhere in the world. There are 28 nuclear reactors currently under construction, evoking concern among some elements in the population regarding safety issues.

Despite these advances, Chinese cities present a contradiction to those working to make them more sustainable. In some areas, such as mass transit and solar they lead the world. In others, they are struggling. For instance, China's air pollution is deadly—literally.

Given the complexity and the size of the challenge, environmental governance and financing need to be improved. As everywhere, there is a need for more horizontal integration across government agencies. Especially important will be closer co-operation between the private sector and government, both to tap capital and expertise. More evidence-based, cost-effective strategies to meet objectives and targets need to be formulated as well. China has taken a step in the right direction moving away from "model" experimental eco-cities to policies that focus on existing metropolitan areas.

As the current, seemingly overwhelming, challenges of air, water, and soil pollution are addressed successfully and China nears its forecast urbanization level of 70%, new challenges will emerge. These challenges may include density and population dynamics, agricultural and food security, aging, and inadequate infrastructure, among others. Climate change and its impacts will present additional challenges. New urban sustainability preferences are expected to evolve as well. These will likely be centered on quality of life and amenity issues. This shift could create overwhelming pressures on tourist and amenity migration locales, while at the same time creating a greater gap between highly successful luxury cities and poorly performing depressed cities.

Last, but not least, Chinese urban sustainability stakeholders need to keep their eye on future trends. China's cities can completely remake themselves in ten to twenty years, as shown by the country's track record in urban economic and physical development. Urban environmental sustainability will enjoy a significant tailwind thanks to an accelerating economic restructuring toward a high value service economy and its factory-of-the-world role becoming relatively less important.

Chapter 17

Sustainability in Indian Country:
A Case of Nation-Building Through
the Development of Adaptive Capacity

Judith Dworkin

Introduction

A sustainable, resilient, and adaptable community is a worthwhile objective for all people. For Indian people and Indian nations located within the US, these concepts are bound up in a 225-year history of inconsistent and often abject treatment by the US government (American Diabetes Association, n.d). Over this timespan, some Indian nations did not survive but most did. But, in order to survive, Indian people were forced to adapt, relinquishing significant portions of their ways of life, livelihoods, cultures, languages and economies. As a result, the definition and meaning of a sustainable Indian community constituted the mere survival of people, government, and culture.

During the last three decades, Indian nations have been engaged in transformative changes of their economic, governmental, social, and cultural capacities. They have begun to emerge from a subsistence mentality to rebuild their communities, restore their Indian identity, and build resilient institutions that strengthen economic, social, and environmental capital to enhance self-governance. Some Indian nations are located in close proximity to large cities and have successful economic enterprises such as casinos, while others have developed significant tourist industries leveraging unique resources or historic sites. Other nations are located in remote locations with limited opportunities. Like many of their indigenous counterparts around the world, some have abundant supplies of natural resources for development; others do not. In each of these situations, Indian nation governments have the task of developing a sustainable and resilient economy that is reflective of their culture and builds a stronger community, and at the same time, adapting to and insulating their people from potentially harmful outside influences.

Today, they are actively planning for a sustainable future that takes into consideration the multiple needs of their members. This includes on-reservation employment opportunities, sufficient housing, reliable transportation systems, safe and adequate water supplies, appropriate wastewater systems, access to electricity, Internet access, and extended on-reservation educational opportunities such as higher education centers. Balanced with planning for the economic needs of

Cultural Activities and Institutions Culture is the set of values and beliefs people have about how the world works as well as the norms of behavior derived from that set of values. Cultural activities and institutions include language, religion, cuisine, social habits, music and arts. It also includes economic decisions such as investment in education and willingness to contribution to the public good.

Nation-building Nation-building refers to the process of constructing or developing the identity of a nation through government processes. Nation-building targets the development of a sense of identity within the territory controlled by the nation. It may involve the use of infrastructure development to foster economic growth.

their communities, Indian nations are committing resources to **cultural activities and institutions** which may include restoration of language, reintroduction of ceremonies, protection of sacred places, cultural storytelling, and education. Despite a palpable legacy of political and economic adversity, Indian nations and Indian people have outlived the tragedies of colonization and marginal existence. They are engaged in **nation-building** to ensure sustainability.

This revitalization requires a definition of sustainability that is more expansive than mere survival, maintenance, or economic development. Studies performed on types of 'ethnic reorganization' indicate that cultural restoration and resurgence is helping tribes to re-engage with their culture. A seven member Board of Directors, whose directors are appointed by the Chairperson of the Nation and confirmed by the Legislative Council, governs TONCA. The Board of Directors is viewed as the policy-making entity and delegates the day-to-day operation of the programs under TONCA to the programs' professional staff Indian nations are more inclined now to practice their language and traditional culture in order to promote a sustainable community culture for the future. This strong sense of belonging and community may even result in positive health and economic effects. As Hill describes it, the "sense of belonging as connectedness occurs through the dynamics of relationships between everything in the creation/universe. It is a deep spiritual connection to family, community, nature, the Creator, land, environment, ancestors and traditional way of life" (Background Paper, 2002).

With the recent emergence of and importance placed on developing resilient communities as part of sustainability (see Chapter 7), this chapter explores the evolution of adaptive capacity building and its role in enhancing resilience in the context of US Indian nations. A detailed case analysis of the Tohono O'odham Nation is included to demonstrate how one Indian nation has developed institutional capacity and promoted nation-building, responding to a culturally driven cry to "bring our elders home." To set the stage for this case, the chapter provides the reader with (i) an overview of the special relationship between the US government and Indian nations and the various federal Indian policies to which Indian nations have been subject during the past 225 years, and (ii) statistics regarding American Indians and Indian nations.

An Overview of Federal Indian Policy

Indian nations have a unique relationship with the US' system of federalism. They were sovereign powers at the time the Europeans "discovered" America. Each Indian nation consisted of a unique group of people with a distinct language, culture, and religious structure. It controlled a specific geographical area and possessed executive powers acknowledged by its people and enforced by governmental authority. These entities negotiated treaties with other Indian nations, foreign governments, and after the break from European colonial powers, the newly formed US. While the earliest treaties were negotiated between two equal sovereigns, the US government quickly came to fully embrace the Doctrine of Discovery by which title to newly discovered lands lay with the government whose subjects discovered new territory (Cajete, 2000). Under this doctrine, Indian nations were considered to be

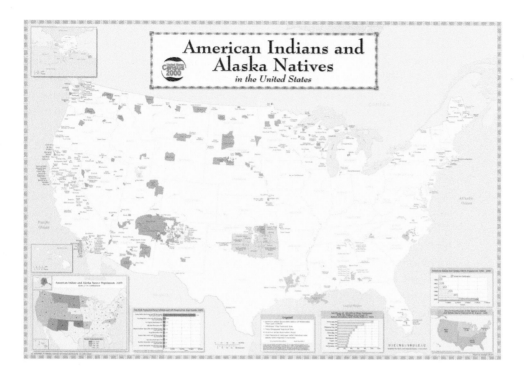

"discovered." Sovereignty and discovery became fundamental features of federal Indian policy and are important for understanding the underlying factors and decisions of Indian nations creating their own sustainable future.

These early expressions of the relationship between the US and the Indian nations were further documented by Justice John Marshall writing for the US Supreme Court. This judicial opinion is referred to as the "Marshall Trilogy." Justice Marshall established three key principles regarding the sovereignty of Indian nations or tribes: (1) "tribes are "domestic, dependent nations," (2) tribes have the right of occupancy but not **fee ownership** in their lands, and (3) the rights of a tribe are always subject to diminishment by Congressional plenary power (CDC Report, 2006). Based upon these concepts, the scope of **tribal sovereignty** has waxed and waned with changes in federal Indian policy over time.

Today there are 566 federally recognized Indian nations or tribes within the US (Chapter 3, infra). These Indian nations have survived a series of federal Indian policies that, at times, sought to eliminate their very existence (Community pursues development at its own pace, 2007). Among the most notable are:

- The Treaty Period (1789–1871). During the Treaty Period, the US negotiated treaties with Indian nations in which the overriding goal was to obtain Indian lands and fix the boundaries of the Indian nation in return for the delivery of goods and services by the US. Common among these goods and services were livestock, farming tools, and health and educational services. A fundamental principle developed during this period called for the US to have a special trust responsibility to Indians.

Fee ownership The right or interest that an individual or group has in lands or private property to the exclusion of all others that entitles them to do what they want with the land or property.

Tribal Sovereignty Tribal sovereignty refers to a tribe or Indian nation's right to govern itself, define its membership, manage its property, both real and personal, and regulate tribal business and domestic relations.

- The Removal Period (1815–1846). The Indian Removal Act of 1830 authorized the President to negotiate with selected tribes, granting them ownership of unsettled lands west of the Mississippi in Oklahoma, Kansas, and Nebraska in exchange for their lands within existing states. With the discovery of gold in California and the need to develop safe routes to western states, reservations were re-established and tribes moved to even more remote locations or confined to smaller reservations.

- The Allotment Period (1887–1934). The General Allotment Act of 1887, known as the Dawes Act, redistributed substantial landholdings to individual Indians rather than to the Indian nations. The stated intent was to protect Indian property rights. The underlying intention was to break up reservations and indoctrinate Indians into the majority culture. Thus, "It would then no longer be necessary for the government to oversee Indian welfare in the paternalistic way it had been obligated to do, or provide meager annuities that seemed to keep the Indian in a subservient and poverty-stricken position" (US National Archives and Records Administration, 2017). This shift in land rights devastated many Indian nations. Allotments were small and often fractioned by familial descent. Additionally, Indian allottees had neither the farming experience nor the capital needed to buy the appropriate equipment and supplies; further, their limited allotments often failed to have access to necessary water supplies. Indian allottees frequently found it necessary to sell the land once it had been converted to fee. Over the next 50-year period, some ninety million tribal acres were acquired by non-Indian parties either directly as **"surplus" land** or indirectly through the acquisition of land from Indian allottees.

Surplus land Reservation land remaining after allotment to individual Indians. The Dawes Act authorized the Secretary of the Interior to acquire this excess, or surplus, land for the US.

- Indian Reorganization (1928–1942). This period was marked by a shift away from assimilation policies and toward the recognition of Indian sovereignty. In 1934, the Indian Reorganization Act was passed by Congress and stopped any further allotment of lands. It established restraints on alienation for trust lands and mechanisms for the organization of tribal governments and tribally owned businesses. It also recommitted the federal government to strengthening tribal sovereignty. Unfortunately, by this time, reservation life had "plunged into a downward spiral of poverty, disease and despondency" Federal recognition of an Indian tribe authorizes that tribe to be eligible to receive services from the U.S. Department of the Interior, Bureau of Indian Affairs. See Federally Recognized Tribes List Act, Pub. L. No. 103–454 1994.; Indian Entities Recognized and Eligible To Receive Services From the United States Bureau of Indian Affairs, 79 Fed. Reg. 4748 (Jan. 29,2014). Some Indian nations have received state recognition but not federal recognition while many Indian nations are unrecognized by either the federal government or a state.

- Termination (1943–1961). A 1943 study entitled *Survey of Conditions Among the Indians of the United States* found that conditions on reservations were deplorable, the Bureau of Indian Affairs bore responsibility for extreme mismanagement, and assimilation of Indian peoples should be expedited. For a summary of the history of Federal Indian policies, *see History &*

Background of Federal Indian Policy, Cohen's Handbook of Federal Indian Law, chapt. 1 2012. Termination of tribes became an express federal Indian policy in 1952 when the House of Representatives directed the Committee on Interior and Insular Affairs to conduct a full investigation into Bureau of Indian Affairs activities and formulate legislative proposals "designed to promote the earliest practicable termination of all federal supervision and control over Indians" (H.R. Rep. No. 82–2503 1952). On August 1, 1953, Congress adopted House Concurrent Resolution 108 which spurred legislation terminating 70 tribes and Indian bands. This legislation ended the special relationship between the US and Indian nations with drastic consequences for the tribes. Federal programs were discontinued so that education, health, welfare, housing assistance, and other social programs were no longer available to Indians. Most tribes ultimately relinquished or lost their land. While the tribal governments of the terminated nations were not expressly extinguished, once the land base was lost, most were unable to exercise governmental power over their tribal members, effectively weakening the sovereignty of the tribes.

- Relocation (1952–1972). After World War II, between 1952 and 1972, approximately 100,000 Indians were "voluntarily" relocated to targeted urban cities such as Minneapolis, Denver, Chicago, San Francisco, and Seattle for vocational training. The Indian Relocation Act of 1956 (Public Law 959) also known as the Adult Vocational Training Program encouraged Indians to move from the reservations to urban areas for educational purposes. It provided for "vocational counseling and guidance, institutional training in any vocation or trade, apprenticeship, and on the job training." It allotted funds for transportation and subsistence for up to 24 months (Public Law 959, 1956). The underlying goal was to continue the assimilation of American Indians into the dominant culture and relieve the government of its responsibilities. Senator Arthur V. Watkins of Utah, chairman of the Senate Committee on Indian Affairs expressed the sentiments of Congress: "The sooner we can get the Indians into the cities, the sooner the government can get out of the Indian business" (Hibbard, Lane, & Rasmussen, 2008). Many of these Indians moved back to the reservations despite the rampant poverty and unemployment there to escape the cultural isolation of the cities.

Notwithstanding the assault on tribal sovereignty by the loss of substantial portions of their land holdings, many tribes managed to survive. After nearly 225 years of land dissipation; ethnocentric regulatory mauling; religious, educational, and economic force-feeding; and destabilization of tribal governments, most reservations and cultures still existed, albeit, with varying degrees of integrity (Hibbard, Lane, & Rasmussen, 2008). Much communal land had been lost; chaotic land holding patterns had been created; poverty had become entrenched; and language, customs, and traditions had been lost. Despite these profound losses and dislocations, the embers of these remarkably resilient tribal cultures still smoldered (Hibbard, Lane, & Rasmussen, 2008).

In 1970, President Richard Nixon declared a new Indian policy of **Indian self-determination** and tribal sovereignty. Subsequent administrations have supported

Indian Self-determination
First used by the National Congress of American Indians in 1966, it refers to three interrelated concepts of tribal self-rule, cultural survival, and economic development.

Nixon's government-to-government policy between the federal government and Indian tribes. President Barack Obama similarly confirmed his administration's commitment "to strengthening and building on the Nation-to-Nation relationship between the United States and tribal nations" (Hibbard, Lane, & Rasmussen, 2008). It is within this government-to-government Indian policy that we consider the ability of Indian nations to promote sustainable community practices and institutions.

An Overview of Indian Demographics

Indians are some of the most disadvantaged people within the US as compared to the general population and the "Non-Hispanic White" population. These statistics are collected by the US Census as well as other federal agencies. According to the 2010 US Census, there were approximately 5.2 million people who identify as American Indians and Alaska Natives ("AI/AN), (Hibbard, Lane, & Rasmussen, 2008), representing 1.7 percent of the U.S. total population (US Census, 2010). Of that population, 22 percent lived on land identified as American Indian reservations, other trust lands, or in Alaska Native Villages. Indian nations have a higher proportion of children than the general population: 30.0 percent of the population is under 17 years of age, compared to 22.9 percent nationwide. The median age of the AI/AN population is 30.8 years, which is younger than the median age of 38 for the US population. Multigenerational homes are common in Indian culture. In 2015, 6.1 percent of grandparents lived with their grandchildren. Of these, 51 percent are solely responsible for them. The median income of AI/AN households in 2015 was $38,367 as compared to $53,889 for the US as a whole. The poverty rate for AI/AN was 25.9% in 2015, compared to 15.3% nationally (US Census, 2011).

Indians are disproportionately afflicted by a number of health issues as well, according to the Center for Disease Control and Prevention (CDC). The Indian Health Service reports that AI/AN are 2.2 times as likely to have diabetes as compared to non-Hispanic Whites. There was a 110 percent increase in diabetes from 1990-2009 in AI/AN youth aged 15-19 years (Hibbard, Lane, & Rasmussen, 2008). AI/AN mothers had the second largest infant death rate compared with other mothers, 48 percent greater than the rate among white mothers (Hibbard, Lane, & Rasmussen, 2008). In 2007, AI/AN population had the highest death rate due to motor vehicle injuries, the second highest death rate due to drugs (including illicit, prescription, and over the counter), and the largest suicide rate compared with other racial/ethnic populations. In 2009, AI/AN adults were among those with the largest prevalence of binge drinking, one of the largest number of binge drinking episodes per individual, and the largest number of drinks consumed during binge drinking compared with other racial ethnic populations. In 2009, only adult (18 years and older) Hispanics in the same age group were less likely to have completed high school than AI/AN adults. A 1990-91 study of 9,000 children aged 5-18 years living on or near an Indian reservation, found 39.3 percent of AI/AN to be overweight or obese (Hibbard, Lane, & Rasmussen, 2008).

Unemployment rates for AI/AN were the highest of any US population in 2015 at 9.9 percent; Blacks are second highest at 9.6 percent. Both were significantly higher than the overall unemployment rate of 5.3 percent. In the aggregate, only 60.6 percent of individuals were employed in 2015, compared to 70 percent nationwide (BLS, 2016). In 2009, the percentage of AI/AN adults living in poverty was among the largest compared with other racial/ethnic populations. Twelve percent more AI/AN adults lived below the federal poverty level as compared to white adults. The three counties with the highest poverty rates in the US are all located in South Dakota and are wholly or predominantly comprised of Indian reservations. Ziebach County's (Cheyenne River Indian Reservation and Standing Rock Indian Reservation) poverty rate is 50.1 percent, Todd County's (Rosebud Indian Reservation) poverty rate is 49.1 percent, and Shannon County's (Pine Ridge Indian Reservation) poverty rate is 47.3 percent (Hibbard, Lane, & Rasmussen, 2008). Residents of Indian reservations utilize food stamps at approximately twice the national rate.

The cumulative effect of federal Indian policies has resulted in the extreme poverty, isolation, and health problems of Indian people. Despite these efforts to assimilate and destroy Indian culture and communities, Indian people have resisted and undertaken efforts to reverse these statistics in order to provide long-term sustainability for their people. The development of community resilience is a cornerstone of these efforts and the basis for this case study.

A Sense of Identity

Resiliency, in the context of Indian nations, can be defined as the "capacity of a complex structured system such as a tribe to survive, grow, adjust, and thrive in the face of unforeseen change and catastrophic events" (Hibbard, Lane, & Rasmussen, 2008). It is also the ability of a system to adapt and cope with long, drawn out events such as climate change or economic restructuring that are not abrupt in nature, but require a community to adapt in order to survive and thrive. Other definitions look at specific aspects of a system.

"Resilient systems expand and contract with variable cycles of growth, accumulation, crisis, and renewal" (Hibbard, Lane, & Rasmussen, 2008). "A resilient social system resists disorder and the unexpected by accepting and planning for episodic catastrophes, crisis and abundance" (Hibbard, Lane, & Rasmussen, 2008). Resilience theory teaches that a resilient community, such as an Indian nation, will embrace the reality of continuous, unpredictable future change, and look for ways to adapt in order to survive the irreversible changes that have already been made or that will occur (Hibbard, Lane, & Rasmussen, 2008). Resilience is the capacity of a complex system (such as an Indian nation) to survive, grow, adjust, and thrive in the face of unforeseen change and catastrophic events, such as colonial contact, placement onto a reservation, or destruction of a community's land base (Hibbard, Lane, & Rasmussen, 2008).

Some US communities simply disappeared when people moved away or became ghost towns, as communities did during the Dust Bowl era and after the Gold Rush (Hibbard, Lane, & Rasmussen, 2008). In contrast, Indian nations are deeply connected to place and "inextricably tied to the sense of community" (Hibbard, Lane, & Rasmussen, 2008) associated with these places. This builds place-based social capital, a key aspect of resiliency. For the past century, and in some instances for time immemorial, that sense of place is most closely connected to land and Indian peoples' "homelands." For some, the reservation may be all or a portion of their traditional homelands, for others it may be located far from their traditional homeland.

"Native people traditionally perceive themselves as embedded in a web of dynamic and mutually respectful relationships among all of the natural features and phenomena of their homelands." (Hibbard, Lane, & Rasmussen, 2008). Many Indian nations perceive their environment as sacred and fundamentally important to their society (Jackson, 1993). Reservation boundaries can be a reminder that the reservation lands "typically represent a fraction of what previously constituted the custodial lands of pre-colonial indigenous populations" (Johnson, 1823). At the same time, members of an Indian nation understand that reservations create a special sense of place and Indian identity as well as conferring intergenerational responsibilities (Jonathan, 2001)

Research is affirming that "the knowledge, practices, values and languages of indigenous people are intimately linked to sustainable living" (King, Leslie, & Hood, 1999). Indigenous people "know a great deal about flora and fauna and they have their own classifications systems and versions of meteorology, physics, chemistry, earth science, astronomy, botany, pharmacology, psychology and the sacred. Sustainability concepts of preserving the environment and saving it for future generations are included and celebrated in their rituals and spirituality"(Kunesh, n.d). The study of ecosystems in relationship to community health has recently been applied to studies of Indigenous peoples. King and Hood (1999) note "Indigenous people have much to teach us about the connection among ecosystems and community health and about integrated ways for managing environments and people" (Mariella & DeWeaver, 2007). Additionally, researchers are working on indicators to determine the overall health of communities and its role within a Reservation ecosystem (Martinez, 1998). This is an important research direction for Indian nations, particularly as they work to create a broader vision of a sustainable future.

As noted above, Indian nation's sustainable future will include myriad components specific to their culture. Revival of language is but one of these critical components. The status of Indian languages is constantly changing. In general, Indian nations are becoming more aware of the need to restore and preserve their languages (Kunesh, 1991). They are recognizing that many of their traditional practices are seeped in language and are striving to reconnect their younger generations with both. Many schools located on reservations are implementing language immersion programs into their curriculum. In addition, many schools, including head start and pre-school, have started to teach native languages to children at a young age to familiarize them with their traditional language.

Establishing a sustainable future for tribal communities requires recognizing and enhancing traditional and customary principles with contemporary legal, economic, and political strategies. In recent years, many Indian nations have established legal and judicial systems under which community discord and commercial disputes are resolved and defined the parameters of their own programs and services (Public Law 959, 1956).

Economic Strategies

For some Indian nations, Americans' apparent passion for gaming activities, the passage of the Indian Gaming Regulatory Act, and the negotiation of gaming compacts between the Indian nations and the states have produced an economic engine not previously seen in Indian country. Gaming revenues have contributed positively toward mitigating the absence of tax revenues as well as the overall lack of funds for basic infrastructure, including funding for roads, housing, public safety, sanitation, education, and health (Ragsdale, 1991). The implementation of gaming activities, however, cannot be the sole basis for a sustainable future. Indian nations control millions of acres of land throughout the US that are rich in natural resources, including reserves of oil, gas, coal, timber, minerals, and renewable resources like wind and solar energy. These resources offer potential opportunities for economic development and employment. This development must incorporate cultural values and traditions, and at the same time, produce development and employment that is sustainable. These seemingly disparate goals are in keeping with Indian nations tradition as conscientious land stewards, based on their deep respect for, and intimate relationship with, the environment (Reyhner, Cantoni, Clair, & Yazzie, 1999). As former Chairman and President of the Navajo Nation, Peterson Zah explained, "on the reservation, land means everything. It touches religious beliefs and spiritual reality" (Reyhner, Cantoni, St. Clair, & Yazzie, 1999).

Political Strategies

Land-use planning is key to developing a sustainable future—for economic development, housing, education, and healthcare. Indian legal principles stemming from the Doctrine of Discovery to allotment and assimilation impact the practical development of land by both an Indian nation and its members. Because most reservation land is held in trust by the federal government for the benefit of the Indian nation and individual Indians, land development is challenging. Tribal trust land generally may not be sold, taxed, or encumbered without the approval of the Secretary of the Interior. Because the land is held in trust, tribal members and the Indian nation cannot use the land for collateral in order to obtain loans for development purposes, even to build much needed housing. In many instances, tribal members inherit fractionated land that was distributed during the allotment period resulting in numerous allottees with ownership interests in a parcel of land. Because the federal government disfavors partitioning of land for the benefit of tribal members, many tribal members are left without options to develop a homesite (Roland, Margaret, & Semali, 2010).

An additional challenge for each Indian nation is determining the vision for its people. As the governing body, the Indian nation is charged with considering how to restore the health of its members, provide opportunities within the reservation for employment, and revitalize the customs and traditions of the Indian nation in accordance with community values. This visioning is an important part of tribal nation building and creating a sustainable future can take many forms. In some instances, Indian nations have successfully established a comprehensive community planning effort. The Spokane Tribe of Indians and the Navajo Nation are two recent examples (S. Rep. No. 78–310 1943). In other instances, Indian nation's have focused attention on specific economic development activities.

Following is a case study of the Tohono O'odham Nation and the institutional capacity building that began with the Nation's response to a need driven by the O'odham people's cultural values to "bring their elders home." This case demonstrates how the expression of a cultural need led to the building of institutional capacity and resiliency through newly developed governance structures.

The Case of the Tohono O'odham Nation: Capacity Building

The Background

The Tohono O'odham Nation is located in the Sonoran Desert of southern Arizona along the US-Mexican border. It is the second largest reservation in Arizona in both population and geographical size. Comprised of three non-contiguous reservations and additional parcels of land, the Tohono O'odham Nation controls a land base of 2.8 million acres or 4,460 square miles of land. Its boundaries begin south of Casa Grande and continue south for about 90 miles to the US-Mexico International border. It runs for 75 miles along the border. The closest urban populations are Tucson (population: 520,116), Casa Grande (population: 49,591), and Gila Bend (population: 1,922).

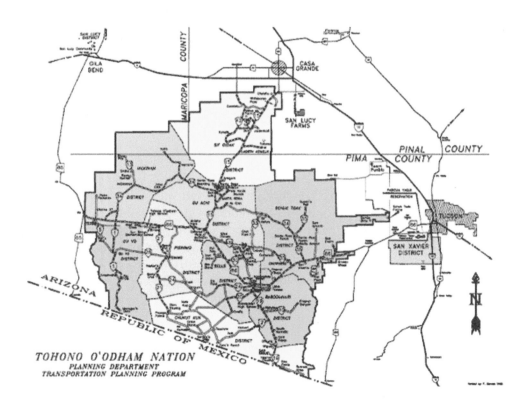

TOHONO O'ODHAM NATION
PLANNING DEPARTMENT
TRANSPORTATION PLANNING PROGRAM

Tohono O'odham members have inhabited portions of the Sonoran Desert for thousands of years. The Spanish established the San Xavier mission in the seventeenth century to minister to the O'odham Indians and convert them to the Roman Catholic religion. Many indigenous traditions remained, adapted into a unique form of "Sonoran Catholicism" (Seivertson, 1999). Over time, the O'odham language and culture adapted and survived. The Tohono O'odham Community Action organization ("TOCA") describes the Tribe's culture and language as "Endangered but Intact" (Semken, 2005).

In 1853, the Gadsden Purchase split the O'odham people between the US and Mexico. In the twentieth century, federal Indian policies brought greater mobility to O'odham people, some voluntary and some forced, including programs involving boarding schools, migrant labor, and World War II. After World War II, numerous O'odham families were moved to cities like Oakland, California and Chicago, Illinois (Smith, 2005).

The rural nature of the Tohono O'odham lands is evident in the number of people per square mile. The Reservation's population density is only 2 persons per square mile. This is significantly lower than the population density of Pima County, in which the majority of reservation lands reside, with 107 persons per square mile, and the State of Arizona with 56 persons per square mile.

The Tohono O'odham Nation is divided into 12 political districts and 83 villages. Nine of these districts are located on the Sells Tohono O'odham Reservation (established in 1917 by Executive Order and expanded in 1931 by a Congressional Act). The community of Sells, centrally located within the Sells Reservation, is the seat of the central government. The remaining districts are the San Xavier District (established by Executive Order in 1874), which covers the entirety of the San Xavier Reservation; the San Lucy District, located on the Gila Bend Indian Reservation (established by Executive Order in 1882 andmodified by Executive Order in 1909); and the Hia-Ced District created in 2012 and located on a parcel of land near Why, Arizona and placed in trust in February 2009. The Sells, San Xavier, Gila Bend Reservations and other parcels of land held by the Nation are referred to as the "Reservation."

According to the 2010 U.S. Census, roughly 22,000 of the 30,000 enrolled tribal members live on the Reservation. More than 3,300 of the enrolled members are 55 years of age or older, 1,800 of whom reside on the Reservation. Multigenerational families are common. Thus, children on the Tohono O'odham reservation are 15 times more likely to live with grandparents as compared to urban Arizona communities.

Health is a major concern on the Reservation. TOCA reports that until the 1960's, no tribal member had ever suffered from type-2 (adult onset) diabetes. Today, more than 50 percent of all Tohono O'odham adults have the disease, which is reported to be the highest rate in the world. Beginning in the 1990's, childhood-age onset began to rise. The percentage of diabetics is expected to exceed 75 percent for O'odham children born after 2002. It is closely correlated with childhood obesity. According to the Arizona Bureau of Public Health Services, the death rate due to type-2 diabetes among Arizona's native population is three times that of the state average.

Tohono O'odham life expectancy is more than six years less than the US average, according to a 2010 study published in *The New England Journal of Medicine.* This study analyzed the longitudinal health data of non-diabetic 4,857 Pima and Tohono O'odham children born between 1945 and 1984. Early data was culled from Indian Health Service records covering 40 years of children's vital signs. The average age of the children was 11 years old when they were first examined for glucose levels, BMI, blood pressure, and cholesterol information. Researchers traced health records of the individuals to see how long they actually lived. The median follow-up was 23.9 years. The researchers documented 166 premature deaths from endogenous (internal) causes for those who died before 55 years of age. The adults who, as children, had been measured at the highest body mass indices (BMI) were 2.3 times more likely to die prematurely. Those with the highest glucose levels were 73 percent more likely (Spokane Tribe of Indians, n.d).

Self-reported health status among Tohono O'odham elders shows that 57 percent rated their health as fair or poor as compared to only 27 percent of non-Native American elders. Those who rated themselves as "fair" or "poor" tended to have one or more illnesses or conditions considered as predictors for future medical intervention and skilled healthcare. These include diabetes (61.9 percent) and obesity (47 percent).

The health of O'odham members is further complicated by disadvantaged living conditions. Many of the villages scattered throughout the Nation lack paved roads, making travel difficult. Housing is limited and often not built according to accepted building codes. According to the Indian Health Service Division of Sanitation, hundreds of homes on the Nation still lack indoor plumbing and telephone service, signifying that often the poorest tribal members live in homes without adequate plumbing, heating, and cooling.

Until recently, health care services on the Tohono O'odham Reservation were quite limited. While limited acute care services were available at the Sells Area Indian Health Service Hospital, most O'odham members did not have access to reservation-based hospital care, post-hospital care, or long-term care services. This situation was particularly problematic for O'odham elders.

Elders admitted to the Sells Area Indian Health Service Hospital for acute care and subsequently requiring long-term skilled nursing care would be discharged to a nursing home facility located in the Tucson, Arizona area. Most of these facilities are more than one and one-half hours away from the majority of O'odham families. In some cases, the ailing elder went directly to a Tucson hospital for his or her initial care and then would be discharged for post-hospital recovery to a Tucson nursing home. The end result, in either case, left O'odham elders away from their families.

Many elders do not speak English, which also complicated their off-Reservation care. Facility caregivers could not speak to these residents in their traditional language resulting in significant challenges for both staff and patients. Other factors such as O'odham residents not being offered traditional foods to eat, not readily being able to seek the assistance of "medicine people," and not being able to spend their remaining days in their homes made long-term care a serious issue. While statistics are not readily available, it is not hard to conclude that placing an ill and frail O'odham elder in a foreign and often time isolated facility hastens both further debilitation and death.In response, the O'odham people asked their government to "bring their elders home."

In 1978, the Tohono O'odham Nation responded to the demands of its members and funded the development of a master plan to build a nursing home within the boundaries of the Nation. Unfortunately, the cost of constructing and operating a skilled nursing facility was not feasible in 1978. But the voices in support of this goal continued for the next 15 years. Those most passionate about the facility formed an advisory committee, which became recognized by the government as the "Tohono O'odham Nation Nursing Home Advisory Committee."

Institutional Capacity Building and Cultural Sustainability

The 1970's brought a new era of self-determination on the part of Indian nations and a new federal Indian policy that addressed Indian nations on a government-to-government relationship. It was, however, the enactment of the Indian Gaming Regulatory Act in 1988, which enabled the Tohono O'odham Nation to amass substantial revenues. The first casino, located in the metropolitan Tucson area, opened in 1993. It was followed by two additional casinos that opened in 1999 and 2003. Due in large part to the financial success of the first casino, the Nation allocated funding to build a skilled nursing facility on the Reservation in 1993.

An important aspect of resiliency is the notion of economic resiliency, which refers to a diversified economic base and the reinvestment of funds back into the community. This not only strengthens a community's economic base, it also strengthens the community from within by buffering it from outside shocks. The Tohono O'odham Nursing Care Facility is an important example of this process in action.

In December 1998, the Nation established the Tohono O'odham Nursing Care Authority ("TONCA") as a tribal enterprise. The Nation could have operated the skilled nursing facility through its Department of Health and Human Services but the formation of TONCA expanded its institutional capacity and promoted nation-building. The Nation's lawmakers intended for TONCA to operate independently of the Nation's government but not be independent of the Nation's government. The Charter clearly states that TONCA controls its own budget, has the authority to open bank accounts, hire professionals including legal counsel, invest funds, establish a capital improvement fund, develop an annual budget, form subsidiary organizations and otherwise operate without requiring consent of the government. Importantly, however, TONCA receives an annual subsidy from the Nation.

This ongoing duality of independence and dependence encourages the maintenance of close, positive relationships among TONCA and the Nation's members, the governing bodies of the 12 districts, the Executive Branch of the Nation's government, and the Legislative Council. These productive relationships have enhanced TONCA's ability to meet its long- and short-term objectives. The financial dependence encourages transparency and regular communications, while the independent control encourages resiliency through adaptability and institutional connections.

The guiding principle of TONCA is defined as "All people deserve to live and die in dignity. Life, death, and dignity are uniquely defined by one's own culture." TONCA's Charter demonstrates the Nation's unique dual objectives in building the facility. First, it operates as a modern twentieth century facility meeting all of the national licensing requirements Secondly, it is guided by **O'odham Himdag.**

O'odham Himdag
Roughly translates as the Desert People's lifeway or way of life. It refers to the traditional cultural skills and knowledge of the O'odham people.

O'odham Himdag, which refers to "a way of life inclusive of terms such as culture, heritage, history, values, traditions, customs, beliefs and language" is an integral part of the O'odham lifestyle (Spokane Tribe of Indians, 2012). Each person is expected to live their life in reverence of the *Himdag*, which promotes cultural sustainability in all aspects of life including tribal government departments. *Himdag* can manifest itself in direct and indirect ways, including "fostering strength and wellness in their community by translating increased economic self-sufficiency and resources derived from gaming into social, health, and educational services which maintain their tribal traditions, thereby providing an effective path toward the maintenance of cultural identity, or *O'odham Himdag*." Thus, *O'odham Himdag* has been a resiliency tool for the Tohono O'odham people.

TONCA was envisioned as a mission-driven institution that would not only successfully develop and operate the skilled nursing facility, it would also advocate for other needed health care programs on the Reservation. The Preamble to the Charter establishes the breadth of the responsibilities delegated to the TONCA Board (TOCA, n.d). "Health is more than the absence of disease. It is a complete combination of intellectual, physical, psychological, social, and spiritual states which form the condition of wellness" (TONCA Charter, n.d). This established the foundation for the Nation's sustainable well-being.

In satisfying its dual objectives, TONCA provides the highest quality in a language that residents understand and in harmony with the cultural values and customs of *O'odham Himdag*. The skilled nursing facility is located in the Sonoran desert and offers residents traditional foods, some purchased from a cooperative farm located on the reservation. Medicine people visit residents and bless the facility. Young people are encouraged to visit to learn traditional stories and improve their O'odham language skills from the elder residents. The facility respects residents' desires to die with family around them in accordance with O'odham tradition. TONCA encourages non-O'odham employees to learn rudimentary O'odham language.

For many Indian nations, the years of abject poverty and the change away from traditional foods have led to some of the worst health statistics in the overall US population. Improvements in health and health services contribute to sustainable development. "The extent to which sustainable development benefits a community is closely tied to its level of health, as health is a product of economic social, political and environmental factors, as well as of health services" (TONCA Preamble to Charter, n.d).

In establishing TONCA, the nation sought to improve the health care services available to O'odham members on the reservation. Effective health services must be accessible and offer good quality care, which requires good organization, sufficient resources, and a capacity for strategic support and effective mobilization of personal action and technological development to improve health (Trosper,

2002). "Poverty leads to ill-health, but possibly ultimately more debilitating is the negative impact of poor health on development" (U.S. Census Bureau, 2010). In contrast, good health enhances development and results in higher productivity among healthier workers and higher rates of savings and investment (US Bureau of the Census, 2010). In summary, effective health services are those that are accessible and offer good quality care, requiring the right focus on health needs, and "equitable distribution, good organization and sufficient resources (human, physical and supplies)" (US Bureau of the Census, 2015).

Recognizing that healthcare is more than just skilled nursing, the Nation charged TONCA with assuming "a leadership role in providing a continuum of care and services designed to enhance the physical, spiritual, emotional, social and intellectual qualities of life for aging O'odham and other members of the Nation" (US Bureau of Labor Statistics, 2016). As such, TONCA seeks to ensure that services are provided without duplication, fragmentation, or gaps—a seamless system integrating health with nation-building. There are substantial barriers to improving the health of O'odham members however. These barriers include:

- *Lack of Financial Resources.* Half of O'odham elders live below the poverty line.
- *Far Distances to Healthcare.* Primary health care services are centralized, requiring patients to spend inordinate amounts of time traveling for treatment. These distances can lead to postponement of care. This often results in treatable and chronic illnesses being first diagnosed at a later stage as compared to non-Natives, increasing the needs and costs of invasive treatments as well as less satisfactory outcomes.
- *Limited and Undependable Transportation.* The Tohono O'odham Nation does not have a public transportation system and many families do not have a personal vehicle.
- *Lack of Readily Available In-home, Locally Based Services.* Elders need help with basic activities of daily living, which is particularly true for the 16% of elders who live alone.
- *Language Barriers.* Language is still a barrier to healthcare for the many O'odham elders for whom the O'odham language is their only language. This requires healthcare workers who can either understand the language or the assistance of an interpreter, either of which is often unavailable.
- *Limited Understanding of Traditional Medicine by Healthcare Team.* Many non-Indian healthcare workers, doctors, and administrators often misunderstand the use and benefit of traditional medicine. Seventy percent of O'odham elders desire to include traditional medicine as a component of their compliment of healthcare resources.
- *Treatment Instead of Prevention.* Treatment of healthcare conditions continues to have priority over prevention and early intervention, resulting in health care needs that progress unchecked until the related treatment is acute and costly or the condition becomes debilitating.
- *Poor Communication and Coordination.* Poor communication and co-ordination of services leads to the fragmentation of healthcare services. In such an environment, the quality of care is compromised.

- *Supportive Housing Options.* Housing stock on the Reservation is insufficient and lacks the quality necessary to support the needs of elders. There is a clear need for a robust program of home repair and home rehabilitation including the modification of homes to meet accessibility standards for changing needs of aging elder residents.

The need for long-term care services and supportive housing facilities continues to rise, resulting in a demand that surpasses the supply of such services and facilities on the Reservation. The ultimate result is that O'odham members who require this help cannot be guaranteed access to on-Reservation options. These options include group homes; independent living where meals, housekeeping, and maintenance are provided; and assisted living that can extend the independent living services to include modest health related services. TONCA seeks to address these barriers as it expands its leadership role in its continuum of care mission.

Growing Institutional Capacity and Adaptability

When the Nation formed TONCA, its long-term mission would rely on developing the capabilities of the institution. In 1998, it would have been difficult to predict the significant success of the organization 16 years later.

The Skilled Nursing Facility

In November 2002, the Archie Hendricks Sr. Skilled Nursing Facility opened to its first four residents. Since then, it has operated at or near its 60-bed capacity providing short and long-term care to O'odham members. The majority of the admissions to the facility come from hospitals in Tucson, Casa Grande, and Sells. As of Spring 2014, 586 O'odham members had been admitted and discharged from the facility.

TONCA has faced many challenges including staffing, physical infrastructure, and training needs. Senior professional staff, including a licensed administrator, Director of Nursing, and a Business Office Manager, had to be recruited from outside of the Nation, as there were no O'odham members with the appropriate qualifications. The lack of employee housing inhibited recruiting efforts as well. The closest housing for employees was either 45 minutes away in Casa Grande or 90 minutes away in Tucson. Additionally, the Skilled Nursing Facility's building had some construction defects. And, finally, some of the O'odham members who were hired in managerial positions were in need of mentorship to be able to meet the challenges of their positions.

TONCA was able to overcome these challenges. It began operations with non-O'odham members in key positions with the understanding that over time it would replace those individuals with O'odham members. To ensure that members were able to assume these positions, TONCA started a mentorship program to facilitate the management training of its O'odham employees. Employee housing was constructed in order to facilitate the task of hiring and retaining both O'odham and non-O'odham staff. Building defects from the initial construction were corrected. This series of initial challenges and practical solutions reflect what would become a defining resiliency characteristic of TONCA—its ability to recognize current or future problems and the adaptability to overcome or prevent such issues.

Staying Relevant, Having the Capacity to be Adaptive, and Being Responsive

Healthcare organizations encounter community needs that change as the health dynamics of the community evolve over time. Such organizations must be alert to these trends so that their services can meet such needs as they arise. In directing TONCA "to take a leadership role," its founding board offered broad flexibility for the institution to stay relevant and be adaptive while staying true to its core value of providing services in accordance with O'odham culture and customs. This is one of several respects in which TONCA may be viewed as a proactive organization.

To ensure that the provision of services meets the needs of the Nation's members and determine if new services need to be included, the Board methodically revisits TONCA's provision of services and other providers on the Reservation. It projects the healthcare needs of the Nation's members and evaluates the need for future endeavors. It also sets relevant goals for its programs, acquires data that relates to the goals and programs, evaluates the data in light of the issue(s) being addressed, and determines a strategic plan of action. After its implementation, the Board evaluates if the results were consistent with the expected outcomes and modifies the course of action when necessary. The intensive levels of attentiveness, feedback, and flexibility ensure that its programs remain relevant to the O'odham community. Furthermore, it insures that TONCA supports a sustainable future for the Tohono O'odham Nation. TONCA's focus is expanding healthcare services and improving O'odham health. The bi-product is the growth of institutional capacity. The outcome is nation building.

Over the course of its first 16 years, TONCA saw the Nation's population demographics change and the healthcare needs shift on the Reservation. It established and collaborated with other institutions in a number of programs in response to these changes. Such programs include pre-hospice, hospice, and assisted living services. Each of these additional services enhances the knowledge base and experience of TONCA, provides additional health services for members of the Tohono O'odham Nation, and contributes to the sustainable development of the Nation.

Hospice Services: Shortly after the Skilled Nursing Facility opened its doors, TONCA became aware of the lack of reliable hospice services on the Reservation. In 2007, the Board expanded TONCA's services to provide hospice care. Its first hospice admission occurred on November 16, 2007. The Tohono O'odham Hospice brings together a team of traditional healers, doctors, nurses, social workers, chaplains, and volunteers to meet the physical, emotional, and spiritual needs of those at the end of life and their families. Hospice patients, and by extension family members, receive services either in their own homes or at the Skilled Nursing Facility, which allows the patient to complete his or her life in the presence of family and community, and in accordance with O'odham traditions.

Desert Pathways: With a small grant from the Robert Wood Johnson Foundation, TONCA began "Desert Pathways" to help O'odham members and their families address the challenges of facing a serious illness. In 2013, Desert Pathways staff made 336 visits to 65 members of the Nation and their families. Grant funds have since been exhausted and no additional funding has been obtained, despite the growing number of O'odham members with long-term conditions, such as diabetes and obesity.

Assisted Living Facility: By 2008, TONCA recognized that many of the residents of the Skilled Nursing Facility no longer needed skilled nursing care but were unable to be discharged from the facility because they could not live independently. Unfortunately, there were no options for them on the Reservation. A new master plan was developed for TONCA's 30-acre site that provided for four assisted living buildings that would accommodate 10-12 residents each. After several years of unsuccessfully seeking to obtain funding to construct the first building, the TONCA Board resolved to commit monies from its Board directed fund to construct a single ten-resident assisted living facility in July 2011. In February 2013, the Assisted Living Facility opened with 4 residents. By the end of 2013, ten O'odham members were living in this residence. Some of these residents transferred from the Skilled Nursing Facility; others came from the O'odham community.

The Elder Care Consortium and Health Care Network: As an advocate for enhancing the quality of life for O'odham elders and to accomplish its mission, TONCA required the collaboration of other service providers. It set out to identify the full spectrum of needed services, identify gaps in service, and advocate for filling the gaps by coordinating with other service providers and create a shared plan for the O'odham Nation. TONCA became a founding member of the Tohono O'odham Elder Care Consortium (the "ECC"), which includes the Nation's Department of Health and Human Services ("HHS"); the Tohono O'odham Community College ("TOCC"), a tribal enterprise of the Nation; and the Tucson Area Indian Health Service ("IHS"), a Federal health care agency.

In April 2009, ECC members executed a Collaborative Agreement in which members agreed to cooperate in grant applications, program development, and the preparation and dissemination of position statements, reports, and other written materials. The ECC enables representatives of these entities and the Community College to meet and discuss issues, breaking down the "silo" behavior of these organizations.

It has also improved the skilled healthcare workforce on the Reservation. Before the ECC was founded, the Nation sent students to off-Reservation programs to receive healthcare training and certification. With the support of TONCA, TOCC now offers an entry-level healthcare course to certify students as caregivers. Graduates are eligible to work in assisted living facilities, including TONCA's facility. Several of the initial graduates are now working in the

Nation's Assisted Living Facility. Today, TONCA is reaching out to other Indian nations to establish an organization that can advocate for needed resources in the US Congress.

The Future

National trends suggest skilled nursing facilities will see (i) a growing requirement for more sophisticated care and treatment, and (ii) an increasing number of requests for short-term rather than long-term skilled nursing care. The Tohono O'odham Nation has developed an institution, TONCA, that has grown, developed, and performed well in addressing larger, ever evolving health care challenges. As such, TONCA is well positioned to meet the projected trends.

TONCA also continues to maintain vigilance over the Nation's current and projected healthcare-related needs. This allows the Board to make informed decisions regarding future services and administrative and funding needs. Currently, TONCA must contend with the changing healthcare landscape, the needs associated with physical infrastructure construction and maintenance, recruiting qualified staff, and securing the funding to continue operating. These issues are addressed as follows.

The Changing Healthcare Landscape: The incidence of chronic health conditions continues to rise on the reservation, which results in elders needing longer and more-intensive care. Given this trend, the roles of skilled nursing and assisted living are expected to change with skilled nursing facilities handling the "sickest" O'odham members in the near future as an alternative to hospital stays. Assisted living facilities will function more like skilled nursing facilities.

Physical Infrastructure: A continuing challenge will entail finding capital funds to complete the planned assisted living buildings and expand the wastewater system to service the additional buildings. The current wastewater system is operating at (or near) maximum capacity. TONCA has been seeking to partner with the Nation's housing authority to collaborate on a regional wastewater facility that would meet TONCA's needs as well as those of a proposed neighboring housing project.

Staffing: The location of TONCA's facilities, about 100 miles from the urban centers of Tucson and Phoenix, continues to be a challenge in hiring and retaining professional staff. TONCA is always looking for ways in which it can improve the quality of staff and increase O'odham employment. The Board remains ready to consider alternative employment arrangements in order to obtain the highest quality of care providers.

Financial Resources: Funding is an ever-present concern for any organization or program. TONCA continually looks for ways to secure financial resources that will keep its programs running. Its primary source of funding comes from the Nation through an annual allocation. It has successfully modified the funding cycle so that it receives funding on a five-year plan whereby the Nation agrees to provide annual support prior to the start of the fiscal year in a single lump sum. Currently, TONCA is in the third year of the latest five-year cycle and has started the process of seeking support for the next five-year resolution well before the current five-year cycle ends.

Conclusion

The Tohono O'odham Nation established TONCA to "enhance the physical, spiritual, emotional, social and intellectual qualities of life for aging O'odham and other members of the Nation in need of skilled nursing care." In achieving its initial objective, it gained the institutional capabilities necessary to expand its services. Healthcare needs, like many other types of needs, operate in a fluid environment in which resilient and adaptable institutions are better able to support the dynamics of a sustainable foundation. The shifting healthcare needs and TONCA's responses are a prime example of the type of institution that can successfully support a sustainable future for an Indian nation.

People, society, regulations, and technology evolve requiring institutions and their leadership teams and administrators to focus on new trends and prepare for the impacts of such trends. TONCA achieves this by (1) constantly seeking data on new healthcare and regulatory needs within and outside of their community, (2) analyzing the data to make appropriate plans, and, (3) evaluating current programs for effectiveness in meeting current goals. This practice ensures that TONCA is able to stay true to its vision while also staying versed in on-the-ground needs and trends of its community.

In pursuing its mission, many organizations find that their existing procedures are ineffective in meeting challenges. This often becomes a stumbling block for organizations that are reluctant to consider new paths and forego outdated plans. The adaptability inherent in TONCA's operating culture means that it is able to quickly respond to an issue in order to meet its ultimate objective. In addition, these improvements in health services contribute to sustainable development of the Nation.

The Tohono O'odham Nation's reliance on *O'odham Himdag* demonstrates how traditional cultural norms can be used to help develop sustainable communities. Integration of O'odham culture and values into its health services ensures that the needs of its members are met. The process by which the skilled nursing facility was established and the ability of those charged with elder care to expand beyond the initial objective of "bringing the Elders home" shows a clear connection to the tenets of resilient thinking. It expands the healthcare system in a holistic manner that engages the community, government, and tribal members. The case of the Tohono O'odham Nation and the Tohono O'odham Nursing Care Authority (TONCA) is being replicated in other locations as Indian nations pursue sustainable futures consistent with their own cultures and customs.

References

Chapter 1

1. CNN Library. (2016). *Haiti Earthquake Fast* Facts. Accessed 5/2/2017 at http://www .cnn.com/2013/12/12/world/haiti-earthquake-fast-facts/.

2. CNN Library (2016a). Five years after Japan's worst nuclear disaster. W. Ripley, J. Ogura, and J. Griffiths. Accessed 5/2/2017 at http://www.cnn.com/2016/03/08 /asia/fukushima-five-year-anniversary/.

3. Diamond, J. (2005). Collapse: How societies choose to fail or succeed. New York: Penguin.

4. Food and Agriculture Organization (FAO) of the United Nations, The International Fund for Agricultural Development (IFAD) and the World Food Programme (WFD). (2013). The State of Food Insecurity in the World 2013: The multiple dimensions of food security. Rome: FAO.

5. Food and Agriculture Organization of the United Nations, The International Fund for Agricultural Development (IFAD) and the World Food Programme (WFD). (2015). *The State of Food Insecurity in the World*. Rome: FAO.

6. Global Footprint Network (GFN). (2017a). *Reserve/Deficit Trends*. Accessed 5/2/2017 at http://data.footprintnetwork.org/countryTrends.html?cn=5001&type=cdPC.

7. Global Footprint Network (GFN). (2017b). *Climate Change*. Accessed 5/2/2017 at http://www.footprintnetwork.org/our-work/climate-change/.

8. Kyoto Protocol. (1997). United Nations framework convention on climate change. *Kyoto Protocol, Kyoto, 19*.

9. Liu, M., Huang, Y., Ma, Z., Jin, Z., Liu, X., Wang, H., . . . & Kinney, P. L. (2017). Spatial and temporal trends in the mortality burden of air pollution in China: 2004–2012. *Environment International*, 98, 75–81.

10. Meadows, D., Randers, J., & Meadows, D. (2004). *Limits to growth: The 30-year update*. Utah: Chelsea Green Publishing.

11. United Nations Department of Economic and Social Affairs (2013). World population prospects: the 2012 revision. Population division of the department of economic and social affairs of the United Nations Secretariat, New York.

12. United Nations Department of Economic and Social Affairs (2014). *World Urbanization Prospects: The 2014 Revision*. Population division of the department of economic and social affairs of the United Nations Secretariat, New York.

13. United Nations, Department of Economic and Social Affairs, Population Division. (2016). *The World's Cities in 2016 – Data Booklet* (ST/ESA/ SER.A/392). Accessed 8/3/2017 at http://www.un.org/en/development/desa/population/publications/pdf /urbanization/the_worlds_cities_in_2016_data_booklet.pdf.

14. United Nations Framework Convention on Climate Change (UNFCCC). (n.d.). *The Paris Agreement*. Accessed 5/8/2017 at http://unfccc.int/paris_agreement/items/9485.php.

15. US Energy Information Administration (EIA). (2017). *What Is U.S. Electricity Generation by Energy Source?* Accessed 5/2/2017 at https://www.eia.gov/tools/faqs /faq.php?id=427&t=3.

16. World Commission on Environment and Development (WCED). 1987. *Our Common Future*. New York: Oxford University Press.

17. World Health Organization (WHO). (2016). *Ebola Virus Disease: Fact Sheet*. World Health Organization. Accessed 4/20/2107 at http://www.who.int/mediacentre /factsheets/fs103/en/.

Chapter 2

1. Adams, W. M. (2006). The future of sustainability: Re-thinking environment and development in the twenty-first century. In *Report of the IUCN Renowned Thinkers Meeting*, 29–31 January 2006.

2. Balboa, M. W. (1973). United Nations Conference on the Human Environment. *Women Law J, 59*, 26.

3. Boone, C., & Modarres, A. (2006). *City and Environment*. Temple University.

4. Brown, L. R. (2011). *World on the Edge: How to Prevent Environmental and Economic Collapse*. New York: Earth Policy Institute.

5. Carson, R. (1951). *The Sea Around Us*. New York: Oxford University.

6. Carson, R. (1962). *Silent Spring*. New York: First Mariner.

7. Clark, W. C. (1999). Sustainability science: a room of its own. *PNAS, 104(6)*, 1737–1738.

8. Commoner, B. (1963). *Science and Survival*. New York: Viking.

9. Commoner, B. (1971). *The Closing Circle: Confronting the Environmental Crisis*. London: Cape.

10. Commoner, B. (1975). How Poverty Breeds Overpopulation (and Not the Other Way Around). *Ramparts, 13*(10), 21–24, 59.

11. Cronon, W. (2011). *The Riddle of Sustainability: A Surprisingly Short History*. Lecture presented at Wrigley Lecture Series. Arizona State University, Tempe, AZ.

12. Diamond, J. (2006). *Collapse: How Societies Choose to Fail or Succeed*. New York: Penguin Group.

13. Du Pisani, J. A. (2006). Sustainable development: Historical roots of the concept. *Environmental Science, 3*(2), 83–96.

14. Egan, M. (2007). *Barry Commoner and the Science of Survival: The Remaking of American Environmentalism*. Cambridge: M.I.T.

15. Ehrlich, P. (1970). The Population Bomb. *New York Times, 47*.

16. Ehrlich, P. (1968). *The Population Bomb*. New York: Ballantine.

17. Grober, U. (2012). *Sustainability: A Cultural History*. (Translated by Ray Cunningham). UIT Cambridge Limited.

18. Guthrie, W., & Guthrie, W. (1988). *Dust Bowl Ballads*. Rounder.

19. Hardin, G. (1998). Extensions of "the tragedy of the commons." *Science, 280*(5364), 682–683.

20. Hay, P. R. (2000). *Main Currents in Western Environmental Thought*. Bloomington: Indiana University.

21. *Hempel, L. C. (2012).* Evolving concepts of sustainability in environmental policy. *The Oxford Handbook of US Environmental Policy, 67.*

22. Hopwood, B., Mellor, M., & O'Brien, G. (2005). Sustainable development: Mapping different approaches. *Sustainable Development, 13*(1), 38–52.

23. The Intergovernmental Panel on Climate Change (IPCC). (2017). Accessed 8/4/2017 at http://www.ipcc.ch/.

24. International Union for Conservation of Nature, Natural Resources, & *World Wildlife Fund. (1980). World Conservation Strategy: Living Resource Conservation for Sustainable Development.* Gland, Switzerland: IUCN.

25. Kates, R. W. (2011). What kind of a science is sustainability science? *Proceedings of the National Academy of Sciences, 108*(49), 19449–19450.

26. Kates, R. W., & Clark, W. C. (1999). *Our Common Journey: A Transition Toward Sustainability.* Washington: National Academy.

27. Kates, R. W., & Parris, T. M. (2003). Long-term trends and a sustainability transition. *Proceedings on the National Academy of Sciences, 100*(14), 8062–8067.

28. Leopold, A. (1949). *A Sand County Almanac: And Sketches Here and There.* Oxford: Oxford University.

29. Malthus, T. R. (1898). *An Essay on the Principle of Population or a View of Its Past and Present Effects on Human Happiness, an Inquiry Into Our Prospects Respecting the Future Removal or Mitigation of the Evils which it Occasions by Rev. TR Malthus.* Reeves and Turner.

30. Meadows, D., Randers, J., & Meadows, D. (2004). *Limits to Growth: The 30-Year Update.* Chelsea Green.

31. Meadows, D. H., Meadows, D. L., & Randers, J. (1972). *Beyond the Limits: Global-Collapse or a Sustainable Future.* Earthscan.

32. Meadows, D. H., Meadows, D. L., Randers, J., & Behrens III, W. W. (1972). *The Limits to Growth: A Report for the Club of Rome's Project on the Predicament of Mankind.* Nova York: New American Library.

33. Miller, C. (2009). The once and future forest service: Land-management policies and politics in contemporary America. *Journal of Policy History, 21*(1), 89–104.

34. *Minteer, B. A. (2006). The Landscape of Reform: Civic Pragmatism and Environmental Thought in America. MIT Press.*

35. Minteer, B. A. (2011). *Refounding Environmental Ethics: Pragmatism, Principle and Practice.* Philadelphia: Temple University.

36. Naess, A., & Sessions, G. (1986). Basic principles of deep ecology. *Trumpeter, 3*(4).

37. Naess, A. (1973). The shallow and the deep, long-range ecology movement. A summary. *Inquiry, 16*(1–4), 95–100.

38. Odum, E. F. (1971). *The Fundamentals of Ecology,* (3rd ed). Philadelphia: Suanders.

39. Orr, D. W. (2002). Four challenges of sustainability. *Conservation Biology, 16*(6), 1457–1460.

40. Osborn, F. (1948). *Our plundered planet.* Boston: Little, Brown.

41. Oxford English Dictionary Online (OED). (2014). Oxford University Press. Accessed 4/6/2014 at www.oed.com.

42. Radkau, J. (2009). *Nature and Power. A Global History of the Environment.* New York: Cambridge University.

43. Redman, C. L., Grove, J. M., & Kuby, L. H. (2002). Integrating social science into the long-term ecological research (LTER) network: social dimensions of ecological change and ecological dimensions of social change. *Ecosystems, 7*(2), 161–171.

44. Rees, W. E., & Wackernagel, M. (1996). Urban ecological footprints: Why cities cannot be sustainable and why they are a key to sustainability. *Environmental Impact Assessment Review, 16*(4), 223–248.

45. Rees, W. E. (1992). Ecological footprints and appropriated carrying capacity: what urban economics leaves out. *Environment and Urbanization, 4*(2), 121–130.

46. Rittel, H. W., & Webber, M. M. (1974). Planning problems are wicked. *Polity, 4,* 155–169.

47. Sale, K. (1993). *The Green Revolution: The Environmental Movement 1962–1992.* New York: Macmillan.

48. Steinbeck, J. (1939). *The Grapes of Wrath.* Penguin.

49. Summit, E. (1992). Agenda 21. *The United Nations Programme for Action from Rio.*

50. Thoreau, H. D. (1949). *On the Duty of Civil Disobedience* [1849, original title: *Resistance to Civil Government*]. Rockville: Manor.

51. Thoreau, H. D. (2006). *Walden.* Yale University Press. [Originally published in New York: Tinker & Co., 1854.]

52. Von Carlowitz, H. C. (1732). *Sylvicultura Oeconomica Oder Haußwirthliche Nachricht und Naturmäßige Anweisung zur Wilden Baum-Zucht Nebst Gründlicher Darstellung Wie . . . dem allenthalben und insgemein einreissenden Grossen Holtz-Mangel, Vermittelst Säe-Pflantz-und Versetzung vielerhand Bäume zu rathen . . . Worbey zugleich eine gründliche Nachricht von dem in Churfl. Sächß. Landen Gefundenen Turff . . . befindlich* (Vol. 1). Bey Johann Friedrich Brauns sel. Erben.

53. Williams, R. (1976). *Keywords: A Vocabulary of Culture and Society.* London: Fontana.

54. Wilson, E. O. (1998). *Consilience: The Unity of Knowledge.* New York: Vintage.

55. Wilson, E. O. (2002). *The Future of Life.* London: W.W. Norton & Co.

56. World Commission on Environment and Development (WCED). (1987). *Our Common Future: Report of the World Commission on Environment and Development.* The Brundt land Commission. Oxford: Oxford University.

57. Worster, D. (1979). *Dust Bowl: The Southern Plains in the 1930s.* Oxford, UK: Oxford University.

58. Worster, D. (1985). *Rivers of Empire: Water, Aridity, and the Growth of the American West.* Oxford: Oxford University.

59. Worster, D. (1994). *Nature's Economy: A History of Ecological Ideas.* Cambridge: Cambridge University.

Chapter 3

1. Diamond, J. 2005. *Collapse: How societies choose to fail or succeed.* New York, NY: Penguin Group.

Chapter 4

1. Ahmadi, F. & Toghyani, S. (2011). The role of urban planning in achieving sustainable urban development. *OIDA International Journal of Sustainable Development, 2*(11), 23–25.

2. Arcadis. (2015). *Improving Quality of Life: Annual Report 2015.* Netherlands. https://www.arcadis.com/media/E/F/4/%7BEF45D8F7-4203-4B22-A343-0311C621061E%7DArcadis%20NV%20Annual%20Report%202015.pdf (accessed 8.3.2017).

3. Arcadis. (2016) *Sustainable Cities Index.* https://www.arcadis.com/en/global/our-perspectives/sustainable-cities-index-2016/ (accessed 8.4.2017)

4. Beatley, T. (2010). *Biophilic Cities: Integrating Nature into Urban design and Planning.* Washington: Island Press.

5. Berggren, C. (2013). *Stockholm, Sweden: An Urban Planner's Dream City.* Accessed 8/12/2017 at http://www.smartcitiesdive.com/ex/sustainablecities collective/stockholmsweden-urban-planners-dream-city/179351/

6. Bündnis 90/Die Grünen. (2014). Accessed 15/8/2014 at http://www.gruene.de/startseite.html.

7. C40. (2014). C40 Cities: Climate Leadership Group. Accessed 15/8/2014 at http://www.c40.org/.

8. Caine, T. (2011). *The Garden City vs the Green City*. Sustainable Cities Collective. Accessed 15/8/2014 at http://sustainablecitiescollective.com/tcaine/21809/garden-city-vs-green-city.

9. CH2M HILL Canada Limited. (2016). *2015 Transportation Panel Survey prepared for City of Vancouver*. Accessed 11/10/2016 at http://vancouver.ca/files/cov/Transportation-panel-survey-2015-final-report.pdf.

10. City Climate Leadership Awards. (2014). *Stockholm: Hammarby Sjöstad*. Accessed 8/12/2014 at http://cityclimateleadershipawards.com/stockholm-hammarby-sjostad/.

11. City of Freiburg. (2014). Accessed 15/8/2014 at http://www.freiburg.de/pb/,Lde/205243.html.

12. City of Stockholm. (2010). *The Walkable City: Stockholm City Plan*. Stockholm City. Accessed 11/11/2016 at http://international.stockholm.se/globalassets/ovriga-bilder-och-filer/the-walkable-city---stockholm-city-plan.pdf.

13. City of Stockholm. (2011). *The Walkable City: Stockholm City Plan*. Accessed 25/1/2012 at http://international.stockholm.se/Future-Stockholm-Stockholm-City-Plan/.

14. City of Stockholm. (2012). *Urban Mobility Strategy*. Stockholm: City of Stockholm, the City of Stock-holm Traffic Administration.

15. City of Stockholm. (2014). *Welcome to the Capital of Scandinavia*. Accessed 8/12/2014 at http://international.stockholm.se/.

16. City of Vancouver. (2011). *Downtown Separated Bicycle Lanes Status Report*. Accessed 15/8/2014 at http://bettercities.net/sites/default/files/penv2DowntownSeparatedBik-eLanesStatusReportSummer2011.pdf.

17. City of Vancouver. (2012). *Greenest City 2020 Action Plan*. Accessed 15/8/2014 at http://vancouver.ca/files/cov/Greenest-city-action-plan.pdf.

18. City of Vancouver. (2014). Accessed 15/8/2014 at http://vancouver.ca/.

19. Demographia. (2014). Accessed 15/8/2014 at http://www.demographia.com/.

20. Economist Intelligence Unit (EIU). (2012a). *A Summary of the Livability Ranking and Overview*. Accessed 15/8/2014 at http://www.tfsa.ca/storage/reports/Liveability_rankings_Promotional_August_2013.pdf.

21. Economist Intelligence Unit. (2012b). *The Green City Index: A summary of the Green City Index Research Series*. Munich: Siemens AG.

22. Floater, Graham and Rode, Philipp and Zenghelis, Dimitri. (LSE 2013). Stockholm: green economy leader report. Economics of Green Cities Programme, LSE Cities, London School of Economics and Political Science, London, UK

23. Fourteenislands.com. (2014). *Bicycle Paths*. Accessed 8/12/2014 at http://www.fourteenislands.com/bicycle-paths/.

24. Future Communities. (2014). *Hammarby Sjostad, Stockholm, Sweden, 1995 to 2015: Building a 'Green' City Extension*. Accessed 8/12/2014 at http://www.futurecommunities.net/case-studies/hammarby-sjostad-stockholm-sweden-1995-2015.

25. FWTM - Freiburg Economic Tourism and Fair. (2017), http://fwtm.freiburg.de/pb/,Lde/451264.html (accessed 8.4.2017)

26. Green City Freiburg. (2014). Freiburg Green City Brochure.

27. Grosvenor. (2014). *Resilient Cities: A Grosvenor Research Report*. Accessed 15/8/2012 at www.grosvenor.com.

28. Howard, E. (1902). *Garden Cities of Tomorrow*. Harvard University: S. Sonneschein & Company, Limited. Accessed 15/8/2014 at http://www.fwtm.freiburg.de/servlet/PB/menu/1182949_l2/index.html.

29. Ignatieva, M. E. & Berg, P. (2014). *Hammarby Sjöstad—A New Generation of Sustainable Urban Eco-Districts*. Accessed 8/12/2014 at http://www.thenatureofcities.com/2014/02/12/hammarby-sjostad-a-new-generation-of-sustainable-urban-eco-districts/.

30. Nelson, A. (2014). *Stockholm, Sweden: City of Water*. Accessed 8/12/2014 at http://depts.washington.edu/open2100/Resources/1_OpenSpaceSystems/Open_Space_Systems/Stockholm_Case_Study.pdf.

31. Newman, P., Beatley, T., & Heather, B. (2009). *Resilient Cities: Responding to Peak Oil and Climate Change*. Washington: Islandpress.

32. OECD. (2014). *Environmental Performance Reviews: Sweden Highlights 2014*. Accessed 11/11/2016 at https://www.oecd.org/environment/country-reviews/Sweden%20Highlights%20web%20pages2.pdf.

33. Organization for Economic Cooperation and Development (OECD). (2013). *Green Growth in Stockholm, Sweden, OECD Green Growth Studies, OECD Publishing*. Accessed 8/12/2014 at http://dx.doi.org/10.1787/9789264195158-en.

34. Parsons, K. J., & Schuyler, D. (2002). *From Garden City to Green City: The Legacy of Ebenezer Howard*. Baltimore: John Hopkins University Press.

35. Port Metro Vancouver website. (2014). Accessed at 15/8/2014 at http://www.portmetrovancouver.com/.

36. Statistics Canada, Census. (2011). Accessed 15/8/2014 at http://www12.statcan.gc.ca/census-recensement/2011/dp-pd/hlt-fst/pd-pl/Table-Tableau.cfm?LANG=Eng&T=303&SR=1&S=51&O=A&RPP=9999&PR=0&CMA=933.

37. Statistikomstockholm. (2013). *Stockholm Gacts & Figures 2013*. Accessed 8/12/2014 at http://www.statistikomstockholm.se/attachments/article/21/facts%20and%20figures%202013_webb.pdf.

38. The EcoTipping Points Project. (2011). *Germany – Freiburg – Green City*. Accessed 15/8/2014 at http://www.ecotippingpoints.org/our-stories/indepth/germany-freiburg-sustainability-transportation-energy-green-economy.html.

39. Transport Styrelesen. (2016). *Congestion Taxes in Stockholm and Gothenburg*. Accessed 11/11/2016 at https://www.transportstyrelsen.se/en/road/Congestion-taxes-in-Stockholm-and-Goteborg/.

40. TreeKeepers. (2012). Accessed 15/8/2014 at http://www.treekeepers.ca/.

41. United Nations (UN) Habitat. (2012). *State of the World's Cities 2012/2013*. Accessed 15/8/2014 at http://mirror.unhabitat.org/pmss/listItemDetails.aspx?publicationID=3387.

42. Vauban. (2014). www.vauban.de: Stadtteil Vauban, Freiburg. Accessed 15/8/2014 at http://www.vauban.de.

43. Vince, G. (2012). *Sustainability in the New Urban Age*. Smart Planet. Accessed 15/8/2014 at http://www.bbc.com/future/story/20121214-greening-the-concrete-jungle.

44. Winters, M., Babul, S., Becker, H. J., Brubacher, J. R., Chipman, M., Cripton, P., . . . & Teschke, K. (2012). Safe cycling: how do risk perceptions compare with observed risk? *Canadian Journal of Public Health, 103*(9 Suppl 3):eS42–eS47.

45. Zottis, L. (2014). *Cidades Planejam-se Para Mudanças Climáticas, Aponta MIT*. Accessed 15/8/2014 at http://thecityfixbrasil.com/2014/06/10/cidades-buscam-resiliencia-mit/.

Chapter 5

1. Allsopp, P. 2015. Interview by author. Scottsdale, AZ. March 12.

2. Allsopp, P. 2016. *Did it work according to plan?* London: Royal Institute of British Architects.

3. American Society of Civil Engineering. 2016. http://www.infrastructurereportcard.org/ (accessed 12.15.2016).

4. Barber, D. 2016. *A House in the Sun*. New York: Oxford University Press: 43.

5. BREEAM. 2017. http://www.breeam.com/ (accessed 2.28.2017).

6. British Council. 2016. https://www.britishcouncil.org/cubed/urban-mind-app (accessed 12.20.2016).

7. Center for Design Excellence. 2016. Accessed 7/28/2017 at http://www.urbandesign.org/home.html.

8. Cervero, R. 1988. Land-use mixing and suburban mobility. *Transportation Quarterly* 42, 3: 429-446.

9. City of Burlingame. 2015. https://www.burlingame.org/index.aspx?page=3498 (accessed 1.11.2017).

10. City of Phoenix. 2012. https://www.phoenix.gov/streetssite/Documents/104445.pdf (accessed 1.11.2017).

11. Ibid.

12. Connors, P. and McDonald, P., 2010. Transitioning communities: community, participation and the transition town movement. *Community development journal*, 46(4), pp.558-572.

13. Cool California. 2016. http://www.coolcalifornia.org/cool-pave-how (accessed 1.19.2017).

14. Coyle, S. 2011. *Sustainable and Resilient Communities*. Hoboken, N.J.: John Wiley & Sons, Inc.: 174.

15. Curtis, 206.

16. Curtis, 54.

17. Curtis, W. 2014. *The Last Great Walk*. New York: Rodale: 89.

18. Form-Based Codes Institute. 2017. http://formbasedcodes.org/definition/ (accessed 1.20.2017).

19. Frank, 180.

20. Frank, L.D. et al. 2003. *Health and Community Design*. Washington: Island Press: 150.

21. Frederick, C.P. 2017. *America's Addition to Automobiles*. Westport, Connecticut: Praeger Publishers.

22. Gaffin, S.R. et al. 2012. http://iopscience.iop.org/article/10.1088/1748-9326/7/1/014029/pdf (accessed 1.20.2017).

23. Glaeser, E. 2011. *Triumph of the City*. New York: The Penguin Press: 223.

24. GreenCity Freiburg. 2014. http://www.freiburg.de/pb/site/Freiburg/get/params_E-680488302/640888/GreenCity_E2017.pdf, (accessed 1.25.2017).

25. Greenroof Projects Database. 2015. http://www.greenroofs.com/projects/pview.php?id=1457 (accessed 1.12.2017).

26. Heffernan, S. 2013. The Ultimate Guide to Living Green Walls. http://www.ambius.com/blog/ultimate-guide-to-living-green-walls/ (accessed 1.12. 2017).

27. Howard, E. 1898. *Tomorrow: A Peaceful Path to Real Reform*. London: Swan Sonnenschein. https://archive.org/details/tomorrowpeaceful00howa (accessed 1.5.2017).

28. Leff, D. 2004. *The Last Undiscovered Place*. Charlottesville: University of Virginia Press.

29. Litman, T. 2017. Evaluating Public Transit Benefits and Costs. British Columbia: Victoria Transport Policy Institute. http://www.vtpi.org/tranben.pdf (accessed 1.11. 2017).

30. McPherson, G. et al. 2005. Municipal Forest Benefits and Costs in Five US Cities. *Journal of Forestry* (December): 411-416.

31. Montgomery, C. 2014. *Happy City*. New York: Farrar, Straus and Giroux.

32. National Aeronautics and Space Administration. 2017. https://www.nasa.gov/press-release/nasa-noaa-data-show-2016-warmest-year-on-record-globally (accessed 1.20.2017).

33. National Institute of Building Science. 2016. http://www.wbdg.org/resources/green-building-standards-and-certification-systems (accessed 2.26.2017).
34. Ibid.
35. National Park Service. 2017. https://www.nps.gov/TPS/how-to-preserve/briefs.htm (accessed 1.23.2017).
36. Newton, N.T. 1971. *Design on the Land.* Cambridge: Harvard University Press.
37. New York City Department of Transportation. 2012. http://www.nyc.gov/html/dot/downloads/pdf/2012-10-measuring-the-street.pdf (accessed 2.28.2017).
38. Ibid.
39. Partnership to Fight Chronic Disease. 2015. http://www.fightchronicdisease.org/latest-news/130-million-americans-chronic-disease-cost-more-25-trillion-annually (accessed 12.18.2016).
40. Pollio, V., 1914. *Vitruvius: The ten books on architecture.* Harvard university press.
41. Radford, W. 1908. *Radford's Artistic Home.* Chicago: The Radford Architectural Company: 4.
42. Roseland, M. 2012. *Toward Sustainable Communities: Solutions for Citizens and Their Governments.* Vol. 6. New Society Publishers, 2012: 287.
43. Royal Society for the Arts. 2017. www.thersa.org (accessed 2.28.2017).
44. Russell, J. 1958. Late Ancient and Medieval Population. *Transactions of the American Philosophical Society, 48*(3): 65-83. doi:10.2307/1005708 (accessed 12.8.2016).
45. Santa Monica Farmers Markets. 2017. https://www.smgov.net/portals/farmersmarket/ (accessed 1.23.2017).
46. Smart Growth America. 2016. https://smartgrowthamerica.org/program/national-complete-streets-coalition/what-are-complete-streets/ (accessed 12.22.2016).
47. Ibid.
48. Smart Growth America. 2016. www.smartgrowthamerica.org/documents/cs/factsheets/cs-safety.pdf, page 1 (accessed 12.16.2016).
49. Smart Growth America. 2016. https://smartgrowthamerica.org/ (accessed 1.20.2017).
50. Speck, J., 2013. *Walkable city: How downtown can save America, one step at a time.* Macmillan.
51. Ibid.
52. Sustainable Santa Monica. 2012. https://www.smgov.net/uploadedFiles/Departments/OSE/Categories/Sustainability/Sustainable_City_Report_Card_2012.pdf (accessed 1.23.2017).
53. US Department of Energy (DOE) 2017. *Cool Roofs.* https://energy.gov/energysaver/cool-roofs Accessed 4.13.2017.
54. US Environmental Protection Agency. 2016. https://www.epa.gov/heat-islands/using-trees-and-vegetation-reduce-heat-islands (accessed 12.21.2016).
55. USDA Forest Service. 2017. *i-Tree.* www.itreetools.org.
56. United Nations. 2014. http://www.un.org/en/development/desa/news/population/world-urbanization-prospects-2014.html (accessed 1.25.2017).
57. United Nations Conference. 2016. http://wcr.unhabitat.org/wp-content/uploads/sites/16/2016/05/Chapter-2-WCR-2016.pdf (accessed 12.8.2016).
58. Vitalyst Health Foundation. 2016. http://vitalysthealth.org/healthy-community-design/ (accessed 12.18.2016) and Robert Wood Johnson Foundation. 2016. http://www.rwjf.org/en/our-focus-areas/topics/built-environment-and-health.html (accessed 12.18.2016).
59. Walk Score. 2016. https://www.walkscore.com/methodology.shtml (accessed 12.21.2016).
60. Wolf, K. 2017. Urban Forestry/Urban Greening Research. 2010. University of Washington, Seattle. http://depts.washington.edu/hhwb/Thm_SafeStreets.html (accessed 12.22.2016).

Chapter 6

1. Asgary, A., Klosterman, R., & Razani, A. (2007). Sustainable urban growth management using what-if? *International Journal of Environmental Research*, 1(3), 218–230.

2. Campbell, S. (1996). Green cities, growing cities, just cities? Urban planning and the contradictions of sustainable development. *Journal of the American Planning Association*, 62(3), 296–312.

3. Crecine, J. P. (1964). *TOMM (Time Oriented Metropolitan Model), Technical Bulletin Number 6*. Pittsburgh: Community Renewal Program.

4. Crecine, J. P. (1968). *A Dynamic Model of Urban Structure*. Santa Monica: Rand Corporation.

5. Feldt, A. G. (1966). *The Cornell Land Use Game*. Ithaca: Center for Housing and Environmental Studies, Cornell University, p. 4.

6. Goldner, W. (1968). *Projective Land Use Model (PLUM): A Model for the Spatial Allocation of Activities and Land Uses in a Metropolitan Region, BATSC Technical Report 219*. Berkeley: Bay Area Transportation Study Commission.

7. Goldner, W. (1971). The Lowry model heritage. *Journal of the American Institute of Planners*, 37(2), 100–110.

8. Goldner, W., & Graybeal R.S. (1965). *The Bay Area Simulation Study: Pilot Model of Santa Clara County and Some Applications*. Berkeley: Center for Real Estate and Urban Economics, University of California.

9. Guhathakurta, S. (1999). Urban modeling and contemporary planning theory: is there a common ground? *Journal of Planning Education and Research*, 18(4), 281.

10. Haase, D., Schwarz, N., & others. (2009). Simulation models on human-nature interactions in urban landscapes: a review including spatial economics, system dynamics, cellular automata and agent-based approaches. *Living Reviews in Landscape Research*, 3(2), 1–45.

11. Joshi, H., Guhathakurta, S., Konjevod, G., Crittenden, J., & Ke Li. (2006). Simulating the effect of light rail on urban growth in Phoenix: An application of the urbansim modeling environment. *Journal of Urban Technology*, 13(2), 91–111.

12. Klosterman, R. E. (2008). A new tool for a new planning: The what if?™ planning support system. In R.K. Brail, (Ed.),*Planning Support Systems for Cities and Regions*. Cambridge: Lincoln Institute of Land Policy, pp. 85–99.

13. Landis, J. D. (1994). The California urban futures model: a new generation of metropolitan simulation models. *Environment and Planning B*, 21, 399–399.

14. Landis, J. D. (1995). Imagining land use futures: applying the California urban futures model. *Journal of the American Planning Association*, 61(4), 438–457.

15. Landis, J., & Zhang, M. (1998). The second generation of the California urban futures model. Part 2: Specification and calibration results of the land-use change submodel. *Environment and Planning B*, 25, 795–824.

16. Lowry, I. S. (1964). *A Model of Metropolis*. Santa Monica: Rand Corporation.

17. Lozano, R. (2008). Envisioning sustainability three-dimensionally. *Journal of Cleaner Production*, 16(17), 1838–1846.

18. McHarg, I. L. (1969). *Design with Nature*. Published for the American Museum of Natural History [by] the Natural History Press. Accessed at http://student.agsci.colostate.edu/rdfiscus/Design%20with%20Nature.pdf.

19. McHarg, I. L., & Steiner, F. R. (1998). *To Heal the Earth: Selected Writings of Ian L. McHarg*. Island Press. Accessed at http://books.google.com.ezproxy1.lib.asu.edu/books?hl=en&lr=&id=7B0EysYcBs0C&oi=fnd&pg=PR9&dq=McHarg+and+stein-er+1998&ots=jyTT0E4SLI&sig=oIxgusieQoMdk62U8GEJtL7PITc.

20. Poveda, C. A. (2017). The theory of dimensional balance of needs. *International Journal of Sustainable Development & World Ecology*, 24(2), 97–119.

21. Putnam, S. H., & Shih-Liang, C. (2001). The METROPILUS planning support system: urban models and GIS. In R. K. Brail and R. E. Klosterman, (Eds) *Planning Support Systems*, Redland: ESRI Press, pp. 99–128.

22. Putnam, S. H. (1983). *Integrated Urban Models: Policy Analysis of Transportation and Land Use*. London: Pion.

23. Santé, I., García, A. M., Miranda, D., & Crecente, R. (2010). Cellular automata models for the simulation of real-world urban processes: A review and analysis. *Landscape and Urban Planning*, 96(2), 108–122. doi:10.1016/j.landurbplan.2010.03.001.

24. Waddell, P. (2000). A behavioral simulation model for metropolitan policy analysis and planning: residential location and housing market components of UrbanSim. *Environment and Planning B*, 27(2), 247–264.

25. Waddell, P. (2002). UrbanSim: Modeling urban development for land use, transportation, and environmental planning. *Journal of the American Planning Association*, 68(3). Accessed at http://lab.geog.ntu.edu.tw/lab/errml/%E5%A4%A7%E5%B0%88%E9%A1%8C/03021%E5%8E%9F%E6%96%87.pdf.

26. Wegener, M. (1994). Operational urban models state of the art. *Journal of the American Planning Association*, 60(1), 17–29.

27. Wegener, M. (2004). Overview of land use transport models. *Handbook of Transport Geography and Spatial Systems*, 5, 127–146.

Chapter 7

1. 100 Resilient Cities. (2016). *100 Resilient Cities: City Resilience*. Accessed 29/12/2016 at http://www.100resilientcities.org/resilience#/-_/.

2. Adger, W. N. (2000). Social and ecological resilience: Are they related? *Progress in Human Geography*, 24(3), 347–364.

3. Assembly, U. G. (2015). *Transforming our world: the 2030 Agenda for Sustainable Development*. New York: United Nations.

4. Barkham, R., Brown, K., Parpa, C., Breen, C., Carver, S., & Hooton, C., (2014). *Resilient Cities-A Grosvenor Research Report*. Grosvenor. Accessed 29/12/2016 at http://www.grosvenor.com/getattachment/194bb2f9-d778-4701-a0ed-5cb451044ab1/ResilientCitiesResearchReport.pdf.

5. Brand, R., & Graffikin, F. (2007). Collaborative planning in an uncollaborative world. *Planning Theory*, 6(3), 282–313.

6. Cutter et al. (2003). Social vulnerability to environmental hazards. *Social Science Quarterly*, 84(1), 242–261.

7. Davidson, J., Jacobson C., Lyth A., Dederkorkut-Howes A., Baldwin C., Ellison J., . . . & Smith, T. (2016). Interrogating Resilience: toward a typology to improve its operationalization. *Ecology and Society*, 21(2), 27–41.

8. Davoudi, S., Shaw K., Haider L. J., Quinlan, A., Peterson, G. Wilkinson, C., & Davoudi, S. (2012). Resilience: a bridging concept or a dead end? 'reframing' resilience: challenges for planning theory and practice integrating traps: resilience assessment of a pasture management systems in Northern Afghanistan Urban Resilience: what does it mean in planning practice? resilience as a useful concept for climate change adaptation? the politics of resilience for planning: a cautionary note. *Planning Theory and Practice*, 13(2), 299–333.

9. Da Silva, J., & Morera, B. (2014). *City Resilience Framework*. Arup & Rockefeller Foundation. Accessed 1/2/2017 at http://publications.arup.com/publications/c/city_resilience_index.

10. Detroit Future City. (2016). *About DFC Implementation Office*. Accessed 10/12/2016 at https://detroitfuturecity.com/about/.

11. Enelow, N. (2013). *The Resilience of Detroit: An Application of the Adaptive Cycle Metaphor to an American Metropolis*. Economics for Equity and the Environment Network.

12. Folke, C. (2006). Resilience: The emergence of a perspective for social–ecological systems analyses. *Global Environmental Change, 16*(3), 253–267.

13. Gallopín, G. C. (2006). Linkages between vulnerability, resilience, and adaptive capacity. *Global Environmental Change, 16*(3), 293–303.

14. Gencer, E. A. (2013). *The Interplay between Urban Development, Vulnerability, and Risk Management: A Case Study of the Istanbul Metropolitan Area* (Vol. 7). Springer Science & Business Media.

15. Hall, P. & Lamont, M. (2013). Introduction. In P. Hall and H. Lamont (Eds) *Social Resilience in the Neoliberal Era*. P. 1–34. New York: Cambridge Press.

16. Holling, C. S. (1973). Resilience and stability of ecological systems. *Annual Review of Ecology and Systematics, 4*, 1–23.

17. Holling, C. S. (1996). Engineering Resilience versus Ecological Resilience. In P.C. Schulze (Ed) *Engineering within Ecological Constraints*. P31-44 Washington: National Academy Press.

18. Jabareem, Y. R. (2006). Sustainable Urban Forms: Their typologies, models, and concepts. *Journal of Planning Education and Research, 26*, 38–52.

19. Local First, 2014. localfirrst.com.

20. Magis, K. (2010). Community resilience: an indicator of social sustainability. *Society and Natural Resources, 23*, 401–406.

21. Maguire, B., & Cartwright, S. (2008). *Assessing a Community's Capacity to Manage Change: A Resilience Approach to Social Assessment*.

22. Margerum, R. (2011). *Beyond Consensus: Improving Collaborative Planning and Management*. Cambridge: MIT Press.

23. McAslan. (2015). Assessing Urban Sustainability: Using indicators to measure progress. In D. Pijawka (Ed.). *Sustainability for the 21st Century: Pathways, programs, and policies*. Dubuque: Kendal Hunt. p. 235–258.

24. Newman, P., Beatley, T., & Boyer, H. (2009). *Resilient Cities: Responding to Peak Oil and Climate Change*. Washington: Island Press.

25. Norris, F., & Stevens S. (2007). Community resilience and the principles of mass trauma intervention. *Psychiatry, 70*(4), 320–328.

26. Olsson, P., Folke, C. & Berkes, F. (2004). Adaptive co-management for building resilience in socio-ecological systems. *Environmental Management, 34*, 75–90.

27. Pelling, M. (2003). *The Vulnerability of Cities. Natural Disasters and Social Resilience*. London, Sterling, VA: Earthscan Publications.

28. Putnam, R. (2001). *Bowling Alone: The Collapse and Revival of American Community*. New York: Simon and Schuster.

29. Quay, R. (2010). Anticipatory Governance. *Journal of the American Planning Association. 76*(4), 496–511.

30. Resilience Alliance. (2007a). *Assessing Resilience in Social-Ecological Systems: A Workbook for Scientists*.

31. Resilience Alliance. (2007b). *Urban Resilience: Research Prospectus*. A Resilience Alliance Initiative for Transitioning Urban Systems towards Sustainable Futures.

32. Resilience Alliance. (2014). *The Adaptive Cycle*. Accessed at *resalliance.org/index. php/adaptive_cycle*.

33. Rockefeller Foundation, The & ARUP. (2015). *The City Resilience Framework: November 2015*. Accessed 1/2/2107 at http://lghttp.60358.nexcesscdn.net/8046264

/images/page/-/100rc/Blue%20City%20Resilience%20Framework%20Full%20 Context%20v1_5.pdf.

34. Sherrieb, K., Norris, F., & Galea, S. (2010). Measuring Capacities for Community Resilience. *Social Indicators Research, 99*, 227–247.

35. Smith, S., Morris, W., & Voorhees, R. W. (Eds). (1998). *New International Webster's Comprehensive Dictionary of the English language*, Trident Press International.

36. Turner, B. L., Kasperson R. E., Matson P. A., McCarthy J. J., Corell R. W., Christensen L., . . . & Polsky, C. (2003). A framework for vulnerability analysis in sustainability science. *Proceedings of the National Academy of Sciences, 100*(14), 8074–8079.

37. Wamsler, C., 2014. *Cities, Disaster Risk, and Adaptation*. New York: Routledge.

38. Wilson, G. (2012). *Community Resilience and Environment Transitions*. London: Routledge.

Chapter 8

1. Archer, D., & Rahmstorf, S. (2010). *The Climate Crisis: An Introductory Guide to Climate Change*. Cambridge: Cambridge University Press.

2. BREEAM-USA. (2016). Accessed 11/4/2106 at http://www.breeamusa.com/.

3. Calthorpe, P. (2011). The Urban Network. In *Urbanism in the Age of Climate Change*. Washington: Island Press/Center for Resource Economics.

4. Center for Transit-Oriented Development. (2014). *Trends in Transit-Oriented Development 2000–2010*. US Department of Transportation Federal Transit Administration. FTA Report No. 0050. Accessed 11/4/2016 at https://www.transit .dot.gov/sites/fta.dot.gov/files/FTA_Report_No._0050_0.pdf.

5. City of Chicago. (2008). *Chicago Climate Action Plan: Our City*. Our Future. Accessed 20/5/2014 at http://www.chicagoclimateaction.org.

6. City of Portland. (2001). *Local Action Plan Global Warming*. Portland: Office of Sustainable Development & Department of Sustainable Community Development.

7. CNN. (2013). *Hurricane Sandy Fast Facts*. Accessed 6/5/2014 at http://www.cnn. com/2013/07/13/world/americas/hurricane-sandy-fast-facts/.

8. CNU. (2014). *Charter of the New Urbanism. Congress for the New Urbanism Online*. Accessed 11/3/2016 at https://www.cnu.org/who-we-are/charter-new-urbanism.

9. Copenhagenize Design Company. (2015). *The Copenhagenize Index 2015*. Accessed 11/4/2016 at http://copenhagenize.eu/index/index.html.

10. Crimmins, A., Balbus J., Gamble J. L., Beard C. B., Bell J. E., Dodgen, R. J. . . . & Ziska L., (Eds). *U.S. Global Change Research Program*, Washington, p. 312. Accessed at http://dx.doi.org/10.7930/J0R49NQX.

11. Denmark.dk. (2016). *Cycle Superhighway*. Accessed 11/4/2016 at http://denmark.dk /en/green-living/bicycle-culture/cycle-super-highway/.

12. EPA. (2016a). *Sources of Greenhouse Gas Emissions: Industry Sector Emissions*. Accessed 11/4/2016 at https://www.epa.gov/ghgemissions/sources-greenhouse -gas-emissions#industry.

13. EPA. (2016b). *Fast Facts: U.S. Transportation Sector Greenhouse Gas Emissions 1990–2014*. United States Environmental Protection Agency. Accessed 11/3/2016 at https://nepis.epa.gov/Exe/ZyPDF.cgi?Dockey=P100ONBL.pdf.

14. Food and Agriculture Organization of the United Nations (FAO). (2008). *Climate Change and Food Security: A Framework Document*. Rome, Italy: Food and Agriculture Organization of the United Nations.

15. Fuerth, L. S. (2009). Foresight and anticipatory governance. *Foresight, 11*(4), 14–32.

16. Global Governance Project (2012). Global Environmental Governance Reconsidered. MIT Press.

17. Global Food Security Index. (2016). *Ranking and Trends. New York: The Economist Intelligence Unit.* Accessed 31/10/2016 at http://foodsecurityindex.eiu.com/Index.

18. Intergovernmental Panel on Climate Change (IPCC). (2007). *Climate Change 2007: The Physical Science Basis.* Contribution of Working Group 1 to the Fourth Assessment Report of the Intergovernmental Panel on Climate Change. Cambridge, UK and New York, NY, USA: Cambridge University Press.

19. Intergovernmental Panel on Climate Change (IPCC). (2013). *Climate change 2013: The Physical Science Basis.* Contribution of Working Group 1 to the fifth Assessment Report of the Intergovernmental Panel on Climate Change. Cambridge, UK and New York, NY, USA: Cambridge University Press.

20. Intergovernmental Panel on Climate Change (IPCC). (2014a). *Climate Change 2014: Impacts, Adaptation, and Vulnerability.* Contribution of the Working Group 2 to the fifth Assessment Report of the Intergovernmental Panel on Climate Change. Cambridge, UK and New York, NY, USA: Cambridge University Press.

21. Intergovernmental Panel on Climate Change (IPCC). (2014b). *Climate Change 2014: Mitigation of Climate Change.* Contribution of the Working Group 2 to the fifth Assessment Report of the Intergovernmental Panel on Climate Change. Cambridge, UK and New York, NY, USA: Cambridge University Press.

22. International Transport Forum. (2010). *Reducing Transport Greenhouse Gas Emissions: Trends & Data.* Leipzig, GER: The Organization for Economic Co-operation and Development.

23. Millard-Ball, A. (2010). Where the Action is Local governments are taking climate action plans to a new level. *Planning*, pp. 16–21.

24. National Aeronautics and Space Administration (NASA). (2010). *NASA Research Finds Last Decade was warmest on Record, 2009 One of Warmest Years.* National Aeronautics and Space Administration. Accessed 9/1/2010 at http://www.nasa.gov /home/hqnews/2010/jan/HQ_10-017_Warmest_temps.html.

25. National Aeronautics and Space Administration (NASA). (2016). *GISS Surface Temperature Analysis.* National Aeronautic and Space Administration. Accessed 10/4/16 at http://data.giss.nasa.gov/gistemp/graphs/.

26. National Oceanic and Atmospheric Administration (NOAA). (2005). *Stratospheric Ozone Layer Depletion and Recovery.* Accessed 6/4/2013 at http://www.esrl.noaa .gov/research/themes/o3/.

27. National Oceanic and Atmospheric Administration (NOAA). (2015). *State of the Climate: Global Analysis - Annual 2015.* Accessed 10/4/2016 at http://www.ncdc .noaa.gov/sotc/global/201513.

28. National Research Council (NRC). (2010). *Advancing the Science of Climate Change.* Washington: The National Academic Press.

29. National Research Council (NRC). (2009). Driving and the Built Environment: The Effects of Compact Development on Motorized Travel, Energy Use, and CO2 Emissions, Special Report 298: Committee for the Study in the Relationships among Development Pattern, Vehicle Miles Traveled, and Energy Consumption.

30. NBC. (2005). *Bush: Kyoto Treaty Would Have Hurt Economy.* Accessed 6/4/2013 at http://www.nbcnews.com/id/8422343/ns/politics/t/bush-kyoto-treaty-would-have-hurt-economy/#.U5CuPXJdV8E.

31. Pittock, B. A. (2009). *Climate Change: The Science, Impacts and Solutions.* London: Earthscan.

32. Quay, R. (2010). Anticipatory governance: A tool for Climate change adaptation. *Journal of the American Planning Association, 76*(4), 496–511.

33. Rosenzweig, C., Solecki, W. D., Hammer, S. A., & Mehrotra, S. (2011). *Climate Change and Cities: First Assessment Report of the Urban Climate Change Research Networks.* Cambridge: Cambridge University Press.

34. Smart Growth America. (2014). *Smart Growth Overview.* Smart Growth Online. Accessed 24/5/2014 at https://smartgrowthamerica.org/about/overview.asp.

35. Smart Growth America. (2016). *National Complete Streets Coalition.* Accessed 11/4/2016 at https://smartgrowthamerica.org/program/national-complete-streets-coalition/.

36. Stern, N. (2006). *The Economics of Climate Change: The Stern Review - Executive Summary.* Cambridge: Cambridge University Press.

37. The Blog by Copenhagenize Design Company. (2016). *The Bicycle Bridges of Copenhagen.* Accessed 11/4/2014 at http://www.copenhagenize.com/2016/08/the-bicycle-bridges-of-copenhagen.html.

38. The United States Conference of Mayors. (2014). *Organization Overview.* Accessed 20/5/2014 at http://www.usmayors.org/about/overview.asp.

39. Transit Oriented Development Institute. (2016). *Components of Transit Oriented Development.* Accessed 11/4/2014 at http://www.tod.org/.

40. United Nations. (1998). *Kyoto Protocol to the United Nations Framework Convention on Climate Change.* Accessed 6/4/2014 at http://unfccc.int/resource/docs/convkp/kpeng.pdf.

41. United Nations Framework Convention on Climate Change (UNFCCC). (1998a). *First steps to a safer future: Introducing The United Nations Framework Convention on Climate Change.* United Nations: UNFCCC Framework Convention on Climate Change. Accessed 11/3/2016 at http://unfccc.int/essential_background/convention/items/6036.php

42. United Nations Framework Convention on Climate Change (UNFCCC). (1998b). *Kyoto Protocol to The United Nations Framework Convention On Climate Change.* United Nations: UNFCCC Framework Convention on Climate Change. Accessed 11/3/2016 at http://unfccc.int/resource/docs/convkp/kpeng.pdf.

43. United Nations Framework Convention on Climate Change (UNFCCC). (2008a). Now, up to and beyond 2012: The Bali Road Map. United Nations Framework Convention on Climate Change. Accessed 2/11/2016 at http://unfccc.int/key_steps/bali_road_map/items/6072.php.

44. United Nations Framework Convention on Climate Change (UNFCCC). (2008b). *Report of the Conference of the Parties on its thirteenth session, held in Bali from 3 to 15 December 2007.* United Nations: UNFCCC Framework Convention on Climate Change. Accessed 20/5/2014 at http://unfccc.int/resource/docs/2007/cop13/eng/06a01.pdf.

45. United Nations Framework Convention on Climate Change (UNFCCC). (2009). *Copenhagen Accord.* United Nations: UNFCCC Framework Convention on Climate Change. Accessed 20/5/2014 at http://unfccc.int/resource/docs/2009/cop15/eng/l07.pdf.

46. United Nations Framework Convention on Climate Change (UNFCCC). (2011a). *Cancun Agreements.* United Nations: UNFCCC Framework Convention on Climate Change. Accessed 2/11/2016 at http://unfccc.int/meetings/cancun_nov_2010/items/6005.php.

47. United Nations Framework Convention on Climate Change (UNFCCC). (2011b). *Report of the Conference of the Parties on its sixteenth session, held in Cancun from 29 November to 10 December 2010.* United Nations: UNFCCC Framework Convention on *Climate Change.* Accessed 20/5/2014 at http://unfccc.int/resource/docs/2010/cop16/eng/07a01.pdf.

48. United Nations Framework Convention on Climate Change (UNFCCC). (2012). *Durban: Towards full implementation of the UN Climate Change Convention.* United

Nations Framework Convention on Climate Change. Accessed 2/11/2016 at http://unfccc.int/key_steps/durban_outcomes/items/6825.php.

49. US Global Change Research Program (USGCRP). (2016). The Impacts of Climate Change on Human Health in the *United States: A Scientific Assessment*.

50. US Green Building Conference (USGBC). (2012). *Green Building Facts*. U.S. Green Building Council. Accessed 6/4/2014 at http://www.usgbc.org/articles/green-building-facts.

51. United Nations Framework Convention on Climate Change (UNFCCC). (2013). *Doha Amendment*. United Nations Framework Convention on Climate Change. Accessed 11/3/2016 at http://unfccc.int/files/kyoto_protocol/application/pdf/kp_doha_amendment_english.pdf.

52. United Nations Framework Convention on Climate Change (UNFCCC). (2013a). *The Doha Climate Gateway. United Nations Framework Convention on Climate Change.* Accessed 11/3/2016 at http://unfccc.int/key_steps/doha_climate_gateway/items/7389.php.

53. United Nations Framework Convention on Climate Change (UNFCCC). (2014). *Warsaw Outcomes*. United Nations Framework Convention on Climate Change. Accessed 11/2/2016 at http://unfccc.int/key_steps/warsaw_outcomes/items/8006.php.

54. United Nations Framework Convention on Climate Change (UNFCCC). (2015). *Lima Call to Action*. United Nations Framework Convention on Climate Change. Accessed 11/3/2016 at http://unfccc.int/resource/docs/2014/cop20/eng/10a01.pdf.

55. United Nations Framework Convention on Climate Change (UNFCCC). (2015a). *Lima Climate Change Conference - December 2014*. United Nations Framework Convention on Climate Change. Accessed 11/3/2016 at http://unfccc.int/meetings/lima_dec_2014/meeting/8141.php.

56. United Nations Framework Convention on Climate Change (UNFCCC). (2016a). *Paris Agreement*. United Nations Framework Convention on Climate Change. Accessed 11/3/2016 at http://unfccc.int/files/essential_background/convention/application/pdf/english_paris_agreement.pdf.

57. United Nations Framework Convention on Climate Change (UNFCCC). (2016b). *The Paris Agreement*. United Nations Framework Convention on Climate Change. Accessed 11/3/2016 at http://unfccc.int/paris_agreement/items/9485.php.

58. United States Department of Commerce. (2013). *Economic Impact of Hurricane Sandy*. Washington: Economics and Statistics Administration. Accessed 20/5/2014 at http://www.esa.doc.gov/Reports/economic-impact-hurricane-sandy.

59. Wonderful Copenhagen. (2016). *Visit Copenhagen*. Accessed 11/4/2016 at http://www.visitcopenhagen.com/copenhagen-tourist.

Chapter 9

1. Allen, P., Guthman, J., & Morris, A. (2006). Research brief #9: Meeting farm and food security needs through community supported agriculture and farmers' markets in California. *Center for Agroecology and Sustainable Food Systems*.

2. American Farmland Trust. (1997). Bengston et al. 2004. *Landscape and Urban Planning, 69*, 271–286.

3. Arnould, E. J., & Thompson, C. J. (2005). Consumer culture theory (CCT): Twenty years of research. *Journal of Consumer Research, 31*(4), 868–882.

4. Barney and Worth 2008. page 12.

5. Barrett, C. B. (2010). Measuring food insecurity. *Science, 327*(5967), 825–828.

6. Born, B., & Purcell, M. (2006). Avoiding the local trap: Scale and food systems in planning research. *Journal of Planning Education and Research, 26*, 195–207.

7. Bengston et al. (2004). Landscape and Urban Planning, 69, 271–286.

8. Bryld, E. (2003). Agriculture in developing countries. *Agriculture and Human Values, 20*(1), 79–86(8).

9. Bryld, E. (2003). Potentials, problems, and policy implications for urban agriculture in developing countries. *Agriculture and Human Values, 20*, 79–86.

10. Bryld, E. Agriculture in developing countries.

11. Canfield, D. E., Glazer, A. N., & Falkowski, P. G. (2010). The evolution and future of earth's nitrogen cycle. *Science, 330*, 192–196.

12. Cofie, O., van Veenhuizen, R., & Drechsel, P. (2003). Contribution of urban and peri-urban agriculture to food security in sub-saharan Africa. Presented at the Africa session of 3rd World Wildlife (WWF), Kyoto.

13. Colding, J. (2007). Ecological land-use complementation for building resilience in urban ecosystems. *Landscape and Urban Planning, 81*, 46–55.

14. Colding, J. Ecological land-use.

15. Conner, D. S., Colesanti, K. J. A., & Smalley, S. B. (2010). Understanding barriers to farmers' market patronage in Michigan: Perspectives from marginalized populations. *Journal of Hunger and Environmental Nutrition*, 5, 316–338.

16. Connor, D. J. (2008). Organic agriculture cannot feed the world. *Field Crops Research, 106*, 187–190.

17. Conway, G. (1997). *The Doubly Green Revolution: Food for All in the Twenty-first Century*. Ithaca: Comstock Publishing Associates.

18. Conway, G. (1998). *The Doubly Green Revolution: Food for All in the 21st Century*. Ithaca: Cornell University Press.

19. Cordell, D., Drangert, J. O., & White, S. (2009). The story of phosphorus. *Global Environmental Change, 19*(2), 292–305. Traditional Peoples and Climate Change.

20. Cordell, D., Drangert, J. O., & White, S. (2009). The story of phosphorus: Global food security and food for thought. *Global Environmental Change, 19*, 292–305.

21. Dale, V. H., & Polansky, S. (2007). Measures of the effects of agricultural practices on ecosystem services. *Ecological Economics, 64*, 286–296.

22. Deelstra, T., & Girardet, H. 2000. Urban agriculture and sustainable cities. *Growing Cities, Growing Food*. Accessed at http://www.trabajopopular.org.ar/material /Theme2.pdf.

23. Deelstra, T., & Girardet, H. Urban Agriculture.

24. Dimitri, C., & Oberholtzer, L. (2009). *Marketing U.S. organic foods: Recent trends from farms to consumers. Economic Research Service*. Accessed at http://www.ers .usda.gov/publications/eib58/eib58.pdf.

25. Dubbeling 2009.

26. Eaton, R. L., Hammond, G. P., & Laurie, J. (2007). Footprints on the landscape: An environmental appraisal of rural and urban living in the developed world. *Landscape and Urban Planning, 83*, 13–28.

27. Economic Research Service (ERS). (2008). U.S. certified organic farmland acreage, livestock numbers, and farm operations. Economic Research Service.

28. Economic Research Service. (2017). *Organic Market Overview*. US Department of Agriculture. Accessed 5/12/2017 at https://www.ers.usda.gov/topics/natural-resources-environment /organic-agriculture/organic-market-overview/.

29. Edwards-Jones, G., Mila I Canals, L., Hounsome, N., Truninger, M., Koerber, G., Hounsome, B. . . . & Jones, D. L. (2008). Testing the assertion that 'local food is best': The challenges of an evidence-based approach. *Trends in Food Science and Technology, 19*, 265–274.

30. FAO. Food security.

31. Feagan, R., & Morris, D. (2009). Consumer quest for embeddedness: A case study of Brantford Farmer' Market. *International Journal of Consumer Studies, 33*(3).

32. Food Agricultural Conservation and Trade Act of 1990, no. 101–624,104 Stat. 3359. 28 November 1990.

33. Food Agriculture Organization (FAO). (2006). Food security: Policy brief, Food and Agriculture Organization of the United Nations no. 2. ftp://ftp.fao.org/es/ESA/policybriefs/pb_02.pdf.

34. Glanz, K., Sallis, J. F., Salens, B. E., & Frank, L. D. (2007). Nutrition environment measures survey in stores (NEMS-S): Development and evaluation. *American Journal of Preventative Medicine, 32*(4), 282–289.

35. Goland, C. (2002). Community supported agriculture, food consumption patterns, and member commitment. *Culture & Agriculture, 24*(1), 14–25.

36. Grace, C., Grace, T., Becker, N., & Lyden, J. (2007). Barriers to using urban farmers' markets: An investigation of food stamp clients' perceptions. *Journal of Hunger & Environmental Nutrition, 2*(1), 55–75.

37. Grey, M. A. (2000). The industrial food stream and its alternatives in the United States: An introduction. *Human Organization, 59*(2), 143–150.

38. Griffin, M. R., & Frongillo, E. A. (2003). Experiences of farmers from Upstate New York farmers' markets. *Agriculture and Human Values, 20*, 189–203.

39. Guthman, J. (2008). Bringing good food to others: Investigating the subjects of alternative food practice. *Cultural Geographies, 15*, 431–447.

40. Hand, M. S., & Martinez, S. (2010). Just what does local mean? *Choices: The Magazine of Food, Farm and Resource Issues, 25*(1).

41. Heckman, J. (2006). A history of organic farming: Transitions from Sir Albert Howard's to war in the soil to USDA national organic program. *Renewable Agriculture and Food Systems, 21*(3), 143–150.

42. Heimlich, R. E., & Anderson, W. D. (2001). *Development at the Urban Fringe and Beyond: Impacts on Agriculture and Rural Land.* United States Department of Agriculture: Economic Research Service, Agricultural Economic Report no. 803. Accessed at http://www.ers.usda.gov/publications/aer803/aer803.pdf.

43. Heller, M. C., & Keolian, G. A. (2003). Assessing the sustainability of the US food system: A life cycle perspective. *Agricultural Systems, 76*, 1007–1041.

44. Henneberry, S. R., & Agustini, H. N. (2004). *An Analysis of Oklahoma Direct Marketing Outlets: Case Study of Produce Farmers' Markets.* Selected Paper prepared for presentation at the Southern Agricultural Economics Association Annual, Tulsa, Oklahoma.

45. Hill, H. (2008). *Food Miles: Background and Marketing.* ATTRA – National Sustainable Agriculture Information Service. Accessed at https://attra.ncat. org/attra-pub/PDF/foodmiles.pdf.

46. Hinrichs, C. C., Gillespie, G. W., & Feenstra, G. W. (2004). Social learning and innovation at retail farmers' markets. *Rural Sociology, 69*(1), 31–58.

47. Hoefkens, C. (2009). A literature-based comparison of nutrient content and contaminant contents between organic and conventional vegetables and potatoes. *British Food Journal, 111*(10), 1078–1097.

48. Hole, D. G., Perkins, A. J., Wilson, J. D., Alexander, I. H., Price, G. V., & Evans, A. D. (2005). Does organic farming benefit biodiversity? *Biological Conservation, 122*, 113–130.

49. Hoppe, R., MacDonald, J., & Korb, P. (2010). *Small farms in the United States; Persistence Under Pressure.* United States Department of agriculture, Economic Research Service. Accessed at http://www.ers.usda.gov/ publications/eib63/.

50. Hoppe, R., MacDonald, J., & Korb, P. Small farms. *Economic Information Bulletin*, 63.

51. Hunt, A. R. (2007). Consumer interactions and influences on farmers' market ven*dors*. *Renewable Agriculture and Food Systems, 22*(1), 54–66.

52. Ibid.

53. Ibid.

54. Jarosz, L. (2008). The city in the country: Growing alternative food networks in Metropolitan areas. *The Journal of Rural Studies, 24*, 231–244.

55. Kirchmann, H., & Others. Price premiums.

56. Kirchmann, H., Bergstrom, L., Katterer, T., Andren, O., & Andersson, R. (2008). Can organic crop production feed the world? In H. Kirchmann, & L. Bergstrom (Eds.), *Organic Crop Production – Ambitions and Limitations*.

57. Kneafsey, M. (2010). The region in food – important or irrelevant? *Cambridge Journal of Regions, Economy and Society, 3*(2), 1–14.

58. Larson, N. I., Story, M. T., & Nelson, M. C. (2009). Neighborhood environments: Disparities in access to healthy foods in the U.S. *American Journal of Preventative Medicine, 36*(1), 74–81.

59. Lin, B. H., Smith, T. A., & Huang, C. L. (2008). Organic premiums of US fresh produce. *Renewable Agriculture and Food Systems, 23*(3), 208–216.

60. Low, S. A., Adalja, A., Beaulieu, E., Key, N., Martinez, S., Melton, A., . . . & Jablonski, B. R. (2015). Trends in U.S. Local and Regional Food Systems, AP-068, U.S. Department of Agriculture, E*conomic Research Service*, January 2015.

61. Lyso, T., & Guptill, A. (2004). Commodity agriculture, civic agriculture and the future of U.S. farming. *Rural Sociology, 69*(3), 370–385.

62. Mariola, M. (2008). The local industrial complex? Questioning the link between local foods and energy use. *Agriculture and Human Values, 25*, 193–196.

63. Martinez, S., Hand, M., De Pra, M., Pollack, S., Ralston, K., Smith, T., . . . & Newman, C. (2010). *Local Food Systems: Concepts, Impacts and Issues*. United States Department of Agriculture, Economic Research Service Report 97. Accessed at http://www.ers .usda.gov/publications/err-economic-researchreport/err97.aspx.

64. Martinez, S. et al. (2010). Local food. Economic Research Report Number 97.

65. McMichael, A., Powles, J., Butler, C., & Uauy, R. (2007). Food, livestock production, energy, climate change, and health. *The Lancet, 370*, 1253–1263.

66. Milestad, R., Westberg, L., Geber, U., & Björklund, J. (2010). Enhancing adaptive capacity in food systems: Learning at farmers' markets in Sweden. *Ecology and Society, 15*(3).

67. Miller, T. G. (2007). *Living in the Environment*, (15th ed.). Belmont: Thomson Learning.

68. Ministry of Agriculture Botswana (website). (2006). Accessed at http://www.moa .gov.bw.

69. National Organic Program (NOP). *National Organic Program: Background Information*. United States Department of Agriculture, NOP, 2008. Accessed at http://www.ams .usda.gov/AMSv1.0/getfile?dDocName=STELDEV3004443.

70. Nord, M., Andrews, M., & Carlson, S. (2003). Household food security in the United States, 2003. Food Assistance and Nutrition Research Report No. 42.

71. Nord, M., Andrews, M., & Carlson, S. (2009). Household food security in the United States 2008. United States Department of Agriculture, Economic Research Service, Economic Research Report no. 83.

72. Oberholzter, L., Dimitri, C., & Greene, C. (2005). *Price Premiums hold on as U.S. Organic Produce Market Expands*. United States Department of Agriculture. Accessed at http://www.ers.usda.gov/publications/vgs/may05/vgs30801/vgs30801.pdf.

73. Peters, C. J., Bills, N. L., Wilkins, J. L., & Fick, G. (2008). Foodshed analysis and its relevance to sustainability. *Renewable Agriculture and Food Systems*, 1–7.

74. Pimentel, D., & Pimentel, M. (2003). Sustainability of meat-based and plant-based diets and the environment. *American Journal of Clinical Nutrition, 78*, 660–663.

75. *Pimentel, D., & Pimentel, M. (2003). Sustainability of meat-based and plantbased diets and the environment. *American Journal of Clinical Nutrition, 78*, 660–663.

76. Pimentel, D., et al. (2005). Environmental comparisons. *BioScience, 55(7)*.

77. Pimentel, D., et al. (2005). Environmental comparisons. *BioScience, 55(7)*.

78. Pimentel, D., Hepperly, P., Hanson, J., Douds, D., & Seidel, R. (2005). Environmental, energetic, and economic comparisons of organic and conventional farming systems. *BioScience, 55(7)*, 573–582.

79. Pirog, R., & Benjamin, A. (2003). *Checking the Food Odometer: Comparing Food Miles for Local verSus Conventional* Produce Sales to Iowa Institutions. Leopold Center for Sustainable Agriculture. Accessed at http://www.leopold.iastate. edu/pubs/staff /files/food_travel072103.pdf.

80. Ploeg, M. V., Breneman, V., Farrigan, T., Hamrick, K., Hopkins, D., Kaufman, P., ... & Tuckermanty E. (2009). *Access to Affordable And Nutritious Food: Measuring and Understanding Food Deserts and their Consequences*. Economic Research Service. Accessed at http://www.ers.usda.gov/Publications/AP/AP036/AP036.pdf.

81. Postel, S. (2000). Entering an era of water scarcity: The challenges ahead. *Ecological Applications, 10*(4), 941–948.

82. Powell, L. M., Slater, S., Mirtcheva, D., Bao, Y., & Chaloupka, F. (2007). Food store availability and neighborhood characteristics in the United States. *Preventive Medicine, 44*, 189–195.

83. Pretty, J. (2008). Agricultural sustainability: Concepts, principles and evidence. *Philosophical transactions of the Royal Society, 363*, 447–465.

84. Pretty, J. Agricultural sustainability.

85. Ricketts, T., & Imhoff, M. (2003). Biodiversity, urban areas and agriculture: Locating priority ecoregions for conservation. *Conservation Ecology, 8*(2).

86. "Sustainable agriculture" was addressed by Congress in the 1990 "Farm Bill" [Food, Agriculture, Conservation, and Trade Act of 1990 (FACTA), Public Law 101-624, Title XVI, Subtitle A, Section 1603 (Government Printing Office, Washington, DC, 1990) NAL Call # KF1692.A31 1990].

87. Swinton, S. M., Lupi, F., Robertson, G. P., & Hamilton, S. (2007). Ecosystem services and agriculture: Cultivating agricultural ecosystems for diverse benefits. *Ecological Economics, 64*, 245–252.

88. Taylor, C., & Aggarwal, R. (2010). Motivations and barriers to stakeholder participation in local food value chains in Phoenix, Arizona. *Urban Agriculture Magazine, 24*, 46–48.

89. Topp, C. F. E., Stockdale, E. A., Watson, C. A., & Rees, R. M. (2007). Estimating resource use efficiencies in organic agriculture: A review of budgeting approaches used. *Journal of the Science of Food and Agriculture, 87*, 2782–2790.

90. Trevwas, A. (2001). Urban myths of organic farming: Organic agriculture began as an ideology, but can it meet today's needs? *Nature, 410*, 409–410.

91. United Nations Environment Progamme (UNEP). (2011). *Phosphorus and Food Production*. UNEP Yearbook 2011, 35–45. Accessed at http://www.unep.org /yearbook/2011/ pdfs/phosphorus_and_food_productioin.pdf.

92. United States Department of Agriculture (USDA). (2009). *2007 Census of Agriculture: United States Summary and State Data*. The Geographic Area Series, vols. 1 part 51. Accessed at http://www.agcensus.usda.gov/Publications/2007/Full_Report/usv1.pdf.

93. Williams, C. M. (2002). Nutritional quality of organic food: Shades of grey or shades of green? *Proceedings of the Nutrition Society, 61,* 19–24.

94. Winter, M. (2003). Embeddedness, the new food economy and defensive localism. *Journal of Rural Studies, 19,* 23–32.

95. Wolf, M. M., Spittler, A., & Ahern, J. (2005). A profile of farmers' market consumers and the perceived advantages of produce sold at farmers' markets. *Journal of Food Distribution Research, 36*(1), 192–201.

Chapter 10

1. Alberti, M. (2005). The effects of urban patterns on ecosystem function. *International Regional Science Review, 28*(2), 168–192.

2. Asbirk, S., & Jensen S. (1984). *An Example of Applied Island Theory and Dispersal Biology.* In P. Agger & V. Nielson (Eds.). *Dispersal Ecology.* Copenhagen: Naturfredningstradet of Fredningsstrylsen, pp. 49–54.

3. Benedict, T., & McMahon R. (2006). *Green Infrastructure.* Washington: Island Press.

4. Boutequila, L., Miller, J. A., & McGarigal K. (2006). *Measuring Landscapes.* Washington: Island Press. pp.118.

5. Bowler, D., Buyung-Ali L., Knight T., & Pullin A. (2010). Urban greening to cool towns and cities: A systematic review of empirical evidence. *Landscape and Urban Planning, 97,* 147–155.

6. Carr, M., Lambert D., & Zwick P. (1994). *Mapping of Continuous Biological Corridor Potential in Central America: Final Report.* Pase Pantera. University of Florida: Florida.

7. Center for Biological Diversity. (n.d.) *The Extinction Crisis.* Accessed 5/5/2017 at http://www.biologicaldiversity.org/programs/biodiversity/elements_of_biodiversity/extinction_crisi/.

8. Cook, E. A. (2002). Landscape structure indices for assessing urban ecological networks. *Landscape and Urban Planning, 58,* 269–280.

9. Cook, E. A. (2007). Green site design: strategies for stormwater management. *Journal of Green Building. 2*(4), 46–56.

10. Cook, E. A., & van Lier, H. N. (Editors). (1994). *Landscape Planning and Ecological Networks.* Elsevier: Amsterdam. p. 354.

11. Daily, G. C. (Editor). (1997). *Nature's Services, Societal Dependence on Natural Ecosystems.* Washington: Island press.

12. Diaz, E. (Editor). (2008). *Microbial Biodegradation: Genomics and Molecular Biology.* Caister Academic Press.

13. Dramstadt, E. W., Olson, J. D., & Forman, R. T. T. (1996). *Landscape Ecology Principles in Landscape Architecture and Land Use Planning.* Washington: Island Press.

14. Ewing, J. (2005). *The Mesoamerican Biological Corridor: A Bridge Across the Americas.* Ecoworld. Accessed at www.ecoworld.com/home/articles2.cfm?tid=377.

15. Forman, R. T. T. (1995). *Land Mosaics.* Cambridge: Cambridge University Press.

16. Forman, R. T. T. (2008). *Urban Regions: Ecology and Planning.* Cambridge: Cambridge University Press.

17. Forman, R. T. T., & Godron, M. (1984). *Landscape Ecology.* New York: John Wiley. 619 pages.

18. Hellmund, P., & Smith D. (2006). *Designing Greenways.* Washington: Island Press.

19. Hough, M. (1984). *City Form and Natural Process.* New York: Routledge.

20. Huling, S., & Pivetz B. (2006). In-situ chemical oxidation. Engineering Issue of EPA, August 2006.

21. Jongman, R., Bouwma, I., Griffioen A., Jones-Walters L., & Doorn A. (2011). The Pan European Ecological Network: PEEN. *Landscape Ecology, 26*(3), 311–326.

22. Jongman, R., & Pungetti G. (Editors). (2004). *Ecological Networks and Greenways: Concept, Design, Implementation.* Cambridge: Cambridge University Press, pp. 345.

23. Lyle, T. (1985). *Design for Human Ecosystems: Landscape and Natural Resources.* Washington: Island Press.

24. Lyle, T. (1995). *Regenerative Design for Sustainable Landscapes.* New York: Wiley.

25. McGarigal, K., & Marks B. J. (1995). FRAGSTATS: Spatial pattern analysis program for quantifying landscape structure. Gen. Tech. Rep. PNW-GTR-351. US Department of Agriculture, Forest Sevice: Portland, OR, 122 pp.

26. McHarg, I. (1969). New York: Design with Nature. Doubleday.

27. McPherson, E. (1992). Accounting for benefits and costs of urban green space. *Landscape and Urban Planning, 22*(1), 41–51.

28. Miller, K. E. C. [&] Johnson N. (2001). *Defining Common Ground for the Mesoamerican Biological Corridor.* Washington: World Resources Institute.

29. Noss, R. F. (2004). Can urban areas have ecological integrity? Proceedings 4th International Urban Wildlife Symposium.

30. Odefy, J., Detwiler, S., Rousseau, K., Trice, A., Blackwell, R., O'Hara, K., . . . & Raviprakash R. (2012). Banking on Green. American Rivers, The Water Environment Federation, The American Society of Landscape Architects and ECONorthwest.

31. Pagano, M., & Bowman, A. (2000). *Vacant Land in Cities: An Urban Resource.* The Brooking Institute survey series. pp. 1–7.

32. Schultz, F. (2005). *Yellowstone to Yukon: Freedom to Roam.* Seattle: Mountaineers books.

33. Steiner, F. (2002). *Human Ecology: Following Nature's Lead.* Washington: Island Press.

34. Steiner, F. (2008). *The Living Landscape: An Ecological Approach to Landscape Planning.* Washington: Island Press.

35. Steiner, F. (2011). *Design for a Vulnerable Planet.* Austin: University of Texas Press.

36. Suresh, B., & Ravishankar, G. (2004). Phytoremediation – a novel and promising approach for environmental cleanup. *Critical Reviews of Biotechnology, 24*(2–3), 97–124.

37. Van der Ryn, S., & Cowan, S. (1996). *Ecological Design.* Washington: Island Press.

38. Werquin, A., Duhem, B., Lindholm, G., Oppermann, B., Pauleit S., & Tjallingii S. (Editors). (2005). *Green Structure and Urban Planning: Final Report.* Brussels: European Commission.

Chapter 11

1. Cho, R. (2011). *From Wastewater to Drinking Water. State of the Planet.* Earth Institute, Columbia University. Accessed 4/2/2017 at http://blogs.ei.columbia.edu/2011/04/04/from-wastewater-to-drinking-water/.

2. City of Phoenix. (2011). *2011 Water Resource Plan.* Accessed 4/3/2017 at https://www.phoenix.gov/waterservicessite/Documents/wsd2011wrp.pdf.

3. Environmental Protection Agency (EPA). (n.d.) Water Recycling and Reuse: The Environmental Benefits. Accessed April 2, 2017 at https://www3.epa.gov/region9/water/recycling/

4. Loftas, T., & Ross, J. (Eds.). (1995). Dimensions of need: an atlas of food and agriculture. Food & Agriculture Org. Accessed 4/2/2017 at http://www.fao.org/docrep/u8480e/U8480E0c.htm

5. McDonald, R. I., Weber, K. F., Padowski, J., Boucher, T., & Shemie, D. (2016). Estimating watershed degradation over the last century and its impact on water-treatment costs

for the world's large cities. *Proceedings of the National Academy of Sciences, 113*(32), 9117–9122.

6. National Research Council. (2012). *Water Reuse: Potential for Expanding the Nation's Water Supply Through Reuse of Municipal Wastewater.* Washington: National Academies Press.

7. Sato, T., Qadir, M., Yamamoto, S., Endo, T., & Zahoor, A. (2013). Global, regional, and country level need for data on wastewater generation, treatment, and use. *Agricultural Water Management, 130*, 1–13.

8. WALB. (2012). *Special Report: South Georgia's Drought.* Accessed at http://www.walb.com/story/16975753/special-report-south-georgias-drought.

9. Water Aid. (2016). *Water: At What Cost? The State of the World's Water* 2016. Accessed at http://www.wateraid.org/uk/what-we-do/policy-practice-and-advocacy/research-and-publications/view-publication?id=3f44e1ad-49a3-425f-a59b-b5f2c1145fd9.

Chapter 12

1. Ardila-Gomez, A. (2004). *Transit Planning in Curitiba and Bogota: Roles in Interaction, Risk and Change* (PhD Thesis). Cambridge: Massachusetts Institute of Technology.

2. Ardila-Gomez, A. Transit planning in Curitiba and Bogota. (PhD Thesis).

3. Cervero, R., Golub, A., & Nee, B. (2007). City carshare: Longer-term travel demand and car ownership impacts. *Transportation Research Record: Journal of the Transportation Research Board, 1992*, 70–80.

4. Delucchi, J., & McCubbin, D. (2010). External cost of transport in U.S. In A. de Palma, R. Lindsey, E. Quinet, & R. Vickerman (Ed.), *Handbook Transport Economics*, UK: Edward Elgar Publishing Ltd.

5. Etkin, D. S. (2001). *Analysis of oil spill trends in the United States and worldwide.* Presentation at the International Oil Spill Conference, Tampa, Florida.

6. Ewing, R., & Others. (2010). *Growing Cooler—the Evidence on Development.* Washington: Urban Land Institute.

7. Ewing, R., Bartholomew, K., Winkelman, S., Walters, J., & Chen, D. (2010). *Growing Cooler—the Evidence on Urban Development and Climate Change.* Washington: Urban Land Institute, p. 7.

8. Fitch, D., Thigpen, C., Cruz, A., & Handy, S. (2016). Bicyclist Behavior in San Francisco: A Before-and-After Study of the Impact of Infrastructure Investments (No. UCD-ITS-RR-16-06).

9. Freund, P. E. S., & Martin, G. T. (1993). *The Ecology of the Automobile* (p. 18). Montreal: Black Rose Books.

10. Golub, A., & Henderson, J. (2011). The greening of mobility in San Francisco. In M. Slavin (Ed.), *Sustainability in America's cities: Creating the green metropolis.* Washington: Island Press.

11. Harry Wray, J. (2008). *Pedal power: The Quiet Rise of the Bicycle in American Public Life.* Boulder: Paradigm Publishers.

12. Ibid, 12.

13. Ibid.

14. Ibid.

15. Ibid.

16. Ibid.

17. Lublow, A. (2007). *The Road to Curitiba.* New York: The New York Times Magazine.

18. Mohl, R. A. (2004). Stop the road: Freeway revolts in American cities. *Journal of Urban History, 30*(5), 674–706.

19. National Highway Traffic Safety Administration. (2017). 2015 motor vehicle crashes: overview. *Traffic safety facts research note, 2016*, 1–9.

20. Oak Ridge National Laboratory. 2010.

21. Office of Response and Restoration. (n.d.) *How Oil Harms Animals and Plants in Marine Environments*. NOAA, Washington. Accessed at http://response.restoration.noaa.gov/oil-and-chemical-spills/oil-spills/how-oil-harms-animals-and-plants.html.

22. Pisarski, A. E. (2006). *Commuting in America III, Transportation Research Board*. (3rd ed.). Washington: Transportation Research Board.

23. Pucher, J., & Dijkstra, L. (2000). Making walking and cycling safer: Lessons from Europe. *Transportation Quarterly, 54*(3), 25–50.

24. San Francisco Municipal Transportation Agency (SFMTA). (2009). 2008 San Francisco State of Cycling Report, San Francisco, CA: San Francisco Bicycle Program, Department of Parking and Traffic, Municipal Transportation Agency.

25. SFMTA, San Francisco Cycling Report, 2009, 22 and SFMTA, 2010 San Francisco State of Cycling Report, San Francisco: San Francisco Bicycle Program, Department of Parking and Traffic, Municipal Transportation Agency, 2010), 22.SFMTA. 2009. 22.

26. SF Municipal Transit Authority. (2009). *The Citizen's Guide to the Bicycle Plan*. Accessed at https://www.sfmta.com/sites/default/files/projects/SFMTA-CitizensGuideBike_000 .pdf.

27. Shaheen, S., Cohen, A., & Chung, M. (2009). North American carsharing: A ten year retrospective. *Journal of the Transportation Research Board, 2110*, 35–44.

28. Switzky, J. (2002). Riding to see. In C. Carlsson (Ed.), *Critical Mass: Bicycling's Defiant Celebration*. Oakland: AK Press, pp. 186–192.

29. The name was inspired by Return of the Scorcher, an independent documentary film that included a narrative observing hundreds of cyclists bunching at an intersection in China and pushing their way into traffic. Bicyclists would wait until they had "critical mass" to push into the stream of cross-traffic, and the phrase was borrowed by local activists organizing the monthly ride.

30. United States Department of Transportation (USDOT). (2010). *The National Bicycling and Walking Study: 15-Year Status Report* (22). Washington: USDOT.

31. United States Environmental Protection Agency (EPA). (2010). *Our Nation's Air-Status and Trends through 2008*. Research Triangle Park: EPA.

32. United States Environmental Protection Agency (EPA). (2012). *Inventory of U.S. Greenhouse Gas Emissions and Sinks: 1990–2010*. Washington: EPA.

Chapter 13

1. Building Research Establishment Ltd (BRE). (n.d.). *BREEAM*. Accessed 1/13/2017 at http://www.breeam.com/.

2. Denmark.dk. (n.d.) *Independent from Fossil Fuels by 2050*. Ministry of Foreign Affairs of Denmark.

3. Emsley, C., Hitchcock T., & Shoemaker, R. "*London History - A Population History of London*", Old Bailey Proceedings Online. Accessed 1/12/2017 at www.oldbaileyonline .org, version 7.0.

4. Energypost.eu. (2015). *Solar Power Passes 1% Global Threshold*. June 11, 2015. Accessed 1/15/2017 at http://energypost.eu/solar-power-passes-1-global-threshold/.

5. EU SME Centre and the China-Britain Business Council. (2015). *The Automotive Market in China*. Beijing. Accessed 3/13/2017 at https://www.ccilc.pt/sites/default /files/eu_sme_centre_sector_report__the_automotive_market_in_china_update_-_ may_2015.pdf.

6. Gibson, C. (1998). *Population of the 100 Largest Cities and Other Urban Places in the United States: 1790 to 1990*. Population Division Working Paper No. 27. Washington: U.S. Bureau of the Census. Accessed 1/12/2017 at https://www.census.gov/population /www/documentation/twps0027/tab05.txt.

7. Glaskin, M. (2013). *Cycling Science: How Rider and Machine Work Together*. Chicago: University of Chicago Press.

8. Guerra, E., & Cervero, R. (2011). Cost of a ride: the effects of densities on fixed-guideway transit ridership and costs. *Journal of the American Planning Association, 77*(3), 267–290.

9. Hawaii State Energy Office. (2016). *Hawaii's Emerging Future: State of Hawaii Energy Resources Coordinator's Annual Report 2016*. State of Hawaii: Honolulu, Hawaii.

10. MTA.info (n.d.). *Facts and Figures*. New York: Metropolitan Transportation Authority. Accessed 1/15/2017 at http://web.mta.info/nyct/facts/ffsubway.htm.

11. REVE. (2016). U.S. number one in the world in wind energy production. *Wind Energy and Electric Vehicle Magazine*. Spanish Wind Energy Association, March 1, 2016. Accessed 1/15/2017 at http://www.evwind.es/2016/03/01/u-s-number-one-in-the-world-in-wind-energy-production/5552.

12. Rybczynski, W. (2010). *Makeshift Metropolis: Ideas About Cities*. New York: Simon and Schuster.

13. United Nations, Department of Economic and Social Affairs, Population Division. (2015). World Urbanization Prospects: The 2014 Revision, (ST/ESA/SER.A/366).

14. United Nations, Department of International Economic and Social Affairs. (1980). Patterns of urban and rural population growth. United Nations Population Studies, no. 68.

15. United States Green Building Council (USGBC). (2016). *USGBC Statistics*. Accessed 1/13/2017 at http://www.usgbc.org/articles/usgbc-statistics.

16. US Department of Energy. (n.d.). *Where the Energy Goes: Gasoline Vehicles*. Accessed 8/6/2017 at https://www.fueleconomy.gov/feg/atv.shtml.

17. US EIA. (2015). *China: International Energy Data and Analysis*. Accessed 3/13/2017 at https://www.eia.gov/beta/international/analysis_includes/countries_long/China /china.pdf.

18. US Energy Information Administration (EIA). (2016a). *Energy in Brief: How Much of U.S. Electricity Supply Comes from Wind, and How Does that Compare with Other Countries?* US Department of Energy: Washington. Accessed 1/15/2017 at https:// www.eia.gov/energy_in_brief/article/wind_power.cfm.

19. US Energy Information Administration (EIA). (2016b). *Frequently Asked Questions: How Much Carbon Dioxide is Produced by Burning Gasoline and Diesel Fuel?* US Department of Energy, Washington. Accessed 1/15/2017 at http://www.eia.gov/tools /faqs/faq.cfm?id=307&t=11.

20. Zhang, J., Mauzerall, D. L., Zhu, T., Liang, S., Ezzati, M., & Remais, J. (2010). Environmental health in China: challenges to achieving clean air and safe water. *Lancet, 375*(9720), 1110–1119. Accessed at http://doi.org/10.1016/S0140-6736(10)60062-1.

Chapter 14

1. Arnfield, A. J. (2003). Two decades of urban climate research: a review of turbulence, exchanges of energy and water, and the urban heat island. *International Journal of Climatology, 23*, 1–26.

2. Besser, L. M., & Dannenberg, A. L. (2005). Walking to public transit: steps to help meet physical activity recommendations. *American Journal of Preventive Medicine, 29*, 273–280.

3. Bohnenstengel, S. I., Evans, S., Clark, P. A., & Belcher, S. E. (2011). Simulations of the London urban heat island. *Quarterly Journal of the Royal Meteorological Society*, *137*(659), 1625-1640.

4. Brazel, A. J., Gober, P. A., Lee, S. J., Grossman-Clarke, S., Zehnder, J., Hedquist, B., & Comparri, E. (2007). Determinants of changes in the regional urban heat island in metropolitan Phoenix (Arizona, USA) between 1990 and 2004. *Climate Research*, *33*, 171–182.

5. Brazel, A. J., Selover, N., Vose, R., & Heisler, G. (2000). The tale of two climates-Baltimore and Phoenix urban LTER sites. *Climate Research*, *15*, 123–135.

6. Chow, W. T. L., Brennan, D., & Brazel, A. J. (2012). Urban Heat Island research in Phoenix, Arizona: Theoretical contributions and policy applications. *Bulletin of the American Meteorological Society, 93*, 517–530.

7. Centers for Disease Control and Prevention (CDC). (2006). *Extreme Heat: A Prevention Guide to Promote Your Health and Safety.* Accessed 4/29/09 at http://www.bt.cdc.gov /disasters/extremeheat/heat_guide.asp.

8. Davis, M. (2006). *Planet of Slums.* London Verso.

9. Gober, P. A. (2006). *Metropolitan Phoenix Place Making and Community Building in the Desert.* Philadelphia: University of Pennsylvania Press.

10. Golden, J. S. (2004). The built environment induced urban heat island effect in rapidly urbanizing arid regions – a sustainable urban engineering complexity. *Environ Sci, 1*(4), 321–349.

11. Guhathakurta, S., & Gober, P. A. (2007). The impact of the Phoenix urban heat island on residential water use. *Journal of the American Planning Association, 73*, 317–329.

12. Harlan, S. L., & Ruddell, D. (2011). Climate change and health in Cities: Impacts of heat and air Pollution and potential co-benefits from mitigation and adaptation. *Current Opinion in Environmental Sustainability, 3*, 126–134.

13. Hedquist, B. C., & Brazel, A. J. (2006). Urban, residential, and rural climate comparisons from mobile transects and fixed stations: Phoenix, Arizona. *Journal of the Arizona-Nevada Academy of Science*, 38(2), 77-87.

14. Howard, L. (1833). *The Climate of London, Volume 1.* Joseph Rickaby, pp. 348.

15. Hu, E., Yang, Y. P., Nishimura, A., Yilmaz, F., & Kouzani, A. (2010). Solar thermal aided power generation. *Applied Energy, 87*, 2881–2885.

16. Jauregui, E. (1997). Heat island development in Mexico City. *Atmospheric Environment, 31*(22), 3821-3831.

17. Kalkstein, A., & Sheridan, S. (2007). The social impacts of the heat–health watch /warning system in Phoenix, Arizona: assessing the perceived risk and response of the public. *International Journal of Biometeorology, 52*, 43–55.

18. Klinenberg, E. (2002). *Heat Wave: A Social Autopsy of Disaster in Chicago.* Chicago: University of Chicago Press.

19. Kunkel, K. E., Pielke, Jr. R. A., & Changnon, S. A. (1999). Temporal fluctuations in weather and climate extremes that cause economic and human health impacts: A review. *American Meteorological Society, 80*, 1077–1098.

20. Landsberg, H. E. (1981). *The Urban Climate.* New York: Academic Press.

21. Larson, J. (2006). *Setting the Record Straight: More than 52,000 Europeans Died from Heat in Summer 2003.* Earth Policy Institute. Accessed 10/19/2009 at http://www .earth-policy.org/plan_b_updates/2006/update56.

22. Masson, V. (2006). Urban surface modeling and the meso-scale impact of cities. *Theoretical and Applied Climatology, 84*(1–3), 35–45, doi: 10.1007/s00704-005-0142-3.

23. Middel, A., Häb, K., Brazel, A. J., Martin, C., & Guhathakurta, S. (2014). Impact of urban form and design on microclimate in Phoenix, AZ. *Landscape and Urban Planning, 122*, 16–28.

24. Nurse, J., Basher, D., Bone, A., & Bird, W. (2010). An ecological approach to promoting population mental health and well-being – a response to the challenge of climate change. *Prospectives in Public Health, 130*, 27–33.

25. Ojeh, V. N., Balogun, A. A., & Okhimamhe, A. A. (2016). Urban-Rural Temperature Differences in Lagos. *Climate*, 4(2), 29.

26. Oke, T. R. (1973). City size and the urban heat island. *Atmospheric Environment, 7*, 769–779.

27. Oke, T. R. (1982). The energetic basis of the urban heat island. *Quarterly Journal of the Royal Meteorological Society, 108*, 1–24.

28. Oke, T. R. (1987). *Boundary Layer Climates*, (2nd ed). Routledge, pp. 435.

29. Oke, T. R. (1995). The heat island of the urban boundary layer: Characteristics, causes and effects. In Cermak, J. E., et al. (Eds.) *Wind Climate in Cities*, Kluwer, pp. 772.

30. Parmesan, C., Root, T. L., & Willig, M. R. (2000). Impacts of extreme weather and climate on terrestrial biota. *Bulletin of American Meteorological Society, 81*, 443–450.

31. Rosenzweig, C., Solecki W. D., Parshall L., Chopping M., Pope G., & Goldberg R. (2005). Characterizing the urban heat island in current and future climates in New Jersey, Global Environmental Change Part B. *Environmental Hazards*, 6(1), 51–62. doi: 10.1016/j.hazards.2004.12.001.

32. Roth, M. (2007). Review of urban climate research in (sub) tropical regions. *International Journal of Climatology, 27*, 1859–1873.

33. Roth, M., & Chow, W. T. (2012). A historical review and assessment of urban heat island research in Singapore. *Singapore Journal of Tropical Geography*, 33(3), 381-397.

34. Ruddell, D. M., Harlan, S. L., Grossman-Clarke, S., & Buyanteyev, A. (2010). Risk and exposure to extreme heat in microclimates of Phoenix, AZ. In P. Showalter & Y. Lu (Eds.), *Geospatial Techniques in Urban Hazard and Disaster Analysis*. Springer, pp. 179–202.

35. Runnalls, K. E., & Oke, T. R. (2000). Dynamics and controls of the near-surface heat island of Vancouver, British Columbia. *Physical Geography, 21*(4), 283–304

36. Semenza, J. C., McCullough, J. E., Flanders, W. D., McGeehin, M. A., & Lumpkin, J. R. (1999). Excess hospital admissions during the July 1995 heat wave in Chicago. *American Journal of Preventive Medicine, 16*, 269–277.

37. Semenza, J. C., Rubin, C. H., Falter, K. H., Selanikio, J. D., Flanders, W. D., How, H. L., & Wilhelm, J. L. (1996). Heat-related deaths during the July 1995 heat wave in Chicago. *American Journal of Preventive Medicine, 16*, 269–277.

38. Shashua-Bar, L., Pearlmutter, D., & Erell, E. (2009). The cooling efficiency of urban landscape strategies in a hot dry climate. *Landscape and Urban Planning, 92*, 179–186.

39. Stewart, I. D., & Oke, T. R. (2012). Local climate zones for urban temperature studies. *Bulletin of the American Meteorological Society, 93*, 1879–1900.

40. US Census Bureau. (2014). Accessed at www.census.gov.

41. United Nations, Department of Economic and Social Affairs, Population Division (2016). The World's Cities in 2016 – Data Booklet (ST/ESA/ SER.A/392).

42. Wentz, E. A., & Gober, P. A. (2007). Determinants of small-area water consumption for the city of Phoenix, Arizona. *Water Resources Management, 21*, 1849–1863.

43. West, J. J., Fiore, A. M., Horowitz, L. W., & Mauzerall, D. L. (2006). Global health benefits of mitigating ozone pollution with methane emission controls. *Proceedings of the National Academy of Sciences, 103*, 3988–3993.

44. Yang, P., Ren, G., & Liu, W. (2013). Spatial and temporal characteristics of Beijing urban heat island intensity. *Journal of Applied Meteorology and Climatology*, 52(8), 1803-1816.

Chapter 15

1. Abbott, L. (2012). Tired of teaching to the test? alternative approaches to assessing student learning. *Society for Range Management Rangelands, 34*(3), 34–38.

2. Boone, C. G. (2014). Justice and equity in the city. In C. E. Colten & G. L. Buckley (Eds.), *North American Odyssey: Historical Geographies for the Twenty-first Century* (pp. 413). Lanham: Rowman and Littlefield.

3. Buissink-Smith, N., Mann, S., & Shephard, K. (2011). How do we measure affective learning in higher education? *Journal of Education for Sustainable Development, 5*(1), 101–114.

4. Colucci-Gray, L., Camino, E., Barbiero, G., & Gray, D. (2006). From scientific literacy to sustainability literacy: An ecological framework for education. *Science Education, 90*(2), 227–252.

5. de Haan, G. (2010). The development of ESD-related competencies in supportive institutional frameworks. *International Review of Education, 56*(2–3), 315–328.

6. Djordjevic, A., & Cotton, D. R. E. (2011). Communicating the sustainability message in higher education institutions. *International Journal of Sustainability in Higher Education, 12*(4), 381–394.

7. Dweck, C. S., Mangels, J. A., & Good, C. (2004). Motivational effects on attention, cognition, and performance. *Motivation, emotion, and cognition.*

8. Jensen, J. (2008). Educating for ignorance. In B. Vitek & W. Jackson (Eds.). *The Virtues of Ignorance–Complexity, Sustainability, and the Limits of Knowledge.* Lexington: University Press of Kentucky, pp. 307–322.

9. Jickling, B., & Wals, A. E. J. (2008). Globalization and Environmental Education: Looking beyond Sustainable Development. *Journal of Curriculum Studies, 40*(1), 1–21.

10. Kollmuss, A., & Agyeman, J. (2002). Mind the gap: why do people act environmentally and what are the barriers to pro-environmental behavior? *Environmental Education Research, 8*(3), 239–260.

11. Kraiger, K., Ford, J. K., & Salas, E. (1993). Application of cognitive, skill-based, and affective theories of learning outcomes to new methods of training evaluation. *Journal of applied psychology, 78*(2), 311.

12. McKeown, R., Hopkins, C. A., Rizi, R., & Chrystalbridge, M. (2002). *Education for Sustainable Development Toolkit.* Knoxville: Energy, Environment and Resources Center, University of Tennessee.

13. Murray, P. E., & Murray, S. A. (2007). Promoting sustainability values within career-oriented degree programmes: A case study analysis. *International Journal of Sustainability in Higher Education, 8*(3), 285–300.

14. Orr, D. W. (1992). *Ecological Literacy: Education and the Transition to a Postmodern World.* Albany: SUNY Press.

15. Pijawka, D., Yabes, R., Frederick, C., & White, P. (2013). Integration of sustainability in planning and design programs in higher education: Evaluating learning outcomes, *Journal of Urbanism*, 1–13.

16. Rowe, D., Svanström, M., & Lozano-García, F. J. (2008). Learning outcomes for sustainable development in higher education. *International Journal of Sustainability in Higher Education, 9*(3), 339–351.

17. Rode, H., & Michelsen, G. (2008). Levels of indicator development for education for sustainable development. *Environmental Education Research, 14*(1), 19–33.

18. Shephard, K. (2008). Higher education for sustainability: seeking affective learning outcomes. *International Journal of Sustainability in Higher Education, 9*(1), 87–98.

19. Sterling, S., & Thomas, I. (2006). Education for sustainability: the role of capabilities in guiding university curricula. *IJISD International Journal of Innovation and Sustainable Development, 1*(4).

20. Tilbury, D. (2004). Environmental education for sustainability: a force for change in higher education. In P. B. Corcoran & A. E. Wals (Eds.), *Higher Education and the Challenge of Sustainability*. Dordrecht: Kluwer Academic Publishers.

21. Veronica, B. M., & Elizabeth D. D. (2007). Targeted assessment of students' interdisciplinary work: An empirically grounded framework proposed. *The Journal of Higher Education, 78*(2), 215–237.

22. Wiek, A., Withycombe, L., & Redman, C. L. (2011). Key competencies in sustainability: a reference framework for academic program development. *Sustainability Science, 6*(2), 203–218.

Chapter 16

1. China Internet Information Center. (2007). *What is 'Green Consumption?'* Accessed 5/9/2017 at http://www.china.org.cn/english/environment/224177.htm.

2. Chinese Ministry of Environmental Protection. (2016). *Mid-term Evaluation of Action Plan on Air Pollution Prevention and Control*, July 6, 2016. Accessed 4/17/2017 at http://www.mep.gov.cn/xxgk/hjyw/201607/t20160706_357205.shtml.

3. Koleski, K. (2017). *The 13th Five-Year Plan: U.S.-China Economic and Security Review Commission Staff Research Report*, February 14, 2017. Accessed 4/17/2017 at https://www.uscc.gov/Research/13th-five-year-plan.

4. People's Republic of China (PRC). (2016). *13th Five-Year Plan on National Economic and Social Development*, March 17, 2016. Accessed 4/17/2017 at http://news.xinhuanet.com/politics/2016lh/2016-03/17/c_1118366322.htm.

5. People's Republic of China (PRC). (2003). *Environmental Impact Assessment Law of the People's Republic of China*. Accessed 5/9/2017 at http://api.commissiemer.nl/docs/os/sea/legislation/china_s_ea_legislation_03.pdf.

6. People's Republic of China (PRC). (2013). *Action Plan on Air Pollution Prevention and Control*, September 10, 2013. Accessed 4/17/2017 at http://www.gov.cn/zwgk/2013-09/12/content_2486773.htm.

7. People's Republic of China (PRC). (2016). *13th Five-Year Plan on National Economic and Social Development*, March 17, 2016. Accessed 4/17/2017 at http://news.xinhuanet.com/politics/2016lh/2016-03/17/c_1118366322.htm.

Chapter 17

1. Holling. (1996) [Spokane Tribe of Indians, Sustainable Community Master Plan, p.2].

2. *American Diabetes Association*. Accessed at http://www.diabetes.org/living-with-diabetes/treatment-and-care/high-risk-populations.

3. A seven member Board of Directors, whose directors are appointed by the Chairperson of the Nation and confirmed by the Legislative Council, governs TONCA. The Board of Directors is viewed as the policy-making entity and delegates the day-to-day operation of the programs under TONCA to the programs' professional staff.

4. Background Paper: "Health and Sustainable Development, Meeting of Senior Officials and Minsters of Health, Johannesburg, South Africa, 19–22, January 2002, Summary Report at 35.

5. Cajete, G. (2000). Native Science, Natural laws of Interdependence. Santa Fe: Clear Light Publishers. p. 178.

6. CDC Report, CHDIR: (2006 data).

7. Chapter 3, *infra.*

8. Community pursues development at its own pace. (2007). *The East Valley Tribune.* Accessed 8/29/2011 at http://www.eastvalleytribune.com/article_0ef8a52b-aa65-5101-bc16-b4f5b32fa343.html.

9. Federal recognition of an Indian tribe authorizes that tribe to be eligible to receive services from the U.S. Department of the Interior, Bureau of Indian Affairs. See Federally Recognized Tribes List Act, Pub. L. No. 103–454 1994.; Indian Entities Recognized and Eligible To Receive Services From the United States Bureau of Indian Affairs, 79 Fed. Reg. 4748 (Jan. 29, 2014). Some Indian nations have received state recognition but not federal recognition while many Indian nations are unrecognized by either the federal government or a state.

10. For a summary of the history of Federal Indian policies, see History & Background of Federal Indian Policy, Cohen's Handbook of Federal Indian Law, chapt. 1, 2012.

11. H.R. Rep. No. 82–2503 1952.

12. Hibbard, L., & Rasmussen, 2008, p. 141.

13. Hibbard, M., Lane, M. & Rasmussen, K. (2008). The split personality of planning: Indigenous peoples and planning for land resource management. *Journal of Planning Literature, 23*(2), 136–151.

14. *Id.*

15. *Id.*

16. *Id.*

17. *Id.*

18. *Id.*

19. *Id.*

20. *Id.*

21. *Id.*

22. *Id.* at 2.

23. *Id.*at 242.

24. *Id.* at 25.

25. *Id.*at *29–40.*

26. *Id.at 31–32.*

27. *Id. at 33.*

28. *Id. at 37–38.*

29. Jackson M. Y. "Height, Weight, and BodyMass Index of American Indian Schoolchildren, 1990–1991." *Journal of the American Dietetic Association, 93*(10), 1136–1140, 1993. The reference information for this study was the Centers for Disease Control and Prevention's 1970 definitions for children who were at or above the 85th percentile of body mass index (BMI) on sex- and age-specific growth charts. www.leadershipforhealthycommunities.org.

30. Johnson v. McIntosh, 21 U.S. 543 1823.; Cherokee Nation v. Georgia, 30 US 1 1831.; and Worcester v. Georgia, 31 U.S. 515 1832.

31. Jonathan Placito, "Native American tribal Rights: How Arizona's Looming Water-Shortage Threatens Tribal Sovereignty." Rebecca Tsosie, Land, Culture, and Community: Envisioning Native American Sovereignty and National Identity in the 21st Century. 2 International Social Science Review. 180 2001.

32. King, L. A., & Hood, V. L. (1999). Ecosystem Health and Sustainable Communities: North and South. *Ecosystems Health,* (1), 49–57, 56.

33. Kunesh, at 39.

34. Mariella, P., & DeWeaver, N. (2007). *Tribes in Arizona: Growth and Land Use. In: Land use: Challenges and choices for the 21st century.* (pp 63–71). Phoenix: Arizona

Town Hall. Accessed 4/10/2011 at http://www.aztownhall.org/pdf/Complete_91st_Report_FINAL.pdf.

35. Martinez, D. (1998). First People—Firsthand Knowledge. *Winds of Change, 13*(3), 1–4.

36. Nagel, J., & Snipp, M. C. (1993). Ethnic reorganization: American Indian social, economic, political and cultural strategies for survival. *Ethnic and Racial Studies, 16*(2), 203–235.

37. Kunesh P. H., "Constant Governments: Tribal Resilience and Regeneration in Changing Times," 19 Fall Kan. J. L. & Pub. Pol'y 8, 18 1991.

38. Portions of this chapter were prepared for Harvard University Honoring Nations Program as a case profile, authored jointly by Judith Dworkin and Frances Stout and edited by Niina Haas.

39. Public Law 959. (1956). Chapter 930. Stat. 70

40. Ragsdale, Jr. Indian Reservations and the Preservation of Tribal Culture: Beyond Wardship to Stewardship," 59 UMKC Law Review 503, 510 1991.

41. Reyhner, Cantoni, Clair & Yazzie 1999.

42. Reyhner, J., Cantoni, G., St. Clair, R.N. & Yazzie, E.P. (Eds.). (1999). *Proceedings from 5th Annual Stabilizing Indigenous Languages Symposium.* KY: Northern Arizona University; McCarty, T.L., Romero-Little, M.E. & Zepeda, O. (2006). Native American Youth Discourses on Language Shift and Retention: Ideological Cross-currents and Their Implications for Language Planning. *The International Journal of Bilingual Education and Bilingualism, 9*(5), 659–677.

43. Roland, M., & Semali, L. (2010). Intersections of Indigenous Knowledge, Language and Sustainable Development. CIES Perspectives Newsletter: Florida International University, p.1.

44. S. Rep. No. 78–310 1943.

45. Seivertson, B. L. (1999). History/Cultural Ecology of the Tohono O'odham Nation. (Unpublished doctoral dissertation). University of Arizona: Tucson.

46. Semken, S. (2005). Sense of place and place-based introductory geoscience teaching for American Indian and alaska native undergraduates. *Journal of Geoscience Education, 53*(2), 149–157.

47. Smith, L. (2008). Indigenous geography, GIS, and land use planning on the Bois Forte Reservation. *American Indian Culture and Research Journal, 32*(3), 139–151.

48. Spokane Tribe of Indians, Sustainable Community Master Plan: the ST01 2012 Comprehensive Plan @ pp. 2–3 (Draft: September 26, 2012)(citing to Hollings, Crawford S., "Engineering Resilience versus Ecological Resilience," In Engineering with Ecological Restraints. National Academy of Sciences, pp. 31–43. Accessed at www.nap.edu/openbook.

49. Spokane Tribe of Indians, Sustainable Community Master Plan: the ST01 2012 Comprehensive Plan @ pp. 2–3 (Draft: September 26, 2012); Gardner, J., D. Pijawka, and E. Trevan. 2014. Recommendations for Updating the Community-Based Chapter Land Use Plans for the Navajo Nation. Final Report. Office of the President. Navajo Nation.

50. There are a number of terms that Indian people have used to describe their political entities. Some, like the Navajo and the Tohono O'odham use the term "nation." Others, particularly where more than one Indian ethnic group occupy the same land area, describe themselves as "community," such as the Gila River Indian Community or the Salt River Indian Community. Many use the term "tribe," such as the Hopi Tribe or the San Carlos Apache Tribe. The term "tribe" has legal as well as ethnic significance. In this chapter the term "Indian nation" is used throughout unless reference is made to a specific tribal entity.

51. The U.S. Census collects data for Americans Indians and Alaska Natives in a single category.

52. TOCA Tohono O'odham Communicty Action; http://www.tocaonline.org/our-community.html.

53. TONCA Charter, Title 17, ch. 2, Preamble.

54. TONCA Preamble to Charter, codified at Tohono O'odham Code, Title 17, Ch. 2.

55. Trosper, Ronald L. (2002). Northwest coast indigenous institutions that supported resilience and sustainability. *Ecological Economics, 41*, 329–344.

56. U.S. Census Bureau 2010 small Area Income and Poverty Estimates.

57. US Bureau of the Census. (2010). *The American Indian and Alaska Native Population: 2010: 2010* Census Brief. Washington: US Government Printing Office.

58. US Bureau of the Census. (2015). *Profile America: Facts for Features: American Indian and Alaska Native Heritage Month: November 2015.* Washington: US Government Printing Office.

59. US Bureau of Labor Statistics. (2016). *Labor Force Characteristics by Race and Ethnicity, 2015.* US Washington: Department of Labor.

60. US National Archives and Records Administration. (2017). *Dawes Act (1887).* Washington. Accessed 5/3/2017 at https://www.ourdocuments.gov/doc.php?flash=true&doc=50.

61. Whitbeck, L. B., Adams, G. W., Hoyt, D. R., & Chen, X. (2004). Connceptualizing and Measuring Historical Trauma Among Indian People. *American Journal of Community Psychology*, (33 Nos. 3/4), 119–130, at 121.

62. Woods, T. K., Blaine, K., & Francisco, L. (2002). O'odham Himdag as a Source of Strength and Wellness among the Tohono O'odham of Sothern Arizona and Northern Sonora, Mexico. *Journal of Sociology and Social Welfare, 29*(1), 35–53, at 36.

63. "In 1823, Chief Justice Marshall of the United States Supreme Court justified the taking lands from Indian nations using the "Doctrine of Discovery." Johnson v. McIntosh, 21 U.S. 543 1823. Under the "doctrine of discovery," *Oneida II*, 470 U. S. 226, 234 1985., "fee title to the lands occupied by Indians when the colonists arrived became vested in the sovereign—first the discovering European nation and later the original States and the United States," Oneida I, 414 U. S. 661, 667 1974. In the original 13 States, "fee title to Indian lands," or "the pre-emptive right to purchase from the Indians, was in the State." Id., at 670; see Oneida Indian Nation of N. Y. v. New York, 860 F. 2d 1145, 1159–1167 (CA2 1988). Both before and after the adoption of the Constitution, New York State acquired vast tracts of land from Indian tribes through treaties it independently negotiated, without National Government participation. See Gunther, Governmental Power and New York Indian Lands—A Reassessment of a Persistent Problem of Federal-State Relations, 8 Buffalo L. Rev. 1, 4–6 (1958–1959) (hereinafter Gunther)." City of Sherrill v. Oneida Indian Nation of New York, 544 U.S. 197, 203–204 at n.1 2005.

Contributor Biographies

Rimjhim Aggarwal

Dr. Aggarwal's research explores the interface between international development and sustainability. Her research has examined global dimensions of sustainability such as the links between globalization, local ecosystems, and poverty in less-developed countries. Recent research projects include analyzing the synergies and trade offs among the Sustainable Development Goals adopted by UN member states, specifically focusing on cross-country comparative studies on urban agriculture and the framing of basic needs (e.g. water and food) as human rights versus as economic goods.

Andrew Berardy

Dr. Berardy is a Postdoctoral Research Associate at Arizona State University's Food Systems Transformation Initiative. His areas of expertise include consumer behavior, expertise assessment, and life cycle analysis, all in the context of sustainable food systems. His research identifies synergies between food choices, perceived dietary sustainability, and happiness. He also evaluates all stages of the life cycle of food and assesses potential alternatives or improvements that reduce vulnerability and improve environmental performance. He previously worked as a postdoctoral researcher assessing agricultural vulnerability in central and southern Arizona in the food-energy-water nexus in response to anticipated effects of climate change and identifying strategies for adaptation. He has authored articles on the food-energy-water nexus, sustainable food systems and consumption, tacit knowledge and international communication, and sustainability education.

Anthony Brazel

Dr. Brazel is Professor Emeritus at Arizona State University's School of Geographical Sciences and Urban Planning and is a Senior Sustainability Scientist in the Global Institute of Sustainability. His areas of expertise include physical geography and urban climatology. He has authored numerous papers on physical geography, arid lands, urban climatology, and vulnerability to climate change. He received the Helmut E. Landsberg Award from the American Meteorological Society and the Luke Howard Award from the International Association on Urban Climate for distinguished work on urban climate in desert environments. In addition, he received the Jeffrey Cook Prize in Desert Architecture from Ben-Gurion University of the Negev, Israel in 2017 for lifetime contributions to understanding urban climates in arid regions.

Stephen Buckman

Dr. Buckman holds a PhD in Urban Geography and an MPA in Public Administration. He is an Assistant Professor of Urban Planning in the School of Public Affairs at the University of South Florida. His work is centered in two areas. The first, is with issues of urban resiliency especially as it pertains to the economic and social dynamics of resiliency. Secondly, is with Canal Oriented Development which examines small neighborhood based developments on canal banks to increase density in and create walkability within polycentric landlocked urban environments through waterfront development.

Jianming Cai

Dr. Cai is Professor in the Institute of Geographical Sciences and Natural Resources Research at the Chinese Academy of Sciences. He holds a PhD in Urban Geography & City Planning from the University of Hong Kong, and has published more than 140 papers & chapters since 2000 both in Chinese and English. He frequently serves as a senior advisor or consultant on urbanization, sustainable regional and urban development, urban agriculture and food security to both international agencies such as World Bank, ADB, EU, Ford, Lincoln Land Institute and Chinese national and local governments.

Winston T.L. Chow

Dr. Chow is an Assistant Professor in the Department of Geography, National University of Singapore. Previously, he completed his Ph.D. in Geography and postdoctoral training in Engineering at Arizona State University. His areas of published research expertise include investigating the urban heat island and in applying sustainable concepts in urban climatology, which include assessing methods and techniques in minimizing detrimental impacts of the heat island towards residents in cities. Apart from being the news editor of Urban Climate News, the quarterly magazine of the International Association for Urban Climate, Dr. Chow has also taught lower- and upper-level undergraduate courses on geography and meteorology, which include courses examining sustainability at the human-environment interface.

Edward A. Cook

Dr. Cook is a Professor of Landscape Architecture in The Design School at Arizona State University where he teaches courses on urban ecological design, landscapes and sustainability, and landscape ecological planning. He has published books and articles focused on his research in urban ecology, green/ecological networks and sustainable urbanism. He has worked on projects on landscape ecological planning throughout the world and was one of the pioneers in developing planning and design strategies for ecological networks in urban landscapes. He has a PhD from Wageningen University in the Netherlands, a Master of Landscape Architecture degree from Utah State University, and a Bachelor of Science in Landscape Architecture from Washington State University.

Judith Dworkin

Judith Dworkin, J.D., is the managing partner of the Scottsdale law firm of Sacks Tierney P.A. Her practice is devoted primarily to Indian law and water resources law issues. Dr. Dworkin has been selected for inclusion in Best Lawyers of America in the fields of water law and Native American law and selected among Arizona's "50 Most Influential Women in Business" by AzBusiness magazine. She received her M.A. and Ph.D. degrees in geography from Clark University and her J.D. degree, cum laude, from Arizona State University. She clerked for the Honorable William C. Canby, Jr. of the Ninth Circuit Court of Appeals. She is admitted to the Arizona and Navajo Nation bars and is admitted to practice in the courts of the Tohono O'odham Nation, Gila River Indian Community, Hopi Tribe and Hualapai Tribe. Dr. Dworkin lectures regularly and publishes on topics relating to water resource management and economic development on Indian Reservations. She is an adjunct professor in the Sandra Day O'Connor College of Law and the School of Geographical Sciences and Urban Planning at Arizona State University where she teaches graduate courses in water and natural resources law and planning.

Meagan Ehlenz

Dr. Ehlenz is Assistant Professor at Arizona State University's School of Geography and Urban Planning as well as a Senior Sustainability Scholar in the Global Institute of Sustainability. Her areas of expertise include urban and regional planning, with an emphasis in the areas of planning theory and practice; urban revitalization; community development; and the role of anchor institutions. She was previously a planning consultant in Southeastern Wisconsin and a senior planner for the City of Milwaukee's Department of City Development. Dr. Ehlenz is a certified planner with the American Institute of Certified Planners (AICP) and earned her PhD in City and Regional Planning from the University of Pennsylvania.

Chad Frederick

Dr. Frederick, is a Senior Research Associate at the University of Louisville's Center for Sustainable Urban Neighborhoods and an instructor of public policy at Sullivan University. He holds a PhD in Urban Affairs, and a masters in Urban and Environmental Planning. He researches, teaches, and publishes in the field of sustainable urban development. His published works can be found in academic journals such as the The International Journal of Sustainable Transportation, Local Environment, Journal of Urbanism, and the International Journal of Sustainability Education. His recent book, *America's Addiction to Automobiles*, delves into the monumental role of automobiles in the development of US cities and shows how they exacerbate urban inequalities and thwart the sustainability of modern society. Through a comprehensive, thoughtful discussion, Dr. Frederick illustrates how the automobile is fundamentally at odds with the very nature of cities and how automobile dependency has put our entire society at risk, offering a radical solution to the problem of the automobile.

Aaron Golub

Dr. Golub is Director and Associate Professor in the Toulan School of Urban Studies and Planning at Portland State University. He teaches courses on urban transportation planning and policy, research methods, and environmental justice. His research focuses on the social contexts of urban transportation systems, explored in three ways: 1) the effects on social equity of current transportation planning practices – how do people participate in planning, and who wins and who loses from transportation plans; 2) planning, research and activism in support of alternatives to the automobile (car-sharing, public transportation and bicycles); and, 3) the historical roots of automobile dependence in the United States. He worked as a researcher, consultant and advocate in California for a decade on projects related to public transit, car-sharing, and social justice issues in transportation finance. He has worked in Brazil, Mexico, and Colombia as a consultant on various transportation advocacy and planning projects. He sits on two national research committees related to environmental justice in transportation and transportation planning in developing countries. Dr. Golub received his Ph.D. in Civil Engineering from UC Berkeley in 2003. His masters and bachelor's degree were earned in mechanical engineering at the Massachusetts Institute of Technology and Virginia Polytechnic Institute, respectively.

Subhrajit Guhathakurta

Dr. Guhathakurta is the Director of the Center for Geographic Information Systems (CGIS) and Professor of City and Regional Planning at the Georgia Institute of Technology. Dr. Guhathakurta has spent twenty years undertaking research on various dimensions of urban sustainability that included housing, energy, transportation, urban climate, and water use. He was among the founding members of the School of Sustainability at Arizona State University where he also served as Associate Director of the School of Geographical Sciences and Urban Planning. Dr. Guhathakurta is Co-Editor of the Journal of Planning Education and Research, which is the flagship journal for the Association of Collegiate Schools of Planning (ACSP). His publications include over 70 journal papers and several edited books including Integrated Land Use and Environmental Models (2003) and Visualizing Sustainable Planning (2009), both published by Springer.

Jason Kelley

Dr. Kelley is Lecturer of Urban Planning at Arizona State University's School of Geographical Sciences and Urban Planning. His areas of expertise include urban transportation planning, environmental justice, and sustainable urban planning and design. Dr. Kelley's most recent research examined proposed road pricing strategies in the San Francisco Bay area and the impact of these strategies on low-income and transportation-disadvantaged groups. Other research has included the use of network optimization modeling and linear programming to determine a cost-effective shipment system for indigenous goods in the Amazon region of Ecuador. His interests also include pedestrian-oriented design and the impacts of urban design on pedestrian behavior. Dr. Kelley regularly teaches courses on transportation planning, urban design, planning history, planning methods, environmental justice, and sustainability. He was recognized for excellence in teaching in 2011 and 2015.

Ariane Middel

Dr. Middel is an Assistant Professor in the Department of Geography and Urban Studies at Temple University. She received her Ph.D. in Computer Science from the University of Kaiserslautern, Germany and holds a B.Sc. and M.Sc. in Geodetic Engineering from the University of Bonn, Germany. Prior to her appointment at Temple, she was an Assistant Research Professor at Arizona State University.

Dr. Middel's research interests are directed toward understanding the dynamics of urban climate to develop adaptation and heat mitigation strategies, specifically addressing the challenges of sustainable urban form, landscapes, and infrastructure in the face of climatic uncertainty in rapidly urbanizing regions. For the past five years, she has advanced the field of urban climatology through applied- and solutions-oriented research employing local and microscale climate modeling and observations to investigate sustainability challenges related to urban heat islands, thermal comfort, water use and quality, energy use, and human-climate interactions in cities. Dr. Middel is currently serving a 4-year term (2016-2020) on the Board of the International Association of Urban Climate (IAUC), the premier international organization for researchers engaged in all aspects of urban climate scholarship.

Martin J. Pasqualetti

Dr. Pasqualetti's work blends 40 years of experience in the academic, administrative, and business sides of the energy industry. He has advised several US government agencies, including the National Research Council, the Department of Energy, and the Nuclear Regulatory Commission. He has conducted energy research in China, the United Kingdom, Mexico, Canada, the Czech Republic, among others. He helped establish two energy firms: EcoGroup and Strategic Solar Energy. Professor Pasqualetti has authored books, chapters, and articles on energy, and is in demand as a public speaker. At Arizona State University, Dr. Pasqualetti regularly offers four energy courses, including The Thread of Energy, Energy and Environment, Solar Energy Policy, and Energy in the Global Arena. A constant theme in all his courses is the form and function of cities as an energy organism.

Ray Quay

Dr. Quay is a Research Professional with the Decision Center for a Desert City, a unit of the Julie Ann Wrigley Global Institute of Sustainability at Arizona State University. Professor Quay is involved in research on urban planning, water resources, climate change, urban heat island, and scenario analysis. Previously he was the Assistant Director of Water Services and Assistant Director of Planning for Phoenix Arizona, and Assistant Director of Planning for Arlington and Galveston Texas. He is a Fellow of the American Institute of Certified Planners and has authored numerous books and articles related to urban planning and water resources including "Anticipatory Governance: A Tool for Climate Change Adaptation." He has also coauthored "Mastering Change" with Bruce McClendon. He holds a BS in Biology and Environmental Studies from Baylor University, a MS in Community and Regional Planning from University of Texas at Austin, and a PhD Environmental Planning and Design from Arizona State University.

Nelya Rakhimova

Dr. Rakhimova is founder and CEO of the Open School of Sustainable Development (www.openshkola.org). She completed her Ph.D. at the Dresden Leibniz Graduate School in Germany within the Technical University of Dresden. She also obtained an Environmental Management degree at Tyumen State University in Russia. In 2011, she received a master's degree in Urban and Environmental Planning at Arizona State University as a Fulbright scholar. She has worked as a consultant at the United Nations Department of Economic and Social Affairs in the Division for Sustainable Development in New York and completed an internship in the regional office of UN-Habitat in Moscow. Dr. Rakhimova has participated in numerous summer events and international conferences on sustainable development, including the World Urban Forum V, COP15, the ECOSOC Youth Forum, HLPF, the Global Festival on Ideas for Sustainable Development, and the UNESCO Week for Peace and Sustainable Development, among others.

Charles Redman

Dr. Redman has been committed to interdisciplinary research since as an archaeology graduate student he worked closely in the field with botanists, zoologists, geologists, art historians, and ethnographers. Redman received his BA from Harvard University, and his MA and PhD in Anthropology from the University of Chicago. He taught at New York University and at SUNY-Binghamton before coming to Arizona State University in 1983. Since then, he served nine years as Chair of the Department of Anthropology, seven years as Director of the Center for Environmental Studies and, in 2004, was chosen to be the Julie Ann Wrigley Director of the newly formed Global Institute of Sustainability. From 2007-2010, Redman was the founding director of ASU's School of Sustainability. Redman's interests include human impacts on the environment, sustainable landscapes, rapidly urbanizing regions, urban ecology, environmental education, and public outreach. He is the author or co-author of 14 books including Explanation in Archaeology, The Rise of Civilization, People of the Tonto Rim, Human Impact on Ancient Environments and, most recently, co-edited four books: The Archaeology of Global Change, Applied Remote Sensing for Urban Planning, Governance and Sustainability, Agrarian Landscapes in Transition, and Polities and Power: Archaeological Perspectives on the Landscapes of Early States. Redman is currently working on building upon the extensive research portfolio of the Global Institute of Sustainability by leading the newly NSF-funded Urban Resilience to Extremes (UREx) Sustainability Research Network. He continues teaching in the School of Sustainability which is educating a new generation of leaders through collaborative learning, transdisciplinary approaches, and problem-oriented training to address the environmental, economic, and social challenges of the 21st Century.

Darren Ruddell

Dr. Ruddell is an Associate Professor and Director of Undergraduate Studies at the Spatial Sciences Institute at the University of Southern California. Ruddell teaches and develops curricula in GeoDesign and advanced online programs in Geographic Information Science and Technology. GeoDesign is a forward-thinking, interdisciplinary framework that pairs planning, design, and environmental systems with geospatial technologies to explore ways to build a better world. Ruddell earned his Ph.D. from the School of Geographical Sciences and Urban Planning at Arizona State University, and his research efforts utilize geospatial technologies to investigate issues of urban sustainability and resiliency.

Carissa Taylor

Carissa Taylor received a B.A. in Environmental Studies and an M.A. and Ph.D. in Sustainability from Arizona State University. She is currently living in Sydney, Australia writing young adult novels. Her expertise includes curriculum design and environmental consulting. She has a number of published articles including "Motivations and Barriers to Stakeholder Participation in Local Food Value Chains in Phoenix, Az" published in *Urban Agriculture Magazine* and "Reclaiming freshwater sustainability in the Cadillac Desert" published in the *Proceedings for the National Academy of Sciences (PNAS) for the US.*

Craig Thomas

Dr. Thomas studied engineering at West Point for two years, graduated cum laude in English literature from Ohio State University (BA), and holds graduate degrees in creative writing from the Naropa Institute (MFA), and in sustainability and environmental management from Harvard University (MLA). He has written articles and given presentations on water resource management, climate adaptation, and the history of sustainability at national conferences; and published climate-change and regional-policy related articles in the Sierra Sun and Tahoe Daily Tribune. He was an editor of the literary journal, Bombay Gin, and has published short stories in Jumbo Shrimp and The Murray Meaning, as well as being a finalist for the Scribner's Best of the Fiction Workshops. From 2000-2007, Craig toured Southeast Asia and taught advanced English and American literature and writing courses at Toko University. While at Harvard, he was the secretary then president of the Harvard Extension Environmental Club, and also worked for Alexandria Cousteau as the Director of Angkor Wat Research for the non-profit group, Blue Legacy: Telling the Story of Our Water Planet featured on the National Geographic Channel. He is currently a teaching fellow and PhD Candidate at the School of Sustainability at Arizona State University.

Douglas Webster

Dr. Webster is Professor in the School of Geographical Sciences and Urban Planning at Arizona State University. He holds a PhD in City Planning from the University of California, Berkeley. As Senior Advisor to the National Planning

Agency, Thailand from 1993–2003, he managed the Asian Development Bank's trend-setting large-scale initiative, Planning for Sustainable Urbanization in Thailand. Since 2000 Professor Webster's work on urban sustainability has focused on China - in partnership with the Institute of Geographical Sciences and Natural Resources Research at the Chinese Academy of Science. Professor Webster is a frequent advisor on smart city building in East Asia to major international development organizations, corporations, and NGOs. He is author of many publications, both academic and practice-oriented, on the urban environment, resilience, and sustainability in East Asia.

Feifei Zhang

Feifei Zhang is a PhD student in Urban Planning at Arizona State University. She holds a dual Masters degree in International Co-operation and Urban Development from the Technical University of Darmstadt, Germany and Rome University, Tor Vergata, Italy. Before joining ASU, she worked as a Senior Researcher in the Chinese Academy for Environmental Planning (CAEP), Chinese Ministry of Environmental Protection. Her work experience includes research on China's National and Regional Medium-to-Long Term Environmental Protection Strategic Planning, and Environmental Master Planning for Cities and their hinterlands, from a sustainable point of view. Ms. Zhang has extensive experience working as a consultant on large-scale urban projects in major cities oriented to improving the amenity of Chinese cities.

Index

A

accessibility
  of healthy food, 173
  urban planning, 245
active management areas (AMAs), 229
acute pesticide poisonings, 183
adaptation
  global climate change, 146, 168–169
  resilience, 133–134
adaption policies, 14
adaptive capacity, 2, 134–135
adaptive co-management, 138
adaptive cycle, 129–130
adaptive governance
  and collaborative planning, 138–139
  linkages, 138
Adult Vocational Training Program, 337
advanced scenario planning, 160
affective competencies, ESD, 296
agent-based modeling, 120
agribusiness, 177–178
agricultural districts or agricultural preserves, 193
agricultural systems, cities. *see* sustainable agricultural
    systems, cities
agricultural zoning, 193
air pollution, 238–239
airsheds, 320
albedo, 149, 283
amenity, 317
American Association for the Advancement of Science
    (AAAS), 27
American Council for an Energy-Efficient Economy
    (ACEEE), 262
American Environmental Movement, 30–31
American Indians and Alaska Natives (AI/AN), 338
"A Model of Metropolis", 116
anthropogenic waste heat, 274
anticipatory governance, 14, 160, 228
aquaculture ponds, 191
aquifers, 214–216
arcology, 268

Arcosanti, 268–269
area pricing, 250
Arizona Bureau of Public Health Services, 343
Arizona Public Service (APS), 96
Arizona State University (ASU), 124, 125, 264
Armani, Georgiou, 112
Arrhenius, Svante, 106
A Sand County Almanac, 37
atmospheric models, 277
automobile dependence, US
  Curitiba, Brazil, 247–249
  economy, 244
  federal policies, 243
  individual and household, 242–243
  Phoenix, Arizona, 246–247
  planners and developers, 243
  Portland, Oregon, 246
  proactive urban planning, 245–249
  state governments, 243

B

"Bali Action Plan", 161, 161
basic job, 115
Bay Area Rapid Transit (BART) light-rail
    system, 267
Beatley, Timothy, 196
Beautiful City strategy, 323–324
Beautiful Hangzhou sustainability strategy
  dimensions, 325–326
  low carbon city, 326
  public bicycle system, 327
  recycling, 327
  rivers, ecological planning, 327
  transport of solid waste, 327
  trash sorting, 327
Beijing
  bicycling and traffic congestion, 260, 261
  mass rail transit system, 314
  UHI patterns and characteristics, 288–289
Bentham, Jeremy, 24
Benz, Karl, 141

Beyond the Limits: global collapse or a sustainable future, 38
bicycle activism, 252–253
biocapacity, 7
biodiversity, 178, 197
biodiversity value, 197
biological capacity, 11
biophilic cities, 65–73
blackwater, 217
Brandis, Dietrich, 24
brownfields, 210–211
Brundtland-based "three-pillar" sustainability
    framework, 5
Brundtland Commission report, 224
Brundtland Report, 17, 33–35
buffer zones, 203
Building Research Establishment's Energy and
    Environmental Assessment Method (BREEAM),
    106–107, 166, 167, 264
building sector, mitigation, 165–167
Bureau of Indian Affairs, 336
Bus Rapid Transit (BRT), 245–246, 248

C
calibration, 122
California Public Utilities Commission (CPUC), 266
California Solar Initiative (CSI), 266
California State Polytechnic University in Pomona
    (CSPU Pomona), 196
California Urban Futures (CUF-1) model, 117–118
California Urban Futures (CUF-2) model, 118–119
"Cancun Adaptation Framework", 161
"Cancun Agreements", 161
carbon dioxide, 150–151
carbon emissions, 91
carbon footprint, 59, 322
carbon sequestration, 201–202
car-free zones, 250
car-sharing system, 244
    electric car-sharing system, 251
    households, 251–252
    reservation, 251
    short-term auto rental program, 250, 251
    windshield-located card reader, 251
    Zipcar network, 251
Carson, Rachel, 30–31, 182
"C40 Cities Climate Leadership Group", 163
cellular automata (CA), 121–123
Center for Disease Control and Prevention (CDC), 338
Central Arizona Project, 216
central business districts, 281

Central Phoenix/East Valley Light Rail project, 247
Cervero, Robert, 109
China
    city size, 256
    energy consumption, 259
    vehicle growth comparison, 260
China Internet Information Center, 319
Chinese urban sustainability
    air pollution, 308, 311
    energy inefficiency, 312
    energy-inefficient production, 308
    environmental trajectory, 308, 309
    foreseeable future, 327–330
    governance hierarchy, 314–315
    government's priority, 311
    high speed rail system, 313–314
    locally driven sustainability planning, 321–325
    middle-income status, 310
    MOHURD, 315–316
    peri-urban industrial peripheries, 307
    population, 308–309
    population density, 313
    river pollution, 308
    rural-urban transition, 307
    service economies, 310
    shale resources, 312
    social media, 310–311
    soil contamination, 313
    solar capacity, 309
    transit-oriented development, 314
    water pollution, 311–312
Chongming island, 322–323
Chongqing municipality, 321
City Energy Efficiency Scorecard, 262
city or community resilience, 133–134
City Plan 99, 81
Clark, William, 19
Clean Air Act, 238, 246
Clean Air Amendments of 1970, 31
climate change, 214
    action plans, 170–171
climate change governance and strategies
    adaptation strategies and their impact, 168–169
    international treaties and frameworks, 160–162
    local climate change action plans, 170–171
    mitigation. *see* mitigation
    urban efforts and networks, 162–163
climate forcing, 14
climate-forcing agents, 150
climate models and scenarios, 148, 157–158
climate regime, 279

cluster housing and densification, 82
cognitive and affective learning outcomes, ESD
    cognitive and affective, 299–300
    competency categories, 297
    concept vocabulary, 298, 301
    environmental consciousness, 299, 305–306
    ethical conceptualization, 298–299, 303–304
    interdisciplinary understanding, 298, 302–303
Colbert, Jean -B., 23
collaborative urban planning process, 139
Collapse: How Societies Choose to Succeed or Fail, 18
Collins, Samuel, 89
Colorado River Compact, 225
combined heat and power plants (CHP), 75
combustion engines, 164
Commoner, Barry, 30–31, 182
Community Supported Agriculture (CSA) programs, 186
complete streets approach, 96–98, 164–165
complex adaptive systems, 224
complexity, 234
concept vocabulary, 298
Conference of Parties (COP), 40, 161
congestion pricing, 249–250
Consilience, 38
continental-scale conservation strategy, 204
cool roofs, 100–101
Coordinated Hunger Program, 181
"Copenhagen Accord", 161
Corbusier, Le, 93
cordon pricing, 249–250
Costa Rica's economic development strategy, 204–205
coupled human–ecological systems, 3
coupled human-nature interactions, 45
covenants, codes, and restrictions (CCRs), 286
critical mass bicycle rides, 252–253
cultural activities and institutions, 334
Curitiba park system, 137

D
Dawes Act, 336
decision tree, 231
deep ecology, 28–29
deep water aquifers, 216
defensive localism, 188
demand-side management (DSM), 263–264
Department of Transportation (DOT), 103
Design with Nature, 116, 117
Detroit, City of, 135
developable land units (DLUs), 117–118
Diamond, Jared, 50

direct-to-consumer (DTC), 186
discrete choice models, 116
Doha Agreement, 161, 162
Doha Climate Gateway, 161
domestic water, 221
Dongtan eco-city, 322–323
drought, 218–219. *see also* precipitation and drought
Dust Bowl, 18, 20–22

E
earliest urbanism
    challenges, 51
    Mesopotamia, 44–49
earth's carrying capacity, 12
Earth Summit, 35–36
Ebola virus, 9
ecocentrism, 28
eco-city, 74, 322
Eco-Civilization Development Index, 316
eco-duct, 209
eco-friendly paints, 101
ecological bridge, 209
ecological design
    rivers and streams re-wilding, 205–207
    transportation and utility corridors, 208–210
    urban nature preserving, 207, 208
ecological footprint (EF), 11, 178
ecological functions, 197
ecological integrity, 197
ecological overshoot, 12
ecological refugees, global climate change, 155–156
ecological resilience, 197
Economic and Technology Development Zones (ETDZs), 319
economic base sub-model, 115
economic capital, 135
economies of scale, 185
Economist Intelligence Unit (EIU), 59
ecosystem services, 3, 191, 197, 200–202
educational value, 3–4
education for sustainable development (ESD)
    affective competencies, 296
    journals, 300–301
    learning outcomes, 297–306
    surveys, 300
    systems thinking, 296
    vs. traditional education, 295
Ehrlich, Anne, 24
Ehrlich, Paul R., 24, 27–28
Elder Care Consortium (ECC), 350–351
electric car-sharing system, 251

electric streetcar, 241
embodied energy, 185
Emergency Food Assistance Program, 181
Endangered Species Act of 1973, 31
energy
    China, 255
    city size, 256–257
    coal, 258
    organically grown cities, 257–258
    transportation, 259–262
    water and wood, 258
energy-plus buildings, 76
energy-sustainable city
    ACEEE, 262
    automobile dependency reduction, 267–268
    efficient design introduction, 267
    energy efficiency, 262
    European Institute for Energy Research, 262–263
    increased energy efficiency usage, 263–264
    increased renewable energy usage, 264–267
    parking maximums, 271
    policies planning, 269–270
    sprawling development patterns, 270
    urban growth boundaries, 271
    urban spaces, 268–269
    weatherization programs and incentives, 271
    zoning codes, 270
environmental capital, 136–138
environmental consciousness, 298, 299
environmental externality, 312
Environmental Impact Assessments (EIA), 117
environmental justice, 29
Environmental Protection (EP), 321
environmental trajectory, 308, 309
Enz river, Pforzheim, 206
equilibrium, 128
ethical conceptualization, 298–299, 303–304
Euclidean zoning, 270
European Institute for Energy Research (EIFER), 262–263
eutrophication, 178
evolutionary resilience, 129
eye candy, 92

F
"Farm Bill", 175–176
federal Indian policy
    Marshall Trilogy, 335
    relocation, 337
    sovereignty and discovery, 334
    termination of tribes, 336–337

The Allotment Period, 336
    The Indian Reorganization Period, 336
    The Removal Period, 336
    The Treaty Period, 335
Federal Water Pollution Control Amendments of 1972, 31
feed-in tariffs (FITs), 266–267
fee ownership, 335
flat alluvial plain, Mesopotamia, 45
food accessibility, 190
Food and Agriculture Organization (FAO), 156, 190
food deserts, 179–181
food insecurity, 12–13, 178
Food Locator Tool, 180
food miles, 178
food security, 156
    local food, 190
    nutrition and health, 178–181
food supply chain, 176
food systems, 173
food waste, 182
Foote, Eunice, 106
formalized model, 112
form-based codes, 270
Forrester, Jay, 113
fossil fuels, 258
Fourier, Joseph, 105–106
Franklin, Benjamin, 105
freeways, 241–242, 246
Freiburg's sustainability
    green energy, 75–76
    green movement, 74
    population, 73
    sustainable living and place-making, 76–77
    transportation, 74–75
freshwater, 213

G
gallons used per capita per day (GPCD), 221
Garden City concept, 53, 54
Garden City movement, 89
General Allotment Act of 1887, 336
Geographic Information System (GIS), 116
glaciers, 152–153
global climate change, 145
    adaptation strategies and their impact, 168–169
    climate models and scenarios, 157–158
    ecological refugees, 155–156
    food insecurity, 155–156
    future global temperatures, 158
    greenhouse gas emissions, 150–151

human health, 155–156
international treaties and frameworks, 160–162
IPCC assessment reports, 149
local climate change action plans, 170–171
mitigation. *see* mitigation
urban efforts and networks, 162–163
observed *vs.* simulated, 147–148
precipitation and drought. *see* precipitation and drought
science behind, 149
sea level rise and ice sheets, 152–153
significant climate anomalies and events, 148
surface temperature anomalies, 146–147
uncertainty, climate change projections, 159–160
water security, 159
global climate models (GCM), 226
Global Environmental Governance Reconsidered, 156
Global Food Security Index, 156
Global Footprint Network (GFN), 11, 12
Global Governance Project, 156
global hectares, 11
global hectares per capita (GHC), 12
global transport concept (GTC), 74
global warming, 146
Goal 11: Empower inclusive, productive and resilient cities, 128
Golden Section, 92–93
good bones, 88
Grade Point Average (GPA), 92
Grand Avenue, 97
Grand River Complete Street project, 97–98
Grapes of Wrath, 20
graphical user interface (GUI), 120
Great Recession, 193
green belt, 89
Green Building, 88
Green Building Movement, 106–107
Green City Index (GCI), 55–57, 64
green employment, 82
Green energy, 75–76
Greenest City 2020 Action Plan, 59
greenhouse effect, 150
greenhouse gases (GHGs), 11, 111, 145, 150–151, 178, 189, 239, 268
green movement, 74
green networks
    brownfields, 210–211
    buffer zones, 203
    carbon sequestration, 201–202
    climate amelioration, 201
    conceptual diagram, 198

corridors, 203
education and human psychology, 202
hydrologic process, 201
increased biodiversity, 201
maintenance costs and aesthetics, 202
natural/semi-natural open space, 202, 203
nested hierarchy, 204–205
recreation, 201
reduced management, 202
synthetic corridors, 203
vacant and underutilized land, 210
Green Revolution, 177
green roofs, 100–101
"Green Streets" program, Portland, 209–210
green walls, 101–102
groundwater
    aquifers, 214–215
    Central Arizona Project, 216
    natural variability, 216
Groundwater Management Act (GMA), 229
Grovesnor Group, 141
growth boundaries, 318

H
halocarbons, 151
Hammarby model, 80–81
Hangzhou
    Beautiful City strategy, 323–324. *see also* Beautiful Hangzhou sustainability strategy
    environmental issues, 324–325
    history, 323
Health and Human Services (HHS), 350
health co-benefits, 284
health impact assessments, 247
Healthy Food Financing Initiative (HFFI), 180
heat trapping gases, 150
hectares, definition, 11
Hempel, Lamont, 24
high levels of uncertainty, 146
high speed rail (HSR) system, 313
homeowner associations (HOAs), 267
horizontal city, 261
horse-drawn streetcars, 240–241
Hough, Michael, 196
House Concurrent Resolution, 336
Howard, Ebenezer, 89–90, 174–175
Howard, Luke, 274
Howard, Sir Albert, 182
human capital, 108
human-dominated ecosystems, 195
human-ecological systems, 234

human-environmental dynamics, 44
human health, global climate change, 155–156
hydropower, 59, 60

I

ice sheets, 152–153
Incremental Approach, 99
Indian demographics, 338–339
Indian Gaming Regulatory Act, 341, 345
Indian Health Service (IHS), 350
Indian Health Service Division of Sanitation, 344
Indian Relocation Act of 1956, 337
Indian Removal Act, 336
Indian Reorganization Act, 336
Indian self-determination, 337–338
indicators, resilience, 140–143
indoor water, 221
industrial agriculture, 177
industrial food system, 177
    economic impacts, 177–178
    environmental impacts and ecological
      footprint, 178
    food security, nutrition and health, 178–181
    food waste, 182
    Green Revolution, 177
    industrial agriculture, 177
    long-term resilience, 182
industrial organics, 185
industry sector, mitigation, 167–168
institutional capacity, 146
Interagency Crosscutting Group on Climate Change
    and Human Health (CCHHG), 155
inter-cropped polyculture, 183
interdisciplinary approach, 1
interdisciplinary teamwork, 297
interdisciplinary understanding, 298, 302–303
Intergovernmental Panel on Climate Change
    (IPCC), 149
Intergovernmental Panel on Climate Change Fourth
    Assessment Report, 226
interim re-vegetation, 210
intermediated marketing channels, 186
international car-sharing movement, 244
International Code for a Sustainable Built
    Environment, 264
International Council for Local Environmental
    Initiatives (ICLEI), 35
International Energy Agency (IEA), 15
International Panel on Climate Change (IPCC), 40
International Union for the Conservation of Nature
    (IUCN), 34

"internet of things" (IoT) sub-systems, 310
Interstate Highway Act, 241, 244, 246
*Inventory of U.S. Greenhouse Gas Emissions and
    Sinks,* 167
irrigation canal, Mesopotamia, 45
"i-Tree" software tool, 95–96

K

katabatic flows, 282–283
Kates, Robert, 19
Klein, Calvin, 112
Klosterman, Richard, 117
Köppen-Geiger Climate Classification, 280
Köppen, Wladimir, 279
kWh/year, definition, 75
kyoto protocol, 157, 161–162

L

Lagos, UHI effect, 289–290
land classification, 116
Land Evaluation Site Assessment (LESA), 117
Landis, John, 117
landscape change, 198
landscape corridors, 203
landscape ecology, 197
landscape function, 198
landscape metrics, 212
landscape structure
    corridors, 198
    deep structure, 200
    edges, 198
    flows, 198–199
    fragmentation, 198
    patches, 198
    spatial and structural characteristics, 197–198
land suitability approach, 117
land use and land cover change (LULCC), 274
Leadership in Energy and Environmental Design
    (LEED), 107, 166, 167, 264
Lee, Douglas, 116
Leopold, Aldo, 37–38
Lerner, Jaime, 248
lifecycle approach, 176
lifecycle assessment, 184
light rail transit system, 123–125
"Lima Call for Action", 161–162
Limits to Growth, 17, 19
linear corridors, 203
Livability Index, 55–58
local climate zones (LCZs), 275, 276
Local First movement, 136

local food
    defensive localism, 188
    definition, 186–187
    environment and, 188–189
    local inequities, 190–191
    people and, 189–190
    reasons, 187
    sustainability, 187–188
    urban and peri-urban agriculture (UPA), 186
local food systems, 186
Local Governments for Sustainability (ICLEI), 163
local inequities, 190–191
London, UHI patterns and characteristics, 291
long-term resilience, 182
Los Angeles, population density, 261, 262
Lovins, Amory, 263
Lowry, Ira, 115
Lowry model, 115–116
Lyle, John, T., 196
Lyme disease, 155

M

Magis, 134
malaria, 155
Malthus, Thomas, 23
market forces, 91
Marshall Trilogy, 335
maximum sustainable yield, 22
McFadden, Daniel, 116
McHarg, Ian, 116, 117, 195–196
megacities, 10, 53, 273
megapolitan regions, 313
mental model, 112
Mesa, City of, 103–104
Mesoamerica Biological Corridor (MBC) project, 204
Mesopotamia
    environmental stage, 45–46
    first cities, 47–48
    legitimization, 48–49
    physical environment, 45
    pre-urban, Neolithic settlement, 44
    processes of change, 46–47
metabolism, 191–192
methane, 150–151
Metropolitan Transportation Authority (MTA), 268
Mexico City, UHI effect, 290
microclimate, 98, 192
Mill, John S., 24
Ministry of Environmental Protection (MEP), 315
    12th FYP policies, 316–320
    13th FYP policies, 320

Ministry of Housing and Urban-Rural Development
    (MOHURD), 313, 315–316
Mississippi river, 216
Mitigating the Impact of Disasters, 130
mitigation, 145
    building sector, 165–167
    industry sector, 167–168
    policies, 14
    transportation sector, 164–165
mixed use, 89
mixed-use development, 78
mobility, 245
modern sustainability, 2
Mohl, Raymond, 246
Muir Glacier, 152–153
Muir, John, 36
multifunctional land use, 193
multinomial logit, 119
multiple sectors, 234
Multiple-Use Sustained Yield Act of 1960, 26
Multnomah County Board of Commissioners, 171

N

Naess, Norwegian A., 28
National Academy of Science (NAS), 36, 149
National Aeronautics and Space Administration
    (NASA), 100, 147
National Ambient Air Quality Standards, 239
National Development and Reform Committee
    (NDRC), 314
National Environmental Protection Act, 246
National Environmental Protection Model City, 321
National Environment Policy Act, 31
National Highway Traffic Safety Administration
    (NHTSA), 235
*National New-type Urbanization Development Plan
    (2014-2020),* 318
National Oceanic and Atmospheric Administration
    (NOAA), 148
National Organic Program (NOP), 182–183
National Park Service, 109
nation-building, 334
natural disasters, 131
natural regeneration, 210
nature's deep structure, 200
negative feedback, 112
Neighborhood Energy Utility (NEU), 60
new urbanism, 166
nitrogen-fixing cover crops, 183
nitrous oxide, 151
Nixon's government-to-government policy, 338

non-anthropocentric values, 26
non-basic job, 115
non-sewage water, 217
North American Industry Classification System
        (NAICS), 119
North Sea, 258
Nutrition Assistance, 181

O

oasis effect, 275
Obama, Barack, 338
Odum, Eugene, 39
Ogallala aquifer, 216
Olmstead, Frederick Law, 106
Olympic Line, 63
O'odham Himdag, 346
Open Platform for Urban Simulation (OPUS), 119
open space preserves, Phoenix, 207
organically grown cities, 257–258
organic food, 177
        definition, 182–183
        development, 182
        energy reduction, 183–184
        high price tag, 185–186
        lower yield, 184–185
        nutrition, 184
        synthetic chemicals reduction, 183
Orr, David, 17
Osborn, Henry F., 24
Otis, Elisha, 92
Our Common Future, 17
*Our Common Journey: a Transition for
        Sustainability,* 36
Our Plundered Planet, 24
Outdoor water, 221
"overlay method", 116–117

P

Palmer Drought Severity Index (PDSI), 154–155
Pan European Ecological Network, 204
Paris Agreement, 162
park cool islands (PCI), 275
Parker, Barry, 89
passive energy, 269
pedestrian-& bicycle-friendly environment, 82
pedestrian-only streets, US, 250
Pennsylvania Fresh Food Financing Initiative, 180
permafrost soils, 153
petroleum
        dependence, 236–238
        environmental impacts, 239–240

phases of the food system, 189
Phi or Golden Section, 92–93
Phoenix
        park system, 207, 208
        UHI effect, 292–293
Phoenix Metropolitan area, 180–181
Phoenix water-sustainability planning model
        active management areas, 229
        foresight planning, 230
        Groundwater Management Act, 229
        robust short-term strategies, 230–231
        Water Resource Plan, 229
        worst-case infrastructure timeline, 231
physics-based numerical models, 277
Piano, Renzo, 94
Pinchot, Gifford, 22, 24, 25
place making, 97
planter zone, 96
$PM_{2.5}$, particulate matter, 9
Pollio, Marcus Vitruvius, 93
population growth rates, 8–9
Portland City Council, 171
positive feedback, 112
post industrial brownfield sites, 210
potable water, 213
precipitation and drought
        agricultural sector, 154
        future precipitation patterns, 158–159
        heavy rainfall changes, 153–154
        impacts, 154–155
        Palmer Drought Severity Index (PDSI), 154–155
        tourism and winter sports, 154
Preservation Briefs, 109
proactive adaptation, 133
Proceedings of the National Academy of
        Sciences (PNAS), 19
Public Bike Share (PBS) system, 63
purchase of development rights (PDR), 193

R

radiative flux density, 280
radiative fluxes, 280
radiative forcing, 150
radiosonde data, 275
rail-like boarding system, 248
random utility theory, 119
reactive adaptation, 133
recharged water, 215
reclaimed water, 217
Rees, William, 39
reflective paving surfaces, 102–104

"Reinvent Phoenix", 247
re-naturalization process, 206–207
renewable energy, 60
renewable energy resources, 264
    in China, 267
    in Denmark, 267
    European countries, 266
    global wind capacity, 265
    homeowner associations, 267
    rooftop solar hinges, 267
    solar PV capacity, 265–266
    in US states, 266
renewable portfolio standards (RPS), 266
renewable resource, 215
"Requiem for Large Scale Models", 116
residential outdoor water, 221
resilience
    adaptive cycle, 129–130
    adaptive governance, 138–139
    city or community, 133–134
    collaborative planning, 138–139
    definition, 127
    ecological perspective, 128
    economic capital, 135
    engineering perspective, 128
    environmental capital, 136–138
    indicators and indices, 140–142
    outcomes, 143–144
    social capital, 136, 137
    social-ecological systems, 129
    social resiliency, 129
    typologies, 142–143
    urban and community, 130–133
    urban risks, 127
Resilience and Stability of Ecological Systems, 128
Resilience Capacity Index (RCI), 140
resiliency, 2, 146
    community health, 340
    definition, 339
    economic strategies, 341
    place-based social capital, 339–340
    political strategies, 341–342
    reservation lands, 340
    revival of language, 340
Resilient cities: a Grosvenor research report, 141
Resilient Cities Ranking system, 55–57
resilient socioeconomic unit, 47
Resource Conservation and Recovery Act of 1976, 31
return on investment (ROI), 96
Rio Earth Summit, 17
Riparian areas, 222

Rittel, Horst, 18
"road diet" approach, 250
road verge, 96
Rockefeller Foundation's 100 Resilient Cities initiative, 141–142
rooftop garden, 186
rooftop solar hinges, 267
Roosevelt, Theodore, 24
rural-urban transition, 307

S
Safe Drinking Water Act of 1974, 31
Saint Luke's Health Initiative, 247
Salt River Project (SRP), 96
San Francisco
    bay area rapid transit light-rail system, 267
    population density, 261, 262
San Francisco Bicycle Coalition (SFBC), 252–253
saturation mass transit system, 314
scenarios and climate models, 157–158
sea level rise and ice sheets, 152–153
sense of place, 91
shallow water aquifers, 216
Shanghai
    Dongtan Eco City, 322
    saturation mass transit system, 314
    Taopu Smart City, 313
shocks, resilience, 131
short-term auto rental program, 250, 251
sidewalks, 94–95
Silent Spring, 30
siloed, 315
silvicultura oeconomica, 17
Singapore, UHI effect, 291–292
SkyTrain, 63
sky view factor, 281, 282
slope, land use, exclusion, urban extent, transportation and hillshade (SLEUTH), 122
smart city, 310
Smart Growth, 88, 166, 313
smog, 259
social capital, 91, 136, 137
social-ecological system, 129
social organization, 45
Social Security, 131
social vulnerability, 132
socio-ecological system (SES), 3
    "Dust Bowl", 20–22
    Holling's work, 19
    resolution, 18
    super wicked problems, 19

transdisciplinary sustainability science, 19
wicked problems, 18–19
soft energy pathways, 263
soil contamination, 313
solar energy, 75
solar photovoltaics (PV), 265–266
solar radiation, 283
solar roofs, 100–101
solar thermal energy (STE) systems, 286
Sonoran Preserve Master Plan, 207, 208
spatial allocation sub-model, 115
spatial interaction, 113
spatial scales, 273
Speck, Jeff, 95
State of Oregon's Department of Transportation, 246
Statewide Integrated Model (SWIM2), 121
statistic-based regression models, 277
Steinbeck, John, 20
Steiner, Frederick, 196
stepping stone corridors, 203
stewardship, 108
Stockholm's sustainability
    integration and preservation of nature, 81
    land use and urban form, 79–81
    population, 77–78
    transportation, 78–79
storm surge barriers, 169
streetcar network, 63
stressed water, 213
subsurface UHI, 275
suitability score, 116
Supplemental Nutrition Assistance Program (SNAP),
    180–181
supply-side management (SSM), 263
surface-energy balance (SEB), 280
surface temperature
    anomalies, 146–147
    future global temperatures, 158
surface UHI, 275
surface water, 214, 216–217
surplus land reservation, 336
Survey of Conditions Among the Indians of the United
    States, 336
sustainability
    American Environmental Movement, 30–31
    American naturalists, 36–38
    Brundtland Report, 33–35
    Commoner's population ecology, 27–28
    conservation *vs.* preservation, 25–26
    deep ecology, 28–29
    definition, 299

disasters and extreme events, 13–15
    Earth Summit, 35–36
    ecological footprint, 11–12
    educational value, 3–4
    Ehrlich's population studies, 27–28
    environmental justice, 29
    food insecurity, 12–13
    future of, 38–41
    greenhouse gases, 11–12
    interdisciplinary approach, 1
    international conferences, 5–7
    Limits to Growth, 32–33
    Malthus's population ecology, 23–24
    origins of, 17
    Pinchot's sustainable yield, 24–25
    population growth rates, 8–9
    social dimensions, 5
    three pillars, 18, 25
    urbanization issues, 9–11
    von Carlowitz's forestry management, 23
sustainability Report Card, 109–110
sustainability science, 2, 35, 283
sustainable agricultural systems, cities
    agricultural preserves, 193
    agricultural zoning, 193
    aquaculture ponds, 191
    challenge of feeding cites, 173–175
    definition, 175–176
    ecosystem services, 191
    food supply chain, 176
    industrial food system, 177. *see also* industrial food
        system
    lifecycle approach, 176
    local food. *see* local food
    microclimate improvements, 192
    multifunctional land use, 193
    organic food, 177. *see also* organic food
    phosphorous depletion, 191
    purchase of development rights, 193
    vacant public and private land, 193
    waste recycling, 191
    wildlife habitat & biodiversity, 192
sustainable cities
    biophilic cities, 65–73
    cluster housing and densification, 82
    economic, ecological, and social dimensions, 83
    Freiburg–Rieselfeld, 83–86
    Garden City concept, 53
    Green City Index, 55–57
    green employment, 82
    Livability Index, 55–58

low ecological–environmental impact, 81
pedestrian-& bicycle-friendly environment, 82
place making, 82
political and policy support, 82–83
resilience, 82
Resilient Cities Ranking system, 55–57
Stockholm, 77–81
Sustainable Cities Index, 56–58
Vancouver, 58–64
Sustainable Cities Index, 56–58
sustainable development, 3, 17
"synthetic" corridors, 203
system dynamics, 113
systems-thinking approach, 14
systems thinking, ESD, 296

T
Taopu Smart City, 313
The City of Vancouver's Healthy City Strategy, 60
The Closing Circle, 28
The Closing Circle: Nature, Man, and Technology, 182
The Detroit Future City Strategic Framework, 130
The Food, Conservation and Energy Act, 180
The Future of Life, 38
The National Development Framework for Urban
        Environment Quality, 317
The Population Bomb, 24
thermal admittance, 281
thermal expansion, 152
The Smart Growth Manual, 95
The Ten Books on Architecture, 93
"The Walkable City" plan, 79
13th Five-Year Plan (FYP), 315
Thoreau, Henry D., 37
Tianjin eco city, 323
Tianyuan municipality, 321
Tohono O'odham Community Action organization
        (TOCA), 342, 343
Tohono O'odham Community College (TOCC), 350
Tohono O'odham Nation
        economic resiliency, 345
        facility caregivers, 344
        health care services, 344
        institutional capacity building, 345–348
        life expectancy, 344
        population and geographical size, 342
        reservation, 343
        rural nature, 343
        self-reported health status, 344
Tohono O'odham Nursing Care Authority (TONCA)
        assisted living facility, 350

barriers, 347–348
cultural sustainability, 345
desert pathways, 350
effective health services, 346
Elder Care Consortium, 350–351
financial resources, 351
guiding principle, 346
Health Care Network, 350–351
hospice services, 349–350
physical infrastructure, 351
skilled nursing facility, 348–349
staffing, 351
Tomorrow: A Peaceful Path to Real Reform, 89
traditional education, 295
traffic-calming strategies, 250
transdisciplinary sustainability, 19
transition towns, 110
transit-oriented development (TOD), 164, 245–246, 314
transportation
        choices, 98–100
        eco-duct, 209
        ecological bridge, 209
        Freiburg's sustainability, 74–75
        mitigation, 163–164
        parking lots in Essen, 210
        Portland Green Street program, 209–210
        Stockholm's sustainability, 78–79
        in US. *see* urban transportation systems, US
        Vancouver's sustainability, 60, 62–63
Treekeepers, 64
tree ring records, 217
tribal sovereignty, 335
trinary road system, 248
turbulent fluxes, 280
typologies, resilience, 140–143

U
uncertainty
        global climate change, 159–160
        resilience, 133
        water supply, 214
United Kingdom group Building Research
        Establishment Ltd (BRE), 264
United Nations Conference on the Human
        Environment (UNCHE), 17, 32
United Nation's Earth Summit in Rio, 17
United Nations Environment Programme (UNEP), 149
United Nations Framework Convention on Climate
        Change (UNFCCC), 6, 40
United States Green Building Council (USGBC), 264
UN's Food and Agriculture Organization, 174

Unwin, Raymond, 89
urban and community resilience. *see* resilience
urban and peri-urban agriculture (UPA), 186
urban boundary layer UHI, 275
urban canopy layer UHI, 274–275
urban canyons, 281, 282
Urban Data Science Toolkit (UDST), 121
urban dynamics, 113
urban ecological design
    ecosystem services, 197, 200–202
    green networks. *see* green networks
    landscape structure, 198–200
    urban planning, 195–196
urban ecosystem, 63
urban-environmental models, 113
    cellular automata (CA), 122–123
    UrbanSim, 119–122
urban fabric, 91
urban fabric modification, 285
urban forest, 64
urban greening, 285
urban growth boundaries (UGB), 271
urban heat island (UHI) effect, 2, 101, 201
    air temperature simulations, 277, 278
    Beijing, 288–289
    daytime and nighttime surface, 277
    economic sector, 287
    factors causing, 280–282
    intensity, 282–283
    inter- and intragenerational equity, 286–287
    Lagos, Nigeria, 289–290
    local climate zones, 275, 276
    London, 291
    Mexico city, 290
    mitigation and adaptation strategies, 284–286
    Phoenix, 292–293
    Phoenix, Arizona Metropolitan, 274, 275
    physical factors, 277
    Singapore, 291–292
    subtypes, 274
    surface and subsurface UHI studies, 275
    thermal discomfort, 283–284
    urban boundary layer, 275
    urban canopy layer, 274–275
    urban environment, 287
    variations and climate regime, 279–280
    warmer temperatures, 284
urbanization, 9–10, 53, 273
urbanization level, 318
urban models, 111, 113
    first generation models, 113–116

future prospects, 125–126
    light rail transit system, 123–125
    models and modeling, 111–113
    second generation models, 116–119
    urban-environmental models. *see* urban-
        environmental models
urban planning and design
    Beatley's organization, 196
    Forman's work, 196
    Hough's work, 196
    Lyle's contribution, 196
    McHarg's approach, 195–196
    Steiner's work, 196
urban revolution process, 43–44
UrbanSim
    developments, 121–122
    flowchart, 120
    innovations, 119–120
    Urban Data Science Toolkit (UDST), 121
urban structure modification, 285
urban sustainable design
    complete streets, 96–98
    cool roofs, 101
    green roofs, 100–101
    green walls, 101–102
    history, 89–90
    human capital, 108
    human scale, 92–94
    ingrained problems, overcoming of, 107–108
    New Urban Agenda, 87–88
    policy impacts, quality of place, 90–91
    post development assessment, 104–105
    Preservation Briefs, 109
    reflective paving surfaces, 102–104
    solar roofs, 101
    Sustainability Report Card, 109–110
    sustainability standards and rating systems, 105–107
    transition towns, 110
    transportation choices, 98–100
    trees and their canopy, 95–96
    walkability and sidewalks, 94–95
urban transportation systems, US
    assembly-line production, 241
    automobile dependence, 242–244
    automobility paradigm, 242
    bicycle activism, 252–253
    car-sharing system, 250–252
    demand management strategies, 249–250
    economic problems, 238–240
    electric streetcar, 241
    freeways, 241–242

horse-drawn streetcars, 240–241
petroleum dependence, 236–238
rail lines, 241
sedentary lifestyles and health impacts, 236
social inequality, 235–236
suburban sprawl, 242
traffic fatalities and injuries, 235
"walkable" places, 240
urban water sustainability
anticipatory governance, 228
Brundtland Commission report, 224
climate change, 225–226
Colorado River Compact, 225
demand, 227
drought response, 227
financial resources, 227
Phoenix water-sustainability planning model, 229–231
social systems, 224, 226
stationarity, 224–225
US Conference of Mayors Climate Protection Agreement, 163
US Department of Agriculture (USDA), 179
US Energy Information Administration (EIA), 14–15, 265
U.S. Environmental Protection Agency(US EPA), 96, 238
US Global Change Research Program's (USGCRP), 155
U.S. Green Building Council (USGBC), 166–167
US Indian nations. *see also* Tohono O'odham Nation
capacity building, 345–348
federal Indian policy, 334–338
Indian demographics, 338–339
resiliency, 339–340

V
Valencia Street project, 253
Valley Metro Light Rail project, 123–124
Vancouver's sustainability
archaeological finds, 59
climate and weather, 63
climate leadership, 60
diversity and multiculturalism, 59
economic sectors, 59
Green City Index, 64
greenest city plan, 59–60
green spaces and trees, 63–64
green transportation, 60, 62–63
population, 58–59
Treekeepers, 64
urban forest strategy, 64

Vauban district master plan, 76
Vertical Garden, 102
Vespasian, 89
vitamin C, 184
Von Carlowitz, Hans C., 17, 23
Von Goethe, Johann W., 92

W
Wackernagel, Mathis, 39
Walden: or Life in the Woods, 37
walkability and sidewalks, 94–95
Walkable City, 95
Walk Score, 95
Warehouse District, 98–99
waste recycling, 191
wastewater treatment, 213
water cycle, 214–215
water demand, 213
water quality standards, 221
water security, 159
watersheds, 213, 320
water supply, 214
demand for, 220–222
drought conditions, 218–219
groundwater, 214–216
reclaimed water, 217
surface water, 216–217
water cycle, 214–215
water treatment, 222–223
Webber, Melvin, 18
Westergasfabriek Culture Park, Amsterdam, 211
Western Michigan, economic impact, 136, 137
White, Stanley Hart, 101
"wicked" problems, 14
wildlife habitat & biodiversity, 192
Wilson, Edward O., 38
World Conservation Strategy, 33–35
World Meteorological Organization (WMO), 149
"worst-case" scenarios, 159
Worster, Donald, 23

X
Xiamen's carbon footprint, 321–322

Y
Yellowstone to Yukon Conservation Initiative, 204

Z
zero-energy buildings, 76
zoning codes, 270
zoning ordinances, 78